World Sustainability Series

Series Editor

Walter Leal Filho, European School of Sustainability Science and Research,
Research and Transfer Centre "Sustainable Development and Climate Change Management"
Hamburg University of Applied Sciences
Hamburg, Germany

Due to its scope and nature, sustainable development is a matter which is very interdisciplinary, and draws from knowledge and inputs from the social sciences and environmental sciences on the one hand, but also from physical sciences and arts on the other. As such, there is a perceived need to foster integrative approaches, whereby the combination of inputs from various fields may contribute to a better understanding of what sustainability is, and means to people. But despite the need for and the relevance of integrative approaches towards sustainable development, there is a paucity of literature which address matters related to sustainability in an integrated way.

Notes on the quality assurance and peer review of this publication

Prior to publication, the works published in this book are initially assessed and reviewed by an in-house editor. If suitable for publication, manuscripts are sent for further review, which includes a combined effort by the editorial board and appointed subject experts, who provide independent peer-review. The feedback obtained in this way was communicated to authors, and with manuscripts checked upon return before finally accepted. The peer-reviewed nature of the books in the "World Sustainability Series" means that contributions to them have, over many years, been officially accepted for tenure and promotion purposes.

Ayyoob Sharifi • John Lee Candelaria
Dahlia Simangan • Shinji Kaneko
Editors

Frameworks and Practices Towards a Sustainable and Peaceful Future

Springer

Editors
Ayyoob Sharifi
The IDEC Institute
Hiroshima University
Higashi-Hiroshima, Hiroshima, Japan

Dahlia Simangan
The IDEC Institute
Hiroshima University
Higashi-Hiroshima, Hiroshima, Japan

John Lee Candelaria
Graduate School of Humanities
and Social Sciences
Hiroshima University
Higashi-Hiroshima, Hiroshima, Japan

Shinji Kaneko
Network for Education and Research
on Peace and Sustainability (NERPS)
Hiroshima University
Higashi-Hiroshima, Hiroshima, Japan

ISSN 2199-7373 ISSN 2199-7381 (electronic)
World Sustainability Series
ISBN 978-981-95-2309-2 ISBN 978-981-95-2310-8 (eBook)
https://doi.org/10.1007/978-981-95-2310-8

This Springer imprint is published by the registered company Springer Nature Singapore Pte Ltd.
The registered company address is: 152 Beach Road, #21-01/04 Gateway East, Singapore 189721, Singapore

Preface

The interconnections between peace and sustainability remain critical in academic and policy arenas, as reaffirmed by the 2030 Agenda for Sustainable Development. Yet, many initiatives still address peace and sustainability in isolation, missing their synergistic potential. In a world grappling with escalating climate crises, geopolitical tensions, and socioeconomic disparities, integrating peacebuilding with sustainable development is vital to tackle complex global challenges across environmental, social, political, and economic domains.

"Frameworks and Practices Towards a Sustainable and Peaceful Future "explores innovative pathways to unify peace and sustainability efforts, building on three prior volumes from 2023 and 2024. These volumes heightened global awareness of the peace-sustainability nexus, setting the stage for deeper, more actionable integration in response to current global dynamics.

This edited book features a collection of papers presented at the NERPS 2025 conference, organized by the Network for Education and Research on Peace and Sustainability (NERPS) (Fig. 1). The conference took place from March 5–7, 2025 at De La Salle University (DLSU) in Manila, with the primary focus on achieving peaceful and sustainable futures. The conference facilitated dialogue and collaboration among diverse stakeholders, stimulating further research and initiatives that investigate the interconnectedness of critical global issues. Approximately 170 participants, including students, researchers, and practitioners from various fields, explored a comprehensive range of topics related to the peace-sustainability nexus.

"Frameworks and Practices Towards a Sustainable and Peaceful Future" highlights the crucial interdependence of environmental sustainability and peace, a relationship that warrants increased focus in academic inquiry and policy formulation. By compiling insights from across the world, the volume presents theoretical and empirical evidence that underscores the superior effectiveness of integrated approaches to peace and sustainability over pursuing these objectives in isolation. Through analyses of diverse initiatives—spanning educational systems, governance structures, community organizations, and technological innovations—contributors identify both the impediments and prospects for advancement. The case studies within this volume deliver valuable lessons concerning both effective

Fig. 1 The opening ceremony of NERPS 2025

interventions and the obstacles that remain. This work equips researchers, policymakers, and practitioners with knowledge instrumental in tackling the intertwined challenges of conflict and environmental change.

This edited volume offers crucial insights into the intricate relationship between peace and sustainability, a connection that becomes increasingly vital as our world grows more complex. It is imperative that we deepen our understanding of how these concepts interact amidst evolving global challenges. We encourage researchers and policymakers to attend NERPS conferences and events, where they can engage in critical discussions and adapt the peace-sustainability nexus to address emerging global issues. The integration of peace and sustainability is not just beneficial; it is essential for navigating the challenges of our interconnected and complex world. Let us commit to this integration as a cornerstone of our global strategy.

Higashi-Hiroshima, Japan — Ayyoob Sharifi
Higashi-Hiroshima, Japan — John Lee Candelaria
Higashi-Hiroshima, Japan — Dahlia Simangan
Higashi-Hiroshima, Japan — Shinji Kaneko

Contents

About the Editors

Ayyoob Sharifi is Professor at the IDEC Institute, Hiroshima University. He is also a core member of the Network for Education and Research on Peace and Sustainability (NERPS). His research is mainly focused on climate change mitigation and adaptation. Ayyoob actively contributes to global change research programs, such as the Future Earth, and has served as a lead author for the Sixth Assessment Report (AR6) of the Intergovernmental Panel on Climate Change (IPCC). The ultimate goal of his education and research activities is to inform actions toward building sustainable and peaceful communities.

John Lee Candelaria is Assistant Professor at the Graduate School of Humanities and Social Sciences, Hiroshima University and a research fellow of the Network for Education and Research on Peace and Sustainability (NERPS). He holds a PhD from Hiroshima University and a bachelor's and master's degrees in History from the University of the Philippines Diliman. He specializes in peace and conflict research, with a scholarly focus encompassing war propaganda, heritage and memory politics, peace processes, and the critical intersection of peace and sustainability, with particular emphasis on the Philippines and Southeast Asia. His most recent publication is the book, *War Memorialization and Nation-Building in Twentieth-Century Southeast Asia* (Routledge, 2025).

Dahlia Simangan is Associate Professor at Hiroshima University's Graduate School of Humanities and Social Sciences. She is a former Kanagawa University JSPS Postdoctoral Research Fellow (nominated by the United Nations University-Centre for Policy Research in Tokyo). She obtained her PhD in International, Political and Strategic Studies from the Australian National University, MA in International Relations from the International University of Japan, and her BA in Sociology from the University of the Philippines, Diliman. Her research interest in peace and conflict studies includes topics on post-conflict peacebuilding, the peace-sustainability nexus, and international relations in the Anthropocene.

Shinji Kaneko is the Executive Vice President for Global Initiatives and a Professor at the Graduate School of Humanities and Social Science at Hiroshima University. He graduated from the School of Engineering at Kyushu University, majoring in water engineering (Dr. of Engineering). Immediately after the completion of his doctoral program, he joined the Institute for Global Environmental Strategies (IGES) in 1999 as a researcher and conducted research on urban climate policy in Asian megacities for 3 years. He also worked at global Change System for Analysis, Research and Training (START) as a Research Fellow for 3 years from 2005. In 2018, he was appointed as the Director of Network for Education and Research on Peace and Sustainability (NERPS) at Hiroshima University. He has conducted numerous policy research on natural resources, energy, and the environment in developing countries.

Chapter 1
Frameworks and Practices Toward a Sustainable and Peaceful Future

Ayyoob Sharifi, John Lee Candelaria, Dahlia Simangan, and Shinji Kaneko

1.1 Introduction

The year 2024 marked a watershed moment in human history: the convergence of armed conflicts, climate extremes, and biodiversity collapse has reached levels that fundamentally challenge humanity's capacity to respond through traditional, compartmentalized approaches. With 61 active state-based armed conflicts, the highest number since systematic records began in 1946 (Uppsala Conflict Data Program 2025), coinciding with global temperatures breaching 1.55°C above pre-industrial levels for the first time (WMO 2025), the interconnected nature of peace and sustainability challenges has become undeniable. These figures represent a staggering toll on human lives and the natural world. Some 123.2 million people have been forcibly displaced as a result of conflict, persecution, violence, and human rights violations (UNHCR 2025), while wildlife populations have fallen by an average of 73 % since 1970 (WWF 2024). The convergence of these crises illustrates a reciprocal relationship where environmental degradation multiplies the threat of conflict, while social and political instability make ecological protection and sustainable development untenable.

A. Sharifi (✉) · D. Simangan
The IDEC Institute, Hiroshima University, Higashi-Hiroshima, Hiroshima, Japan
e-mail: sharifi@hiroshima-u.ac.jp

J. L. Candelaria
Graduate School of Humanities and Social Science, Hiroshima University, Higashi-Hiroshima, Hiroshima, Japan

S. Kaneko
Network for Education and Research on Peace and Sustainability (NERPS), Hiroshima University, Higashi-Hiroshima, Hiroshima, Japan

A. Sharifi et al. (eds.), *Frameworks and Practices Towards a Sustainable and Peaceful Future*, World Sustainability Series,
https://doi.org/10.1007/978-981-95-2310-8_1

The threats humanity faces have been referred to as a "polycrisis," a buzzword of the 2020s that highlights the interconnected nature of the challenges (Ang 2024). These crises are causally entangled, affecting and significantly degrading humanity's prospects (Lawrence et al. 2024). The worsening crises expose the fundamental inadequacy of approaches that treat peace and sustainability as separate policy domains. Traditional security frameworks focused primarily on military solutions prove insufficient when environmental stress acts as a threat multiplier (Vogler 2023), while conventional sustainability efforts falter in contexts of violence and weak governance (Ide 2020). The Sahel region exemplifies these reinforcing dynamics, where desertification and resource scarcity have contributed to the region accounting for 51% of global terrorism deaths in 2024 (Bojang Jr. 2025), forcing nearly 5 million people into displacement, a 25% increase since 2020 (UNHCR 2025). Similarly, climate-induced sea level rise in the Pacific, occurring at nearly twice the global rate, threatens the physical existence of island nations and their sovereignty, cultural identity, and peaceful relations over maritime boundaries (WMO 2023). These examples demonstrate how the absence of positive peace, understood as social justice, inclusive institutions, and equitable resource distribution, undermines sustainability efforts, while environmental degradation erodes the foundations necessary for peaceful societies.

The unanimous adoption of the 2030 Agenda for Sustainable Development represented a historic recognition of these interconnections, declaring that "there can be no sustainable development without peace and no peace without sustainable development" (United Nations 2015, p. 2). SDG 16 on Peace, Justice, and Strong Institutions was explicitly identified as an enabler of all other goals, acknowledging that health, education, and environmental protection development outcomes remain fragile without peaceful conditions and just institutions (UN DESA 2025b). Yet as we approach the agenda's deadline, the 2025 Sustainable Development Goals Report exposes a sobering reality: only 35% of SDG targets remain on track or are making moderate progress, and nearly half are moving too slowly; most alarmingly, 18% of goals are in reverse (UN DESA 2025a). The reported $4 trillion annual investment gap in developing countries, combined with a 40% increase in armed conflict casualties in 2024, demonstrates how the failure to integrate peace and sustainability approaches undermines progress across all dimensions of human development. These implementation failures demand not incremental adjustments but fundamental reconceptualization of how we understand and pursue the interconnected goals of peace and sustainability in an era of accelerating global transformations.

1.2 Envisioning the Peace-Sustainability Nexus for Future Generations

In response to the escalating polycrisis, the peace-sustainability nexus offers a transformative and essential pathway forward. It presents a vision where thriving communities can live in harmony with regenerating ecosystems, social justice prevails across societies, and future generations inherit a world of expanding rather than diminishing possibilities. This vision is not a utopian fantasy but an achievable horizon, built on a deeper understanding of the integrated solutions required to navigate our increasingly complex world (Sharifi et al. 2024). Amidst the increasing geopolitical and climate-induced risks and uncertainties (Sharifi et al. 2021a, c), this nexus provides the necessary framework for turning the destabilizing global crises outlined previously into opportunities for profound and positive transformation.

Positive peace is central to this envisioned future, ensuring the presence of flourishing societies. This moves beyond the limited scope of "negative peace," the simple cessation of war and direct physical harm, to embrace a more holistic and ambitious goal founded on the pursuit of social justice (Galtung 1969). It is defined by the attitudes, institutions, and structures that actively create and sustain peaceful conditions (IEP n.d.), including inclusive governance, the equitable distribution of resources, a vibrant cultural life, and robust institutions that nurture human potential. Albeit inexhaustive, these pillars are the foundational building blocks of a society capable of navigating challenges constructively and justly (Simangan et al. 2022). Likewise, sustainability for future generations transcends the narrow goal of maintaining current systems: it calls for an evolution from mere protection to proactive restoration, embracing regenerative practices that heal damaged ecosystems, create circular economies, and strengthen community resilience against social and ecological shocks (Folke et al. 2002; Simangan et al. 2025). This comprehensive view, rooted in the principle of meeting "the needs of the present without compromising the ability of future generations to meet their own needs" (WCED 1987, para. 27), integrates the social, economic, and institutional dimensions of development (UN DESA 2001). Such an approach ensures that human activities enhance, rather than degrade, the conditions necessary for all life to thrive.

In this future vision, peace and sustainability function as mutually reinforcing dynamics. The relationship is not incidental but causal and cyclical. Global data demonstrates a powerful correlation between the pillars of positive peace and strong environmental performance; specifically, societies with more equitable resource distribution and higher levels of human capital are significantly more successful at ensuring better air quality and providing safe drinking water and sanitation (Simangan et al. 2022). Free from the immense costs of conflict, peaceful societies are better equipped to invest in long-term environmental stewardship. In turn, healthy ecosystems provide the resources and stability that prevent resource-based competition, a key driver of conflict (Homer-Dixon 2010).

Contemporary initiatives around the world demonstrate glimpses of this integrated future. We see its potential in transboundary "peace parks" that foster both ecological conservation and diplomatic cooperation (Ndimele et al. 2024), in urban planning that prioritizes green space and social equity to reduce conflict (Rambhia et al. 2024), and in agricultural systems that use regenerative practices to ensure food security and build community resilience (Strauss 2024). The transformative potential of the nexus becomes clearest when pursuing peace and sustainability together creates multiple co-benefits that siloed approaches cannot achieve (Simangan et al. 2025). For example, a project to restore a degraded watershed can simultaneously improve water security, create green jobs, enhance biodiversity, and reduce tensions between upstream and downstream communities.

This integrated vision provides a necessary and hopeful counter-narrative to the grim realities of our time. While existing analyses highlight the failures of fragmented policies and the documented inadequacies of our global assessment tools (Candelaria et al. 2023), the peace-sustainability nexus illuminates a viable path forward. However, translating this powerful vision into reality is the central challenge of our era. It requires moving beyond conceptual recognition to confront the deep-seated institutional, financial, and intellectual barriers perpetuating siloed thinking. To bridge this critical implementation gap, we must explore how we can advance the integrated frameworks and practices needed to turn the aspirational goals of the nexus into tangible, on-the-ground action.

1.3 Advancing Frameworks and Practices: Vision to Reality

Bridging the chasm between the aspirational vision of the peace-sustainability nexus and its tangible realization requires a concerted effort to develop, test, and scale new operational frameworks and concrete practices. The vision articulated previously provides the "why"; we must now explore the critical "how." Doing so requires what UN Secretary-General António Guterres describes as a fundamental renewal of multilateralism itself, adapting our systems of global cooperation to be more networked, inclusive, and effective for the challenges of a new era (United Nations 2021). The evidence is stark: nine out of ten countries receiving the most climate-security funding between 2017–2021 were among the most vulnerable to climate change, yet implementation remains fragmented and insufficient (Chambers and Kyed 2024; Gaston et al. 2023). Merely acknowledging the connections between peace and sustainability is insufficient. The central challenge of our time is to actively dismantle the silos that perpetuate fragmented responses and to build, in their place, integrated approaches capable of navigating the non-linear dynamics of a world in polycrisis.

Advancing new frameworks is an essential step. The dominant models for development and peacebuilding have historically been grounded in separate, often contradictory, assumptions. Environmental interventions, designed with a conflict-blind lens, can inadvertently exacerbate social tensions over resources (Ahmadnia et al. 2022), while peace agreements that overlook the long-term environmental drivers of conflict may sow the seeds of future instability (Raymond 2024). Recent reviews of the research landscape have highlighted persistent gaps and a lack of genuine interdisciplinarity in many publications, despite the field's conceptual goals (Sharifi et al. 2021b), while attention to the nexus between peace and sustainability remains marginal (Simangan et al. 2025). To move forward, we must go beyond simple interdisciplinary collaboration toward a genuine analytical fusion, creating hybrid concepts like "environmental peacebuilding" (Ide et al. 2021), "ecological security" (McDonald 2021), and "planetary health" (Myers 2017)—goals that align with the broader push to implement the Humanitarian-Development-Peace nexus (OECD 2022). This evolution is exemplified by the UN Climate Security Mechanism, which builds on the political mandate established in landmark Security Council resolutions connecting climate change to instability (e.g., UNSC 2017) and now provides integrated analysis across multiple field missions, with eleven UN Security Council members having signed Joint Pledges on Climate, Peace and Security (Security Council Report 2023).

An effective nexus framework must be inherently transdisciplinary, politically astute, and culturally sensitive. This requires multi-scalar frameworks that can skillfully navigate the frequent tensions between global norms, such as those embedded in human rights law or climate agreements, and local ontologies, including customary governance systems and Indigenous worldviews (Nardi et al. 2024). Such frameworks must be adaptive by design, built to function amid the deep uncertainty and unpredictable feedback loops that characterize complex socio-ecological systems (Folke et al. 2002), rather than imposing rigid, top-down solutions.

Beyond conceptual frameworks, translating the peace-sustainability nexus into tangible outcomes hinges upon pioneering and scaling innovative practices, a challenge often impeded by significant real-world obstacles. Even the most meticulously designed frameworks require adequately skilled practitioners and functioning institutions prepared to facilitate meaningful implementation. Institutional inertia, frequently driven by entrenched professional identities, existing power structures, and political economies that defend disciplinary boundaries, remains a formidable barrier. Climate adaptation finance is often prevented from reaching conflict-affected states precisely because of institutional risk aversion and fragmented funding mechanisms (Cao 2023). Sectoral disconnects persist from siloed planning and decision-making processes across crucial domains like mobility, energy, water, housing, and urban development. Overcoming these barriers necessitates the adoption of innovative financing strategies. Cao (2023) proposes using blended financial mechanisms such as "peace bonds," instruments explicitly designed to leverage private capital markets toward achieving the dual goals of climate resilience and peacebuilding. These bonds directly counteract traditional funding silos by integrating ecological and social outcomes, creating incentives for cross-sectoral collaboration, and facilitating coordinated responses.

Simultaneously, addressing critical capacity gaps involves nurturing a new generation of "nexus professionals" capable of operationalizing the transdisciplinary approach that Simangan et al. (2021) recommend. This extends far beyond traditional collaboration, demanding the use of methods that incorporate and empower local perspectives, such as deep immersion in living labs and participatory citizen science. By embedding these practices from design through to dissemination, these professionals can effectively narrow the gap between top-down planning and the aspirations of communities disproportionately affected by conflict and environmental change, ensuring solutions are both relevant and legitimate. Research on the nexus also needs to go beyond simplistic quantitative measures (Sharifi et al. 2021b), as sophisticated mixed-method approaches are more capable of capturing complex, emergent phenomena such as community resilience, social cohesion, and governance legitimacy.

However, deeper, more intractable political dynamics lie beyond improved frameworks and enhanced practices. The question of who holds the power to define problems and shape solutions within the nexus discourse remains crucial. Sultana (2022) warns of the risk that nexus-driven policies, if not grounded firmly in equity and justice, could evolve into new forms of "green" colonialism, where environmental agendas from the Global North are imposed upon Global South communities without authentic local participation. Additionally, nexus-oriented interventions must grapple with inherent temporal tensions: the immediate pressures of humanitarian interventions and ceasefire negotiations often contrast starkly with the generational timelines required for environmental restoration and sustainable development. This temporal misalignment is a core implementation gap that requires adaptations in funding timelines and project designs to better harmonize immediate peacebuilding imperatives with long-term climate resilience objectives (Bedoya Taborda et al. 2025).

1.4 The Focus of the Volume and the Way Forward

The chapters in this volume respond directly to these imperatives, emerging from the rich dialogues of the Network for Education and Research on Peace and Sustainability (NERPS) 2025 conference in Manila. Selected from 147 presentations exploring the conference theme "Achieving Peaceful and Sustainable Futures," these chapters collectively chart diverse pathways through the peace-sustainability nexus.

This volume builds on earlier volumes featuring research presented during NERPS 2022 (Sharifi et al. 2023a), NERPS 2023 (Sharifi et al. 2023b), and NERPS 2024 (Sharifi et al. 2024) conferences. The first volume of this series was mainly focused on issues related to politics and governance, refugees, media, women, culture, and religion. These themes reflect an initial emphasis on socio-political dimensions and human rights issues, particularly in post-conflict and fragile contexts. In the second volume, the peace-sustainability nexus was explored with more attention to themes such as conflict resolution, peacebuilding, justice, youth and women's empowerment, social dignity, and resource management, among other topics. This shift indicated a move toward conceptual integration between peace and sustainability, with growing attention to social equity, institutional development, and transformative justice.

The third volume further expanded the coverage of enabling factors along the peace-sustainability pathway, introducing a more inclusive approach that incorporated both positive and negative dimensions of peace, along with broad sustainability concerns. Prominent themes such as environmental change, governance, migration, and agency emerged more strongly, signaling deeper engagement with structural drivers of conflict and sustainability challenges across scales.

This fourth volume reveals a maturing discourse, with dominant themes including peace, sustainability, climate, governance, local, development, conflict, adaptation, and transformation (Fig. 1.1). Spanning scales from grassroots community action to the complexities of global governance, the chapters move beyond abstract frameworks to demonstrate how peace and sustainability are negotiated, contested, and reimagined in practice. As will be detailed, these pathways offer both practical insights and critical perspectives that provide crucial guidance for those working to bridge the persistent gap between vision and implementation.

Compared to previous volumes, there is a greater emphasis on operationalizing the nexus, reflected in the prevalence of practical and action-oriented terms such as "policy," "community," "solutions," "local," and "resilience." This suggests a shift from primarily conceptual and diagnostic explorations to solution-oriented and participatory approaches, integrating local knowledge and contextual adaptation into peacebuilding and sustainability strategies. Overall, it can be seen that, over time, the conference has been successful in expanding the boundaries of the nexus and exploring its multiple dimensions. The progression of themes illustrates an evolving narrative—from foundational socio-political concerns to complex system-level analyses and, most recently, to implementation and praxis. In doing so, it has contributed to addressing multiple priorities outlined in Sharifi et al. (2021b), including the need for holistic, integrated, and actionable approaches to peace and sustainability.

Drawing from the insights presented in this volume, we outline some key recommendations for research and practice. Advancing research and practice in the peace-sustainability nexus requires a localized and context-sensitive approach, particularly in subnational and marginalized regions. Research efforts should prioritize understanding the dynamics of aid effectiveness and peacebuilding through longitudinal studies that evaluate the sustainability and institutional impact of aid programs over

Fig. 1.1 The overall thematic focus of this volume

time. This includes addressing governance structures, resource accountability, and anti-corruption measures to ensure equitable development. Interdisciplinary collaboration among governments, NGOs, academic institutions, and local communities is essential for addressing complex challenges and fostering innovative solutions. Special attention should be given to intersectional analysis to address the needs of vulnerable groups, such as women, displaced populations, and economically disadvantaged communities, ensuring inclusivity and equity in both research and practice.

In practice, programs aimed at promoting the nexus must integrate context-sensitive frameworks to address localized needs and foster trust among communities. Empowering local ownership through community participation in planning, implementation, and monitoring processes ensures that interventions are relevant and sustainable. Sustainability planning should include clear strategies and capacity-building measures to reduce dependency on external aid while strengthening local governance structures. Transparency and accountability mechanisms must be embedded to prevent mismanagement and corruption, ensuring that resources are utilized effectively. Corporate social responsibility and ethical branding can also play a pivotal role in rebuilding trust and aligning business practices with peacebuilding goals, contributing to collective efforts for sustainable development.

Environmental peacebuilding offers a unique opportunity to use ecological cooperation as an entry point for dialogue in conflict-prone settings. By reframing environmental challenges as shared issues, stakeholders can build trust and foster collaboration even in diplomatically stalled relationships. Climate adaptation strategies should be strengthened by studying the impacts of climate change on ecosystem services and implementing measures that support food security and sustainable livelihoods. Integrating environmental education and conflict management into curricula can promote awareness and equip future generations with the tools to address these challenges. Additionally, promoting environmental stewardship and sustainability within aid programs can align development efforts with broader peacebuilding goals, fostering reconciliation and long-term stability.

Promoting justice must be prioritized as a cornerstone for achieving sustainable development and human rights. Public policies should focus on reducing barriers to justice, such as economic constraints, geographic disparities, and judicial inefficiencies, while empowering vulnerable populations to exercise their rights effectively. By promoting inclusive governance, strengthening local capacity, and ensuring equitable access to resources, these efforts can contribute to a more just, sustainable, and peaceful society.

References

Ahmadnia, S., Christien, A. M., Spencer, P., Hart, T., & De Araujo Barbosa, C. C. (2022). *Defueling Conflict: Environment and Natural Resource Management as a Pathway to Peace*. World Bank Group. https://documents1.worldbank.org/curated/en/099520010272224660/pdf/P1771510b38fda01e0afec01edd810d8cde.pdf

Ang, Y. Y. (2024, December 18). *Doing development in the polycrisis*. UNDP. https://www.undp.org/blog/doing-development-polycrisis

Bedoya Taborda, L. F., Barnes, M. L., & Morrison, T. H. (2025). Adaptation and Peace: Extending the Agenda for Capacity-Building in Climate and Conflict-Affected Communities. *WIREs Climate Change*, *16*(1). https://doi.org/10.1002/wcc.921

Bojang, S. Jr. (2025, March 5). *Burkina Faso, Mali, Niger: The Sahel accounts for 51% of global terrorism deaths in 2024 – report*. The Africa Report. https://www.theafricareport.com/378239/burkina-faso-mali-niger-report-shows-sahel-accounted-for-51-of-global-terrorism-deaths-in-2024/

Candelaria, J. L., Sharifi, A., Simangan, D., & Tabosa, R. M. R. (2023). A critical analysis of selected global sustainability assessment frameworks: Toward integrated approaches to peace and sustainability. *World Development Perspectives*, *32*, 100539. https://doi.org/10.1016/j.wdp.2023.100539

Cao, Y. (2023). *Breaking the cycle: Practical solutions to unlock climate finance for fragile states* [Peace Bonds Case Study]. Mercy Corps. https://financeforpeace.org/wp-content/uploads/2025/03/Peace-Bonds-Climate-Finance-Case-Study-2.pdf

Chambers, J., & Kyed, H. M. (2024). *Bridging the gap in climate change financing to violent conflict affected areas* (DIIS Policy Brief). Danish Institute for International Studies. https://www.diis.dk/en/research/bridging-the-gap-in-climate-change-financing-to-violent-conflict-affected-areas

Folke, C., Carpenter, S., Elmqvist, T., Gunderson, L., Holling, C. S., & Walker, B. (2002). Resilience and Sustainable Development: Building Adaptive Capacity in a World of Transformations. *AMBIO: A Journal of the Human Environment*, *31*(5), 437–440. https://doi.org/10.1579/0044-7447-31.5.437

Galtung, J. (1969). Violence, peace, and peace research. *Journal of Peace Research*, *6*(3), 167–191. https://doi.org/10.1177/002234336900600301

Gaston, E., Brown, O., Al Dawsari, N., Downing, C., Day, A., & Bodewig, R. (2023). *Climate-Security and Peacebuilding: Thematic Review*. United Nations University-Centre for Policy Research.

Homer-Dixon, T. (2010). *Environment, Scarcity, and Violence*. Princeton University Press.

Ide, T. (2020). The dark side of environmental peacebuilding. *World Development*, *127*, 104777. https://doi.org/10.1016/j.worlddev.2019.104777

Ide, T., Bruch, C., Carius, A., Conca, K., Dabelko, G. D., Matthew, R., & Weinthal, E. (2021). The past and future(s) of environmental peacebuilding. *International Affairs*, *97*(1), 1–16. https://doi.org/10.1093/ia/iiaa177

IEP. (n.d.). Positive Peace Report. *Institute for Economics & Peace*. Retrieved July 21, 2025, from https://www.economicsandpeace.org/research/positive-peace-report/

Lawrence, M., Homer-Dixon, T., Janzwood, S., Rockstöm, J., Renn, O., & Donges, J. F. (2024). Global polycrisis: The causal mechanisms of crisis entanglement. *Global Sustainability*, *7*, e6. https://doi.org/10.1017/sus.2024.1

McDonald, M. (2021). *Ecological Security: Climate Change and the Construction of Security*. Cambridge University Press. https://doi.org/10.1017/9781009024495

Myers, S. S. (2017). Planetary health: Protecting human health on a rapidly changing planet. *The Lancet*, *390*(10114), 2860–2868. https://doi.org/10.1016/S0140-6736(17)32846-5

Nardi, M. A., Krause, T., & Zelli, F. (2024). Diverse understandings and values of nature at the peace–environment nexus: A critical analysis and policy implications towards decolonial peace. *Ecology and Society*, *29*(4). https://doi.org/10.5751/es-15198-290441

Ndimele, P. E., Akanni, A., Ositimehin, K. M., Shittu, J. A., Ojewole, A. E., & Ayeni, Y. Z. (2024). Fostering international and trans-boundary cooperation in the management of Lake Chad fisheries, wildlife and flora: The role of a trans-boundary Ramsar conservation area. *Aquatic Living Resources*, *37*, 6. https://doi.org/10.1051/alr/2024004

OECD. (2022). *The Humanitarian-Development-Peace Nexus Interim Progress Review*. OECD. https://doi.org/10.1787/2f620ca5-en

Rambhia, M., Volk, R., Rismanchi, B., Winter, S., & Schultmann, F. (2024). Prioritizing urban green spaces in resource constrained scenarios. *Resources, Environment and Sustainability*, *16*, 100150. https://doi.org/10.1016/j.resenv.2024.100150

Raymond, A. (2024, November 14). The Climate-Conflict Nexus: What Research and Peace Agreements Reveal. *PeaceRep:Rethinking Peace & Transition Processes in a Changing Conflict Landscape*. https://peacerep.org/2024/11/14/the-climate-conflict-nexus-what-research-and-peace-agreements-reveal/

Security Council Report. (2023, May 31). *Climate, Peace and Security: June 2023 Monthly Forecast*. Security Council Report. https://www.securitycouncilreport.org/monthly-forecast/2023-06/climate-peace-and-security.php

Sharifi, A., Simangan, D., & Kaneko, S. (2021a). Three decades of research on climate change and peace: a bibliometrics analysis. Sustainability Science, 16(4), 1079-1095. https://doi.org/10.1007/s11625-020-00853-3

Sharifi, A., Simangan, D., & Kaneko, S. (2021b). The literature landscape on peace–sustainability nexus: A scientometric analysis. *Ambio*, *50*(3), 661–678. https://doi.org/10.1007/s13280-020-01388-8

Sharifi, A., Simangan, D., Lee, C. Y., Reyes, S. R., Katramiz, T., Josol, J. C., Islam, M. (2021c). Climate-induced stressors to peace: a review of recent literature. Environmental Research Letters, 16(7), 073006. https://doi.org/10.1088/1748-9326/abfc08

Sharifi, A., Simangan, D., & Kaneko, S. (2023a). Integrated Approaches to Peace and Sustainability. Springer Nature Singapore. https://doi.org/10.1007/978-981-19-7295-9_1

Sharifi, A., Simangan, D., & Kaneko, S. (2023b). Bridging Peace and Sustainability Amidst Global Transformations (1st 2023. ed.). Springer Nature Singapore: Imprint: Springer.

Sharifi, A., Candelaria, J. L., Simangan, D., & Kaneko, S. (2024). *Navigating Peace and Sustainability in an Increasingly Complex World*. Springer Nature Singapore. https://doi.org/10.1007/978-981-97-8772-2

Simangan, D., Virji, H., Hendrix, C. S., Islam, M., Kaneko, S., Ma, Y., Mechler, R., Pangotra, P., Peters, K., Sharifi, A., & Shams, S. (2021). A co-designed heuristic guide for investigating the peace-sustainability nexus in the context of global change. *Sustainability Science*, *16*, 1097–1109. https://doi.org/10.1007/s11625-021-00970-7

Simangan, D., Lee, C. Y., Sharifi, A., Candelaria, J. L., & Kaneko, S. (2022). A global analysis of interactions between peace and environmental sustainability. *Earth System Governance*. https://doi.org/10.1016/j.esg.2022.100152

Simangan, D., Fisher, J., Ide, T., Koubi, V., Sharifi, A., Alfredo, K., Candelaria, J. L., Dalby, S., Hendrix, C., Kharrazi, A., Oswald-Spring, Ú., & Roy, J. (2025). Twelve research agendas for advancing the peace-sustainability nexus. *Peace and Sustainability*, *1*(1), 100008. https://doi.org/10.1016/j.nerpsj.2025.100008

Strauss, T. (2024, November 18). *How regenerative agriculture builds resilient climate solutions*. World Economic Forum. https://www.weforum.org/stories/2024/11/regenerative-agriculture-climate-solutions-resilient/

Sultana, F. (2022). The unbearable heaviness of climate coloniality. *Political Geography*, *99*, 102638. https://doi.org/10.1016/j.polgeo.2022.102638

UN DESA. (2001). *Indicators of sustainable development: Guidelines and methodologies* (2nd ed.). United Nations Department of Economic and Social Affairs. https://sustainabledevelopment.un.org/content/documents/indisd-mg2001.pdf

UN DESA. (2025a). *The Sustainable Development Goals Report 2025*. United Nations Department of Economic and Social Affairs. https://unstats.un.org/sdgs/report/2025/The-Sustainable-Development-Goals-Report-2025.pdf

UN DESA. (2025b, May). Advancing peace, justice and strengthening institutions to accelerate progress on the 2030 Agenda | UN DESA Voice. *UN DESA Voice*, *29*(5). https://desapublications.un.org/un-desa-voice/more-from-undesa/may-2025/advancing-peace-justice-and-strengthening-institutions

UNHCR. (2025). *Global Trends: Forced Displacement in 2024*. United Nations High Commissioner for Refugees. https://www.unhcr.org/global-trends-report-2024

United Nations. (2015). *Transforming our world: The 2030 Agenda for Sustainable Development* (No. A/RES/70/1). https://sdgs.un.org/2030agenda

United Nations. (2021). *Our Common Agenda—Report of the Secretary-General*. United Nations.

UNSC. (2017). *Resolution 2349 (2017) [on peace and security in the Lake Chad Basin region]* (No. S/RES/2349 (2017)). United Nations. https://digitallibrary.un.org/record/863830

Uppsala Conflict Data Program. (2025). *UCDP Conflict Encyclopedia*. https://ucdp.uu.se/

Vogler, A. (2023). Barking up the tree wrongly? How national security strategies frame climate and other environmental change as security issues. *Political Geography*, *105*, 102893. https://doi.org/10.1016/j.polgeo.2023.102893

WCED. (1987). *Our common future*. World Commission on Environment and Development. https://sustainabledevelopment.un.org/content/documents/5987our-common-future.pdf

WMO. (2023). *State of the Climate in the South-West Pacific 2022* (No. 1324). World Meteorological Organization. https://wmo.int/publication-series/state-of-climate-south-west-pacific-2022

WMO. (2025, January 10). *WMO confirms 2024 as warmest year on record at about 1.55°C above pre-industrial level*. World Meteorological Organization. https://wmo.int/news/media-centre/wmo-confirms-2024-warmest-year-record-about-155degc-above-pre-industrial-level

WWF. (2024). *2024 Living Planet Report: A System in Peril*. World Wide Fund for Nature. https://livingplanet.panda.org/

Part I
Knowledge and Learning Pathways

Part I
Methods and Learning Platforms

Chapter 2
Collective Learning in ASEAN's Climate Governance Networks: The Adoption of the 2018 ASEAN Guidelines for Agroforestry Development

Virgemarie Salazar

2.1 Introduction

Learning is a political process wherein consensual knowledge is used to inform the understanding of the causal relationship between the problem and its solutions. Learning, if it happens at all, does not automatically lead to changes in policy or state behavior. Instead of learning, actors would often adopt solutions that, on the surface, looked like a change in behavior but did not necessarily alter their normative and causal beliefs about how a certain issue should be addressed. The paper looks at how learning occurs in ASEAN, given its state-centric institutional design, which offers limited engagement with non-state actors. It aims to address the question of how the participation of transnational networks in ASEAN's climate governance efforts enables collective learning.

Non-state actors such as experts, academics, and advocacy groups have sought to influence policymaking and governance by imparting knowledge. It is assumed that their interactions with decision-makers can bring about new meanings and interpretations of issues, which in turn may lead to courses of action (Adler and Haas 1992). Knowledge introduces ideational elements to discourses in international politics, and actors hinge on these ideas in further defining their interests (Béland and Cox 2010; Haas 1992). Ideas, in the form of world views, principled beliefs, and causal beliefs, affect policies and outcomes by providing roadmaps that guide political actors during uncertain times, solving coordination problems when evaluating viable outcomes, and by reinforcing rules and norms that influence public policy once institutionalized (Goldstein and Keohane 1993). The connection between ideas and

V. Salazar (✉)
Department of Political Science, University of the Philippines Diliman, Quezon City, Philippines
e-mail: vasalazar@up.edu.ph

A. Sharifi et al. (eds.), *Frameworks and Practices Towards a Sustainable and Peaceful Future*, World Sustainability Series,
https://doi.org/10.1007/978-981-95-2310-8_2

policies is commonly illustrated in the concept of epistemic community, which is often a transnational network of experts with an authoritative claim to knowledge that is relevant to policymaking within a particular issue area (Haas 1992).

Studies on epistemic communities, however, remain closely associated with scientific knowledge or the scientific way of knowing despite assertions by Haas (1992) that the term also applies to social scientists or individuals who have a strong claim to policy-relevant knowledge. In this study, the definition of knowledge is not limited to formal science, for it includes local knowledge, which is a legitimate source of context-based information and expertise (Ascher et al. 2010). The study also recognizes the view from the Science, Technology, and Society (STS) approach that knowledge brokers provide advice based on evidence as well as their social and political judgment (Jasanoff 2005).

The shared practices of transnational actors comprise the basic unit of analysis of this chapter. Practices are socially meaningful and recognizable patterns of performative acts informed by background knowledge or common understanding of certain ways of doing (Adler and Pouliot 2011). Transnational actors operating in networks are viewed as communities of practice, which are social structures where meanings are co-created, learning occurs, and practices evolve (Adler 2004). As an analytical construct, the communities of practice theory rest on the intersection of learning, knowledge exchange, and participation. These communities are transnational entities brought together by a common domain and like-mindedness that impart their knowledge through practices (Adler 2019). By operating across territorial boundaries in their interactions with states, firms, and individuals, these communities are mutually constituting agents and structures in the international sphere through their shared practices.

Furthermore, this chapter extends Sondarjee's (2021) collective learning approach to ASEAN's case. Using the three processes of change (i.e., boundary encounters, knowledge brokerage, and the use of epistemic boundary objects), this study explains how communities of practice negotiate, translate, and contest knowledge to govern climate change issues. In ASEAN, an intergovernmental organization of 10 Southeast Asian member states, non-state actors have been participating in climate change governance by sharing information and expertise and engaging in capacity-building activities, among others. This chapter examines the shared practices of state and non-state actors in ASEAN evident in the transnational governance of climate change to determine whether their practices lead to collective learning.

It argues that their practices can constitute climate change governance through collective learning, which is both a process and an outcome. As a process, participation in governance allows members to learn from each other as they engage in joint activities. As an outcome, it is assumed that collective learning can be reflected in policies and programs both at the regional and national levels. The study focuses on the interactions of state and non-state actors operating in transnational climate governance networks and their shared practices in terms of knowledge sharing, capacity building, and standard setting, among others. By doing so, the study demonstrates that collective learning in communities of practice can serve as a building block of regional climate governance.

In theorizing collective learning within the context of ASEAN climate change governance, the chapter examines the shared practices that led to the creation of the 2018 ASEAN Guidelines for Agroforestry Development using seven semi-structured interviews conducted from February to May 2022 and February to April 2023, as well as documentary analysis of relevant published materials. The key informants who participated in this study represent both state and non-state actors associated with the regional climate governance network by taking part in ASEAN-led activities. Some acted as focal persons of ASEAN member states in working group meetings, while some are members of transnational networks of civil society organizations, research institutions, and experts serving as technical partners. Data from relevant documents taken from the websites of ASEAN and the technical partners were also collected to analyze evidence of learning from joint activities, policies, and programs.

The chapter proceeds with a section on the conceptual framework explaining collective learning through the perspective of communities of practice theory to discuss the theoretical underpinnings of transnational climate change governance in ASEAN. To better understand social forestry as a significant aspect of climate change mitigation and adaptation efforts, the third section presents the context of social forestry in Southeast Asia. It is followed by the section on ASEAN's agroforestry guidelines as a case study on collective learning. The fifth section focuses on the analysis of the findings using the practice-based approach. The concluding section reiterates the main findings and the limitations of the study.

2.2 Collective Learning

The concept of learning has been described as a minefield, for it is "difficult to define, isolate, measure, and apply empirically" (Levy 1994, p. 280). In international relations, learning in the context of international organizations has been theorized as a process wherein consensual knowledge is used to inform the understanding of causal relationships between problems and solutions (Haas 1990). To learn means that an international organization can convert information into new activities and programs; hence, its ability to foster learning depends on persuading actors to absorb the lessons and change their behavior in light of new information (Haas 2000). The purveyors of knowledge are often epistemic communities that seek to influence powerful member states to accept and advocate for their ideas. Once their ideas are validated, the international organization acts as the transmitter of consensual knowledge among member governments. While there are many pathways to learning, the outcomes of learning can be categorized as either the adoption of new practices or new goals (Adler and Haas 1992).

Since learning is a political process, it does not automatically lead to policy change due to constraints that hinder actors from implementing what they learned, if they learned anything at all (Levy 1994). It is assumed that true learning among members of an international organization rarely happens. Adaptation is more common wherein actors may change their behavior but not their understanding of the causal paradigm. Moreover, learning in international organizations entails a collective behavioral change from international coalitions that are accountable to domestic groups (Haas 1990). It requires a complex mechanism composed of many parts, enabling the provision of non-partisan information, regular feedback about the activities of governments and firms, capacity building, application of capacities to the domestic policy process, and monitoring of the enforcement of international commitments (Haas and Haas 1995). Thus, learning becomes evident when lessons from international organizations are adopted in the form of national strategies and policies.

From a constructivist perspective, social learning is proposed as an alternative causal pathway to compliance through interactions that lead to the transformation of identities and preferences of decision-makers. Compliance, as explained by rationalists, happens because of coercion, cost–benefit calculations, and incentives. While these rational mechanisms remain valid, state compliance viewed from a constructivist lens can also be explained by persuasion and learning through interactions that mutually constitute the identities and preferences of actors (Checkel 2001). Building on the practice turn in international relations, constructivists suggest that shared practices can lead to learning since interactions are determined not only by strategic calculations and diffusion of ideas but also by regular patterns of actions that gain meaning and become institutionalized in communities of practice (Adler 2013).

In policy studies, several theories explain the processes and outcomes of learning. Policy learning involves the use of knowledge to transform beliefs that matter in making policy decisions. It has different dimensions and can be classified into various modes (i.e., epistemic, reflexive, bargaining, and hierarchical) depending on knowledge use, causal mechanisms, modes of interaction, attention of decision-makers, benefits, and pathologies resulting from poor learning (Dunlop and Radaelli 2018). Learning can also be categorized based on the types of learning products, which can involve either cognitive changes in the form of ideas, beliefs, or values, or changes in collective behaviors ranging from informal routines, programs, formal rules, institutional arrangements, and policies (Heikkila and Gerlak 2013).

The advocacy coalition framework (ACF) defines policy-oriented learning as changes in thoughts or behaviors due to past experiences or new information in relation to attaining or revising certain policy goals. The ACF hypothesizes that deep core beliefs are less receptive to policy-oriented learning since these are often very resistant to change compared to secondary beliefs, which are narrower in scope and require less evidence (Sabatier and Weible 2019). Likewise, social learning is understood as a deliberate act of adjusting policies in response to new information, which results in policy changes categorized into subtypes as first order (i.e., incremental change through routinized decision-making), second order (i.e., change through developing new policy instruments), and third order (i.e., a policy paradigm shift involving a wider policy community) to emphasize the variations (Hall 1993). By providing a theory of policy change, Hall's (1993) notion of social learning underscores the role of ideas and situates the policy process within a system of ideas accepted by a policymaking community (Evans 2010).

On governing environmental issues, identifying the causal relationship between social learning and environmental governance outcomes poses a challenge since it requires accounting for changes in values and perceptions of individuals or groups, as well as improvement in environmental conditions and changes in norms and beliefs of societies (Ernst 2019). Moreover, there is a need to delineate the process of social learning, which is often understood as participatory interaction, from the outcomes of social learning. Further studies on research techniques and methodologies will have to clearly establish the contours of social learning in participatory environmental governance. On climate change, a study of international development practices on adaptation finds that skilled facilitation is central to social learning, for it builds trust, addresses power imbalances, and bridges diverging goals and interests (Ensor and Harvey 2015).

In this chapter, learning is analyzed using the theory of communities of practice, which captures the interrelations of participation through mutual engagement, collaborative sense-making, and collective learning (Wenger 1998). It is assumed that knowledge not only exists in its cognitive form but is also observable through actions. Thus, knowledge is shared through practices or regular patterns of meaningful actions, which can constitute meanings and shape identities over time (Adler 2019). Building on this argument, communities of practice can then be understood as social structures that promote collective learning based on competence in a common domain of interest, sustained interactions through joint activities and discussions, along with shared practices such as knowledge co-production.

Related studies on learning in the context of international organizations as communities of practice point out that international organizations such as the North Atlantic Treaty Organization, European Union, and World Bank provide opportunities for shared practices to bring about learning among their members (Bremberg et al. 2019; Græger 2018; Sondarjee 2021). On transnational climate governance networks in ASEAN, findings show that networks in which state and non-state actors participate function as communities of practice that facilitate learning through the production, exchange, and use of knowledge necessary in developing regional governance processes (Salazar 2023).

One way to demonstrate how learning occurs in communities of practice is to highlight that learning entails acquiring new knowledge along with challenging existing knowledge, not just with members of one community but also with members of other communities through practices that lead to the co-creation of meaning and shared understanding of ways to address important issues (Sondarjee 2021). The collective learning process is explained using the concepts of boundary encounters, where members of different communities converge and interact to share knowledge in the form of practices toward consensus formation; brokerage, defined as a set of practices that are mutually constituting the agents and structure of knowledge thereby enabling members to translate, negotiate, and contest knowledge with other communities; and epistemic boundary objects, or informative documents produced by members of a community to facilitate the exchange of knowledge with other communities (Sondarjee 2021).

Furthermore, this study defines collective learning both as a process and an outcome. Learning as a process is defined as participation in communities of practice wherein members learn by engaging in joint activities or routinized practices, enabling them to negotiate meanings, form identities, and construct social reality (Adler 2008; Lave and Wenger 1991). Consistent with Lave and Wenger's (1991) situated learning concept, learning, as understood in this chapter, is deeply embedded in the social and cultural context in which it takes place. This means that learning in communities of practice is shaped by the shared norms, values, and knowledge among members. Learning not only involves accepting knowledge that leads to changes in beliefs or actions. It goes beyond a shared sense of knowing that brings about the internalization of knowledge by individuals or groups, for it is conceptualized as a process that entails membership in communities of practice wherein participation in collective pursuits constitutes cognition, agency, structure, and their interrelations (Lave 1991). Furthermore, learning is examined as an outcome of interactions within and among communities of practice that are taking part in ASEAN climate change governance. Learning outcomes included in this study are similar to Heikkila and Gerlak's (2013) notion of changes in collective behaviors in the form of programs, rules, institutional arrangements, and policies, among others.

2.3 Social Forestry and Climate Change in ASEAN

Governments in Southeast Asia have acknowledged the importance of social forestry programs since the 1970s in response to the depletion of forest resources due to unsustainable levels of timber extraction. Social forestry in ASEAN involves forest management policies or programs that engage communities living in and around forests by empowering them through awareness raising, capacity building, involvement in policy development, and recognition of their rights and knowledge systems, along with providing them with benefits and access to forest resources for them to participate in sustainable forest management (Ramirez et al. 2019). A respondent explains that social forestry is about addressing two goals: sustainable forest management and sustainable livelihood development. "We have to understand what the local agricultural trends are about and understand the human dimensions of the local livelihood development," Respondent G says. Thus, social forestry programs in ASEAN include both formal government policies and strategies and informal approaches involving communities, which are often comprised of indigenous and other marginalized groups, in the sustainable management and use of forests and forest resources situated in state-owned or public lands.

Social forestry contributes to community-based initiatives on addressing reforestation, reducing deforestation and forest degradation, managing the spread of forest fires, measuring and monitoring of forest carbon stocks, along with increasing food security and income generation (RECOFTC and AWG-SF 2017). There are many modalities of social forestry, and it is referred to using different terms such as community forestry in Cambodia, village forestry in Laos, and community-based forest management in the Philippines. At its core, social forestry is promoted to "enhance the power of local people because they are the ones who maintain the forests for either products or services," Respondent G notes. Over the years, Cambodia, Indonesia, Myanmar, the Philippines, Thailand, and Vietnam have established national goals to allocate land to local people and implement social forestry policies. If these goals are fully achieved, it would result in about 20 million hectares of forest land managed under social forestry programs by 2030 (RECOFTC and AWG-SF 2017). Presently, eight ASEAN member states have official social forestry or community-based forest management programs.

For the period 2016–2025, ASEAN member states formulated the Strategic Plan of Action for the ASEAN Cooperation on Forestry indicating action programs to enhance forest management involving forest-dependent communities by designing and implementing strategies that contribute toward improving the well-being of local people, sustainable management of forests, climate change adaptation and mitigation, rural development, and poverty alleviation. As the body responsible for coordinating regional actions on social forestry, the ASEAN Working Group on Social Forestry (AWG-SF) oversees and facilitates the conduct of activities under the Plan of Action. It is evident from the Plan of Action that ASEAN aims to uphold the rights of local communities to access and use forest resources following social forestry principles. For one, the Plan of Action enjoins member states to review customary and statutory tenure arrangements to ensure that the rights of indigenous peoples and local communities are recognized and protected. Likewise, the adoption of sustainable management of non-timber forest products (e.g., medicinal plants, bamboo, and rattan) is promoted through documentation of sustainable practices of forest communities and cross-country learning among member states through workshops, seminars, and webinars, among others. In terms of strengthening research and development, the AWG-SF is tasked to support the networking of forest research institutions and exchanges of experts.

On climate change, the contributions of social forestry or community-based forest management to mitigation and adaptation efforts have been recognized. Respondent A claims that "we do not only talk about carbon sequestration and land cover in the forests, but also how to adapt since local communities are one of the vulnerable groups due to climate change." However, Respondent G observes that the Strategic Plan of Action "is quite heavy on mitigation and less on adaptation." Similarly, the nationally determined contributions of member states under the 2015 Paris Agreement on Climate Change are more focused on mitigation than adaptation. Since mitigation is often measured through national targets, member state governments tend to harness the mitigation potential of forests. Through this lens, forests are narrowly viewed as either emitting or storing greenhouse gases. Deforestation and land use change are often identified as sources of emissions, while forests are considered a carbon sink since they absorb more carbon from the atmosphere than they release.

Despite the purported gains of social forestry, program implementation across the region is constrained by administrative issues, high costs, complicated rules, and weak capacity at the local level coupled with other factors such as the low benefits communities get from forest lands allocated to them and certain restrictions on the sale of timber that are disincentivizing local people from supporting social forestry initiatives (Broadhead and Izquierdo 2010; Gritten et al. 2015; RECOFTC and AWG-SF 2017). Except for Vietnam, member states with social forestry programs are not meeting their targets, while some cannot even measure their progress due to a lack of data (Ramirez et al. 2019). Others require policy or legal frameworks to fully implement social forestry programs. Similarly, case studies done in Indonesia, Vietnam, and Sabah in Malaysia reveal that states make efforts to include non-state actors in program implementation by forging partnerships with civil society groups and international donors; however, social forestry programs continue to favor states' interests and market-based discourses that limit the agency of local communities and hinder sustainable management of forests (Wong et al. 2020).

In ASEAN, linking social forestry and climate change entails introducing a climate-friendly technical approach and opening the governance space for wider participation. Respondent J affirms that in addressing climate change, "social forestry, community forestry, agroforestry provided a much more participatory, much more inclusive, much more livelihood-responsive approach as opposed to a command and control, top-down, bureaucratic, regulatory approach to forestry, which was the prevailing norm in many member states." Social forestry also underscores the importance of adaptation given its attention to local livelihood development, but promoting adaptation in the forestry sector is challenging due to weak institutional or legal frameworks in member states, Respondent J asserts.

The ASEAN-Swiss Partnership on Social Forestry and Climate Change (ASFCC) is a partnership between the 10 ASEAN member states and the Swiss government, through funding assistance from the Swiss Agency for Development and Cooperation (SDC). The program ran from 2010 to 2020. It aimed to address the issues of food security, poverty, and climate change affecting the region. As the flagship program of the ASEAN Social Forestry Network (ASFN), the ASFCC developed social forestry approaches to integrate those into the climate change adaptation and mitigation strategies of member states. Prior to the establishment of the ASFCC, the ASFN focal persons underscored the need to effectively address climate change through sustainable forest management, multistakeholder collaboration, sharing of local or indigenous practices, and community carbon accounting (Durst and Soriaga 2020). "The Swiss [government] reached out to the ASEAN Social Forestry Network and conceived of a big project to support social forestry development in ASEAN countries," Respondent J explains. Likewise, Respondent G claims that the SDC fund assisted ASEAN in making a "leap forward" by "allowing for more learning structure" through partnerships with non-state actors deviating from the usual formal process in ASEAN.

Under the ASFCC, national focal points of ASEAN member states engaged with implementing partners such as the Center for International Forestry Research (CIFOR), Non-Timber Forest Products-Exchange Programme (NTFP-EP),

International Council for Research in Agroforestry (ICRAF) or the World Agroforestry Centre, Southeast Asian Regional Center for Graduate Study and Research in Agriculture (SEARCA), and The Center for People and Forests (RECOFTC). Working with partners enabled the program to reach selected forest-dependent communities and indigenous groups and bring together government representatives and stakeholders such as civil society groups, experts, and the private sector. The ASFCC also supported the plans and activities of the AWG-SF, which is tasked to facilitate regional-level coordination and knowledge sharing on social forestry.

2.4 Adoption of Agroforestry Guidelines as Collective Learning

2.4.1 Background and Development of the Agroforestry Guidelines

To further illustrate the process and outcomes of collective learning, this section discusses the efforts of ASFCC members in relation to the formulation and adoption of the agroforestry guidelines. One of the action programs of the Vision and Strategic Plan of the ASEAN Food, Agriculture, and Forestry (2016–2025) declares that ASEAN should enhance agroforestry systems given their ecological and economic benefits in improving the region's resilience to climate change, disasters due to natural hazards, and other shocks. ASEAN defines agroforestry as the interaction of agriculture and trees, including trees on farms, farming in forests and at forest margins, and tree-crop production, which embraces an approach that emphasizes the multi-functionality and management of complex systems (Catacutan et al. 2018).

Agroforestry, as traditionally practiced by farmers in the region, can help in managing economic and environmental risks due to climate change by diversifying the income sources of rural farmers to lessen their reliance on monoculture (Catacutan et al. 2019). In recent years, agroforestry has been recognized as a means to achieve climate mitigation and adaptation goals through carbon sequestration as well as the UN Sustainable Development Goals given its contributions in eradicating hunger, reducing poverty, supporting gender equity and social inclusion, providing affordable and cleaner energy, protecting life on land, reversing land degradation, and combating climate change.

At the 39th meeting of the ASEAN Ministers of Agriculture and Forestry (AMAF) in September 2017, ministers adopted the recommendation to develop the ASEAN guidelines on agroforestry to promote the role of agroforestry in achieving economic, environmental, and social outcomes. Respondent J explains that the guidelines "provide assistance to member states to think about how these guidelines can be implemented on a larger scale and how these can be incorporated in policies." While the implementation of the guidelines is voluntary, having one demonstrates the commitment of member states and ASEAN itself to strengthen regional cooperation on agroforestry.

According to Respondent L, the AWG-SF prioritized the development of the agroforestry guidelines because of two things: one, ASEAN's vision and strategic plans identified the promotion of resilient agroforestry systems in the region as a means to address climate change; and two, focal points acknowledged the "sheer lack of capacity" of national governments to support agroforestry programs. The AWG-SF focal points are also aware that individuals involved in social forestry often lack the necessary capacity and technical guidance to establish agroforestry systems that address the concerns of forest-dependent communities. "In the discussions, ASEAN recognized that agroforestry is a technology or an approach that will respond to climate change through environmental protection and provide environmental benefits to people," Respondent L adds. Hence, member states agreed that regional guidelines would help countries to develop their own agroforestry programs. This decision was also backed by the observation that the development and promotion of agroforestry efforts in the region have been slow compared to similar interventions. Thus, the AWG-SF prioritized the drafting of the guidelines and sought assistance from the ASFCC technical partners.

2.4.2 Epistemic Boundary Objects

Prior to the development of the agroforestry guidelines, ICRAF, with the help of other technical partners, published knowledge products such as policy briefs and a white paper on agroforestry to promote a better understanding of the issue. Respondent L explains that their credible knowledge products have contributed to efforts in socializing the importance of having agroforestry guidelines among member states. "There was a period of getting the science to them to make them understand," Respondent L adds. By the time the ASFCC embarked on developing the guidelines, member states had already indicated some level of support, if not buy-in, to the idea. This demonstrates the importance of epistemic boundary objects in the learning process of communities of practice. Through the production of boundary objects, technical partners were able to share knowledge with other stakeholders, especially member state representatives, to promote better understanding and meaning making between communities.

The ASFCC technical partners, led by ICRAF, worked on developing the guidelines. A multi-disciplinary team from ICRAF produced the first draft of the guidelines, which was reviewed by the ASFCC technical partners. At the 2017 annual AWG-SF conference in Chiang Mai, Thailand, more than 100 participants from across the Southeast Asian region attended, including policymakers, farmers' groups and cooperatives, community-based forest management groups, practitioners, technical experts and researchers, to participate in the initial consultation focused on the purpose and general contents of the guidelines, setting the ground for subsequent development (Catacutan et al. 2018). ICRAF facilitated a workshop session at the conference to solicit ideas on the outline and substance of the proposed ASEAN agroforestry guidelines. The workshop served as the first-level consultation to generate inputs on the focus and issues to be considered. "We got beautiful and helpful ideas and inputs toward developing the agroforestry guidelines," according to Respondent L. Feedback from workshop participants was incorporated in the subsequent revisions.

2.4.3 Boundary Encounters

The draft version was also presented several times at the AWG-SF meetings. The guidelines went through several rounds of deliberations before it was approved at the working group level, underscoring the role of ASFCC members in boosting engagement with ASEAN decision-makers. ICRAF then sent the draft to member states through the ASEAN Secretariat for review and comments. Each member state reviewed the document following ASEAN's standard procedure and gave comments. The lead author of the guidelines responded to the comments and questions from government representatives. "We really make sure that they (i.e., governments) are part of it by ensuring that their inputs are respected," Respondent L notes. Hence, boundary encounters in the form of interactions between decision-makers and stakeholders enabled an active and inclusive participation of communities of practice in negotiating the agroforestry guidelines.

After which, the guidelines were submitted to the ministers of agriculture and forestry for adoption. During the deliberation, Respondent L claims that the ASEAN senior officials on forestry were all "very cooperative" and "willing to provide inputs." Some minor issues arose regarding terminology, for example, but by and large, the officials supported the initiative. "There were no major conflicts on ideas really; it was just more clarificatory than anything else," Respondent L adds. A contributing factor to this was the meticulous efforts of ICRAF, along with the other ASFCC partners, to inform the ASEAN senior officials on the significance of agroforestry. In October 2018, the AMAF endorsed the guidelines during its 40th meeting.

The guidelines embodied 14 principles, covering institutional, environmental, socio-cultural, technical design, communication, and scaling, which represent a broad philosophy that member states can apply in developing national agroforestry interventions. For instance, institutional principles propose the creation of an enabling environment to facilitate the development of policies and programs that are consistent with international and regional frameworks. Institutional principles also advise ensuring the organizational capacity of the government agency or ministry tasked with handling agroforestry issues. Moreover, governments are encouraged to support participatory decision-making to allow the inclusion of all stakeholders and the integration of traditional knowledge systems in evidence-based policymaking. Along with the principles, the guidelines also incorporate implementation considerations for governments to deliberate on, such as institutional roles and arrangements, planning and financing, research and continuous learning, monitoring and evaluation, and knowledge management.

2.4.4 Brokerage Activities

The formulation of the agroforestry guidelines was also in line with capacity-building efforts of the ASFCC. "The ASEAN Working Group on Social Forestry, facilitated by the ASEAN Secretariat, agreed or recommended that there should be agroforestry guidelines at the ASEAN level to guide countries in developing, implementing, and putting investments in agroforestry," Respondent L claims. Furthermore, Respondent J explains that agroforestry is a technical solution to enhance adaptive systems necessary in achieving climate resilience and food security, among others. "There is a very important implication for the rights and livelihood of the people who are normally marginalized," such as swidden farmers, upland dwellers, and indigenous groups, according to Respondent J. These marginalized groups are often excluded from formal governance processes that affect their welfare and livelihood.

Despite the high-level endorsement from ASEAN ministers, national governments in the region are still in the process of advancing their own agroforestry policies and programs based on the ASEAN guidelines. Among the 10 member states, only Laos, Myanmar, the Philippines, and Vietnam have agroforestry-related targets in their national development programs. The guidelines are meant to steer governments, with the help of stakeholders, toward enacting policies and strategies that will provide institutional support to agroforestry programs. The working group expects member states to act on the guidelines "because within ASEAN, once you have a guideline at the regional level, there is already buy-in because it will not be approved if not all countries agree to it," Respondent L adds. Member states are "nudged" into implementing the guidelines because they are "duty-bound in a soft way" to report back to the AWG-SF about their progress, Respondent J explains.

After the adoption of the guidelines, ICRAF facilitated the preparation of the agroforestry roadmaps in Myanmar, Cambodia, and Laos as pilot countries in partnership with the United Nations Food and Agriculture Organization (UN FAO) through its technical cooperation program. The UN FAO offered funding and technical support to assist Mekong sub-region countries in implementing the guidelines. Respondent L explains that Cambodia started "piloting agroforestry projects in seven provinces to try and test the application of the guidelines, and Laos did the same." As for Myanmar, it was making progress on agroforestry, but the military coup in 2021 put on hold the developments that were underway. ICRAF also proposed a system to monitor the progress of member states in adopting the guidelines; however, the monitoring and evaluation design was not approved at the working group level, given the short period of the technical cooperation program. The AWG-SF decided to postpone the decision for 2022, but the grant program ended in 2021. "That is one thing that is needed now if the ASEAN Secretariat will follow up on the progress of the adoption of the agroforestry guidelines," Respondent L explains.

Brokerage practices demonstrate that coordination, consensus building, and capacity building are important mechanisms fostering collective learning among

ASFCC members, government representatives, and ASEAN officials that paved the way for the adoption of the guidelines at the regional level. Without ICRAF's leadership and the support from other ASFCC members, the regional guidelines on agroforestry would not have been approved by ASEAN decision-makers. However, more work is expected to successfully implement the guidelines. Translating the regional guidelines into concrete national policies and programs entails multiple levels of collective learning occurring at the regional, national, and sub-national levels. This shows that the effective implementation of regional guidelines also requires momentum at the national level and is contingent upon the domestic context, which varies from one member state to another. Thus, collective learning process and outcomes are also shaped by domestic conditions given the multi-level nature of climate change governance.

2.5 Participation in Governance Through Practices

Collective learning in the case of the adoption of the ASEAN agroforestry guidelines emphasizes the participation of non-state actors in climate change governance, which can be further analyzed using the practice-based approach (Arts et al. 2014, 2013; Behagel et al. 2019). This approach, which was originally developed to study practices related to community forest management, is applied to examine participation in social forestry and climate change governance. In doing so, participation is explained using the concepts of situated agency, logic of practice, and performativity to expound on the implications of collective learning for governance. Thus, this analysis theorizes how collective learning contributes to climate change governance in ASEAN.

The integration of a practice-based approach in governance brings forth a perspective that reconciles the false dichotomy between individualistic rationality and structuralism in the form of institutions and offers some space for the interaction of humans and nature, especially in knowledge production. These three concepts serve as guiding principles in examining governance and knowledge processes by underscoring the following: "(1) being sensitive to local and situated contexts, (2) keeping an open vision for alternative explanatory factors than the ones that previous research results and/or hypotheses would suggest, and (3) explicitly recognizing the heterogeneity of practice" (Arts et al. 2014, p. 10). This approach, which proposes that actions and meanings are intertwined, is particularly useful in understanding transnational governance of climate change, which is both a local and regional concern, in the context of ASEAN with its own institutions as a regional organization vis-à-vis the prevailing domestic conditions and governance systems in member states.

The concept of situated agency emphasizes the social dimension of agency along with its discursive and material aspects, which means that the source of agency is not solely from individual actions but also from shared practices. Hence, it decenters the focus on agency from individuals toward communities of practice. While actors, be they individuals or organizations, often follow routines, they also have the capacity to improvise or act differently, particularly during disruptive or catastrophic events that can lead to change. Likewise, this principle argues that the interests, identities, and ideas of actors are shaped by the natural environment, history, traditions, institutions, and discourses that constitute a specific context. Actions and behavior are based on certain interpretations of rules, discourses, and knowledge. This makes situated agency closely related to the idea of situated learning, which suggests that knowledge and understanding are deeply embedded in the shared practices, norms, and values of communities engaged in governance.

Situated agency can explain collective learning among those involved in the formulation and adoption of the agroforestry guidelines. For one, it assumes that engagement among members of communities is a precondition for learning to occur. Collective learning practices bring together state and non-state actors to interact by exchanging information, negotiating differing interests and knowledge claims, and co-creating knowledge among others. Through these interactions, technical partners, with the support of focal persons from member states, were able to produce boundary objects such as knowledge products, build consensus on regional guidelines through boundary encounters, and act as knowledge brokers. Their shared practices illustrate that communities of practice have the ability to produce, reproduce, and transform knowledge that can lead to collective learning. Thus, it can be said that state and non-state actors were able to exercise their agency through their participation in governance mechanisms in ASEAN.

However, their participation and the impact of their shared practices are constrained by certain realities at the regional and national levels, which point out the dynamics between practices at the micro level and the underlying conditions at the macro level that impinge on the agency of state and non-state actors. At the macro level, the norm of non-interference, along with other norms referred to collectively as the ASEAN Way, promotes a shared understanding of appropriate behavior among actors participating in ASEAN processes. ASEAN's firm commitment to these norms is often perceived as an obstacle to effective cooperation since the norms prioritize state interests and sovereignty over regional concerns. Moreover, the state-centric institutional design of ASEAN limits the engagement of non-state actors, which also impedes learning.

At the micro level, the implementation of ASEAN guidelines at the national level requires the involvement of forest-dependent communities whose everyday experiences are vastly different from the reality experienced by knowledge brokers and decision-makers. Transforming regional guidelines into national policies also requires adapting certain provisions to fit domestic considerations. According to the practice-based approach, change cannot be reduced to the decisions and actions of individuals or organizations, nor can it be solely attributed to institutional effects, because change through practices is constituted by the interactions between agency and structure. These forces at play determine the landscape of ASEAN's governance of climate change.

Logic of practice challenges the assumption that social structures like incentives, rules, norms, and procedures are the only principal drivers that steer human behavior and produce change. Logic is understood as an intrinsic characteristic of shared practices. It is often implicit and taken for granted after having been internalized by members of communities of practice, such as practical knowledge or routine behavior. However, it is not impervious to social and historical materialities where practices take place. While this logic acknowledges that shared practices are stable and enduring, thereby allowing actors to coordinate and navigate complex situations, it is also responsive to the actual experiences of those belonging to communities of practice. Furthermore, it is assumed that as practices evolve, the structures that enable such practices are also transformed due to the emergence of new rules and norms and the modification of old ones, since institutions are produced and reproduced in communities of practice. Another assumption is that actors adopt rules when doing so makes sense in their everyday lives. Hence, rules have to be in accordance with the realities on the ground to be effective. By understanding the logic of practice, one can better explain the occurrence of change, particularly those that happen contrary to the intended effects of social structures.

The logic of practice that accounts for cooperation among members of the ASFCC cannot be fully ascribed to ASEAN institutions alone, nor the strategic calculations of member states. This study demonstrates that technical partners and member state representatives had a significant role in promoting shared practices that enabled collective learning. Their efforts in encouraging participation in activities and dialogue granted access to civil society and local groups to interact with ASEAN decision-makers, albeit within the limits of ASEAN norms and structure. Moreover, collective learning became possible given the relevance of social forestry and climate change issues to member states and their societies. Learning became possible with the inclusion of representative stakeholders in boundary encounters, production of boundary objects, and facilitation of brokerage practices. While decision-making is still confined to ASEAN focal points and senior officials, non-state actors such as experts from research institutions and advocates from civil society groups were able to participate in the formulation of the agroforestry guidelines.

Performativity assumes that knowledge is performative and not just cognitive; thus, it is evident through practices, which include discursive and social activities that promote collective learning. Performativity underscores the role of knowledge in generating and regenerating interactions and interpretations in context-specific fields. Just as situated agency is contingent on a particular milieu, so is performativity. Hence, the significance of knowledge is influenced by social factors along with historical and political settings. Since knowledge is embedded in practices, it can, in turn, constitute social reality. Therefore, social change can be achieved through the active performance of knowledge by interpreting, negotiating, and contesting meanings.

The ASFCC members applied performativity through collaborative and inclusive initiatives that opened pathways for knowledge exchanges and integration of local traditions and wisdom in governance practices. As a result of collective learning, most ASEAN member states have incorporated social forestry approaches in their policies and programs. Under the social forestry framework, the welfare and livelihood of communities became central to forest governance. Forest communities have also been recognized as important actors in governance, thereby empowering them. Although the capacity of member states to transform social forestry goals into actual gains remains limited, having policies in place can be considered a first step in fulfilling their commitments to ASEAN. Performativity was also evident in the soft approaches to learning, which helped foster trust and identity building among ASFCC members. Forging identities among members enhances the prospect of establishing full-fledged communities of practice in ASEAN.

2.6 Collective Learning and Regional Governance of Climate Change

By examining collective learning through the lens of communities of practice, this chapter demonstrates that shared practices in the form of boundary encounters, knowledge brokerage, and the use of epistemic boundary objects can promote participation in regional governance efforts on climate change. This study highlights that the intersection of learning and governance can be explained through the shared practices of actors participating in ASEAN climate change governance mechanisms. Through shared practices, collective learning becomes possible as shown in ASEAN's governance processes on climate change and in the associated outcomes, such as the regional guidelines on agroforestry and national policies and programs on agroforestry.

On collective learning, the study illustrates that learning and the regional political order, under which climate change governance occurs, are co-produced through the practices of transnational actors operating in networks serving as communities of practice. Within these networks, meanings are co-created, knowledge is performed, and learning occurs through the participation of transnational actors. Therefore, the analysis and findings of the study suggest that collective learning, both as an outcome and a process, is an important aspect of climate change governance at the regional and multilateral level.

In this study, collective learning is mainly examined through participation in transnational climate governance networks where state and non-state actors interact by exercising their situated agency and exchanging knowledge through shared practices. It does not determine whether collective learning has changed the principles and causal beliefs of actors. Nevertheless, evidence points out that shared practices enabled actors to learn from engaging with others by participating in governance mechanisms on climate change. The different activities involving ASEAN focal persons from member states and regional stakeholders indicate the processes that facilitated learning. The adoption of the regional guidelines on agroforestry manifests the learning process. More importantly, the guidelines per se exemplify a learning outcome since the shared practices of boundary encounters, knowledge brokerage, and the use of epistemic boundary objects led ASEAN member states to adopt the guidelines. However, learning does not end at the regional level. The guidelines have to be transformed into policies and programs at the national level. While some member states have made headway, successful national implementation requires overcoming governance constraints.

Although this study analyzed a single case, the findings apply to other transnational governance issues in ASEAN. For one, ASEAN mechanisms follow a similar structure. Further studies can look at other regional guidelines in ASEAN to identify parallelisms and variations in how collective learning takes place. In addition, the participation of non-state actors in other ASEAN mechanisms can also be scrutinized to find evidence linking learning in communities of practice with governance outcomes. Finally, research on collective learning in ASEAN contributes to understanding the role of knowledge in governance and shared practices as building blocks of regionalism and international order.

Funding The author acknowledges the Office of the Chancellor of the University of the Philippines Diliman, through the Office of the Vice Chancellor for Research and Development, for funding support through the PhD Incentive Grant 242412 PhDIA YEAR 1.

References

Adler, E. (2004). *Communitarian international relations: The epistemic foundations of international relations* (1st ed.). Routledge.

Adler, E. (2008). The spread of security communities: Communities of practice, self-restraint, and NATO's post–Cold War transformation. *European Journal of International Relations, 14*(2), 195–230. https://doi.org/10.1177/1354066108089241.

Adler, E. (2013). Constructivism in international relations: Sources, contributions, and debates. In W. Carlsnaes, T. Risse, & B. Simmons (Eds.), *Handbook of international relations* (2nd ed., pp. 112–144). Sage.

Adler, E. (2019). *World ordering: A social theory of cognitive evolution*. Cambridge University Press.

Adler, E., & Haas, P. (1992). Conclusion: Epistemic communities, world order, and the creation of a reflective research program. *International Organization, 46*(1), 367–390. http://www.jstor.org/stable/2706960.

Adler, E., & Pouliot, V. (2011). International practices. *International Theory, 3*(1), 1–36. https://doi.org/10.1017/S175297191000031X.

Arts, B., Behagel, J., van Bommel, S., de Koning, J., & Turnhout, E. (2013). Prelude to practice: Introducing a practice based approach to forest and nature governance. In B. Arts, J. Behagel, S. van Bommel, J. de Koning, & E. Turnhout (Eds.), *Forest and nature governance: A practice based approach* (pp. 3–21). Springer.

Arts, B., Behagel, J., Turnhout, E., de Koning, J., & van Bommel, S. (2014). A practice-based approach to forest governance. *Forest Policy and Economics, 49*, 4–11. https://doi.org/10.1016/j.forpol.2014.04.001.

Ascher, W., Steelman, T. A., & Healy, R. G. (2010). *Knowledge and environmental policy: Re-imagining the boundaries of science and politics*. MIT Press.

Behagel, J. H., Arts, B., & Turnhout, E. (2019). Beyond argumentation: A practice-based approach to environmental policy. *Journal of Environmental Policy & Planning, 21*(5), 479–491. https://doi.org/10.1080/1523908X.2017.1295841.

Béland, D., & Cox, R. H. (2010). Introduction: Ideas and politics. In D. Béland & R. H. Cox (Eds.), *Ideas and politics in social science research* (pp. 3–20). Oxford University Press.

Bremberg, N., Sonnsjö, H., & Mobjörk, M. (2019). The EU and climate-related security risks: A community of practice in the making? *Journal of European Integration, 41*(5), 623–639. https://doi.org/10.1080/07036337.2018.1546301.

Broadhead, J., & Izquierdo, R. (2010). Assessment of land use, forest policy and governance in Cambodia. Food and Agriculture Organization, Regional Office for Asia and the Pacific. https://www.un-redd.org/document-library/assessment-land-use-forest-policy-and-governance-cambodia.

Catacutan, D., Finlayson, R., Gassner, A., Perdana, A., Lusiana, B., Leimona, B., Simelton, E., Öborn, I., Galudra, G., Roshetko, J., Vaast, P., Mulia, R., Lasco, R., Dewi, S., Borelli, S., & Yasmi, Y. (2018). *ASEAN guidelines for agroforestry development*. ASEAN Secretariat. https://asean-crn.org/wp-content/uploads/2019/09/2018-ASEANGuideline-agroforestry.pdf.

Catacutan, D., Finlayson, R., Perdana, A., Lusiana, B., Leimona, B., Simelton, E., Öborn, I., Galudra, G., Roshetko, J., Vaast, P., Mulia, R., Lasco, R., Dewi, S., Borelli, S., & Yasmi, Y. (2019). Policy guidelines for agroforestry development adopted by ASEAN. In M. van Noordwijk (Ed.), *Sustainable development through trees on farms: Agroforestry in its fifth decade* (pp. 337–359). World Agroforestry (ICRAF) Southeast Asia Regional Program. https://www.cifor-icraf.org/knowledge/publication/14138/.

Checkel, J. T. (2001). Why comply? Social learning and European identity change. *International Organization, 55*(3), 553–588. https://doi.org/10.1162/00208180152507551.

Dunlop, C., & Radaelli, C. (2018). Does policy learning meet the standards of an analytical framework of the policy process? *Policy Studies Journal, 46*(Suppl 1), S48–S68. https://doi.org/10.1111/psj.12250.

Durst, P., & Soriaga, R. (2020). *Final evaluation of the ASEAN-Swiss Partnership on Social Forestry and Climate Change (ASFCC)*. Swiss Agency for Development and Co-operation (SDC). https://www.aramis.admin.ch/Default?DocumentID=67101&Load=true.

Ensor, J., & Harvey, B. (2015). Social learning and climate change adaptation: Evidence for international development practice. *Wiley Interdisciplinary Reviews: Climate Change, 6*(5), 509–522. https://doi.org/10.1002/wcc.348.

Ernst, A. (2019). Research techniques and methodologies to assess social learning in participatory environmental governance. *Learning, Culture and Social Interaction, 23*, 100331. https://doi.org/10.1016/j.lcsi.2019.100331.

Evans, M. (2010). *New directions in the study of policy transfer*. Routledge.

Goldstein, J., & Keohane, R. (1993). Ideas and foreign policy: An analytical framework. In J. Goldstein & R. Keohane (Eds.), *Ideas and foreign policy: Beliefs, institutions, and political change* (pp. 3–30). Cornell University Press.

Græger, N. (2018). European security as practice: EU–NATO communities of practice in the making? In I. B. Neumann, H. Leira, & I. Sending (Eds.), *European diplomacy in practice* (pp. 88–111). Routledge.

Gritten, D., Greijmans, M., Lewis, S. R., Sokchea, T., Atkinson, J., Nguyen Quang, T., Poudyal, B., Chapagain, B., Sapkota, L. M., & Mohns, B. (2015). An uneven playing field: Regulatory barriers to communities making a living from timber from their forests—Examples from Cambodia, Nepal and Vietnam. *Forests, 6*(10), 3433–3451. https://doi.org/10.3390/f6103433.

Haas, E. (1990). *When knowledge is power: Three models of change in international organizations*. University of California Press.

Haas, P. (1992). Introduction: Epistemic communities and international policy coordination. *International Organization, 46*(1), 1–35. https://doi.org/10.1017/S0020818300001442.

Haas, P. (2000). International institutions and social learning in the management of global environmental risks. *Policy Studies Journal, 28*(3), 558–575. https://doi.org/10.1111/j.1541-0072.2000.tb02048.x.

Haas, P., & Haas, E. (1995). Learning to learn: Improving international governance. *Global Governance, 1*, 255–285. https://www.jstor.org/stable/27800115.

Hall, P. A. (1993). Policy paradigms, social learning, and the state: The case of economic policymaking in Britain. *Comparative Politics, 25*(3), 275–296. https://doi.org/10.2307/422246.

Heikkila, T., & Gerlak, A. K. (2013). Building a conceptual approach to collective learning: Lessons for public policy scholars. *Policy Studies Journal, 41*(3), 484–512. https://doi.org/10.1111/psj.12026.

Jasanoff, S. (2005). Judgment under siege: The three-body problem of expert legitimacy. In S. Maasen & P. Weingart (Eds.), *Democratization of expertise* (pp. 209–224). Springer.

Lave, J. (1991). Situating learning in communities of practice. In L. B. Resnick, J. M. Levine, & S. D. Teasley (Eds.), *Perspectives on socially shared cognition* (pp. 63–82). American Psychological Association.

Lave, J., & Wenger, E. (1991). *Situated learning: Legitimate peripheral participation*. Cambridge University Press.

Levy, J. S. (1994). Learning and foreign policy: Sweeping a conceptual minefield. *International Organization, 48*(2), 279–312. https://doi.org/10.1017/S0020818300028198.

Ramirez, M. A. M., Lecciones, A. M., & Capiña, X. G. B. (2019). *Social forestry in the ASEAN region: Gaps and strategic interventions*. Southeast Asian Regional Center for Graduate Study and Research in Agriculture. https://www.searca.org/pubs/monographs?pid=445.

RECOFTC, & AWG-SF. (2017). *Social forestry and climate change in the ASEAN region: Situational analysis 2016*. RECOFTC – The Center for People and Forests. https://www.recoftc.org/sites/default/files/publications/resources/recoftc-0000156-0001-en.pdf.

Sabatier, P. A., & Weible, C. M. (2019). The advocacy coalition framework: Innovations and clarifications. In C. M. Weible & P. A. Sabatier (Eds.), *Theories of the policy process* (4th ed., pp. 189–220). Routledge.

Salazar, V. (2023). From practices to praxis: ASEAN's transnational climate governance networks as communities of practice. *Journal of Current Southeast Asian Affairs, 42*(2), 190–215. https://doi.org/10.1177/18681034231167443.

Sondarjee, M. (2021). Collective learning at the boundaries of communities of practice: Inclusive policymaking at the World Bank. *Global Society, 35*(3), 307–326. https://doi.org/10.1080/13600826.2020.1835833.

Wenger, E. (1998). *Communities of practice: Learning, meaning, and identity*. Cambridge University Press.

Wong, G. Y., Moeliono, M., Bong, I. W., Pham, T. T., Sahide, M. A. K., Naito, D., & Brockhaus, M. (2020). Social forestry in Southeast Asia: Evolving interests, discourses and the many notions of equity. *Geoforum, 117*, 246–258. https://doi.org/10.1016/j.geoforum.2020.10.010.

Chapter 3
Mapping Global Trends in Creative Imagination Research for Quality Education: A Bibliometric Analysis

Desfa Yusmaliana, Pg Dr Hjh Nur-Ashikin binti Pg Hj Petra, Suyadi, Ahmad Zaldi, and Gamal Abdul Nasir Zakaria

3.1 Introduction

Education, recognized as a process of teaching and learning that can occur anytime and anywhere, plays a pivotal role in an individual's life. In Islam, the first revelation, iqra', meaning "read," in the Qur'an (Surah al-'Alaq, 96:1–5) serves as the highest imperative for humanity to engage in learning and continual intellectual growth. The concept of reading in this context transcends the literal act, encompassing both comprehension and deeper meaning. Consequently, efforts to enhance the quality of education remain a central focus in national development. In the context of Sustainable Development, education is identified as one of the key targets for 2030, recognized as a primary catalyst for achieving all 17 Sustainable Development Goals (SDGs) (Rieckmann 2017). The essence of the SDGs agenda lies in ensuring the long-term sustainability of human life and promoting sustainable living practices.

One of the key strategies for achieving educational targets in alignment with the Sustainable Development Goals (SDGs) is the integration of creative imagination into education. In this context, creative imagination in education can be defined as the mental ability to process and synthesize sensory experiences into new,

D. Yusmaliana (✉)
Universitas Muhammadiyah Bangka Belitung, Belitung, Indonesia

Universiti Brunei Darussalam, Gadong, Brunei
e-mail: desfa.yusmaliana@unmuhbabel.ac.id

Pg Dr Hjh Nur-Ashikin binti Pg Hj Petra · A. Zaldi · G. A. N. Zakaria
Universiti Brunei Darussalam, Gadong, Brunei

Suyadi
Universitas Ahmad Dahlan, Yogyakarta, Indonesia

A. Sharifi et al. (eds.), *Frameworks and Practices Towards a Sustainable and Peaceful Future*, World Sustainability Series,
https://doi.org/10.1007/978-981-95-2310-8_3

meaningful representations (Al-Farabi 1993; Widiawati 2019), while simultaneously linking past experiences with creativity in learning activities, thereby encouraging the creation of innovative and adaptive solutions tailored to the developmental needs of learners (Vygotsky 2004). This definition emphasizes that creative imagination is not only important for constructing new knowledge but is also essential in fostering critical thinking, problem-solving, and preparedness to face future challenges. OECD (2018) explicitly emphasizes that creative imagination plays a crucial role in realizing SDG 4 (Quality Education). Education that prepares individuals for jobs that have not yet been created, technologies that have not yet been discovered, and challenges that have yet to be anticipated requires students who foster imagination and curiosity. Additionally, it is essential that students learn to respect and value the ideas, perspectives, and values of others, while cultivating lifelong learning, critical thinking, and problem-solving skills.

To understand how imagination can play a role in education, it is essential to examine the philosophical foundations developed by thinkers such as Al-Farabi. Al-Farabi posited that knowledge begins with the senses and then becomes an intellectual conception through imagination, as everything understood by the soul contains an element of imagination (Al-Farabi 1993). According to Al-Farabi (1968) in Widiawati (2019), the acquisition of knowledge occurs through three faculties: the sensory faculty, which captures material objects; the imaginative faculty, which organizes and synthesizes stored images into new representations; and the rational faculty, which makes judgments about the truth of these representations. This philosophical perspective underscores the importance of imagination as a vital component in the process of learning, aligning with the view that fostering creative imagination in education is crucial for preparing students to navigate the complexities of the future, as outlined in the Sustainable Development Goals (SDGs).

Furthermore, imagination is not only conceptual in nature but also has a practical dimension in human activities, as explained by Wilner (1975) in Farklı et al. (2013) and Vygotsky (2004). Both researchers argue that human activity can be divided into two types: reproductive and creative. Reproductive activity involves repeating something that already exists, while creative activity involves combining elements from past experiences in new or different ways. Several studies have demonstrated how creative imagination is implemented in various educational contexts. Dias (2013), through his research on twenty-first-century learning, developed a pragmatic imagination model emphasizing the role of the creative-adaptive teacher in building students' resilience and positive self-vision (Hidayat and Syafe'i 2018; Kasmar et al. 2019; Zuhaeriah et al. 2020). Furthermore, Shank (2016) explored the integration of imaginative pedagogy in mainstream schools in Kenya, highlighting the power of imagination in the learning process. Meanwhile, Ariffin (2018) developed a learning model based on traditional Malay songs to enhance the creative imagination of preschool children. Additionally, Dandarova-Robert et al. (2023) explored children's creativity in portraying the divine across diverse religious backgrounds. These findings confirm that creative imagination can be optimized through innovative and contextual learning strategies, whether by strengthening the role of teachers, developing cognitive tools, or integrating interdisciplinary values within the curriculum.

The role of creative imagination is not limited to the artistic realm but also encompasses scientific fields and decision-making, as Oktaviani (2022) demonstrates that meaningful work is more likely to emerge when artistic and scientific aspects are combined. Moreover, the role of creative imagination is not limited to the artistic realm but also encompasses scientific fields and decision-making, as demonstrated by recent research. In the process of creating a work, whether it is writing or filmmaking, one can use active and limitless creative imagination, while integrating logical thinking to bring the work into reality. Recent studies also show that the rational and intuitive ways of thinking cooperate in forming more integrative patterns in decision-making (Wang et al. 2017).

In addition, understanding creative imagination requires a holistic approach that involves the interaction between the mind, body, and soul, as explained by Steiner (2019). The harmonious interaction between these three aspects—mind (thinking), body (feeling), and soul (will)—shapes human development. Therefore, the learning process must align with the developmental stages of the individual. For example, during preschool years, play activities serve as an effective means to foster development by linking the imagined states of children with the rules of the game (Lima 2018; Schousboe and Winther-Lindqvist 2013). Through play, children acquire significant experiences that contribute to the formation of creative imagination (Pelaprat and Cole 2011; Sombrio et al. 2014).

By examining the role and application of creative imagination in education, it is important to systematically explore the development of research in this field. Therefore, this article aims to explain and analyze the evolution of research on creative imagination in education through a bibliometric approach. Bibliometric analysis was chosen as a scientific mapping technique employing a quantitative method, as proposed by Zupic and Čater (2015). They emphasize that bibliometric studies are increasingly used to map the structure and development of various fields and disciplines. Based on the search results, the growing popularity of research related to imagination is illustrated in Fig. 3.1.

Through bibliometric analysis, trends in the development of research on creative imagination can be observed more clearly, as shown in Fig. 3.1. Over the past decade (2015–2024), the number of publications related to imagination has increased annually, with an 87% growth from 2015 to 2023. This data indicates that creative imagination is becoming an increasingly popular topic in research. Furthermore, the analysis reveals that publications on creative imagination in the context of education are not limited to a single field of study but also span across various disciplines.

Research by Karpyuk and Davydenko (2022) revealed that well-developed spatial awareness and imagination are prerequisites for successful learning in many academic disciplines. For instance, Rolling (2016) highlighted the importance of imagination in STEAM education (Science, Technology, Engineering, Arts, Mathematics) as a tool for individual learning, fostering social responsibility, and solving problems creatively. However, on the other hand, factors such as anxiety

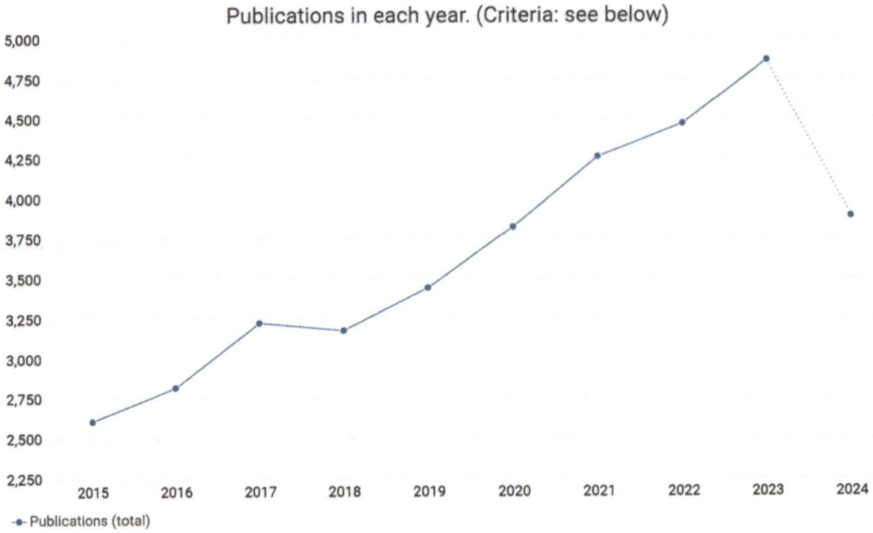

Fig. 3.1 Publications on imagination from 2015 to 2024. Source: https://app.dimensions.ai

and boredom can limit learners' imaginative abilities and hinder creativity, which is a critical skill for the twenty-first century (Cheng 2023).

Given the rapid advancements in research on creative imagination in education, conducting more comprehensive studies has become increasingly important. In the context of an ever-evolving educational landscape, understanding how creative imagination can be integrated into curricula and teaching methods is essential for developing a relevant and effective education system. With the continuous progression of technology and societal dynamics, this area of research holds significant potential to contribute to preparing future generations to tackle complex challenges across various domains. Therefore, this study highlights the importance of understanding the development of creative imagination in education. As trends and discoveries continue to evolve, it is essential to actively monitor these dynamics, address research gaps, respond to emerging themes, and deepen our understanding of creative imagination's potential. Such an approach will help ensure that educational practices remain adaptable and aligned with the demands of a rapidly changing world. This review aims to map and visualize the development of research on creative imagination in education, emphasizing its relevance to the targets of Sustainable Development Goal 4, which focuses on ensuring inclusive and quality education for all.

3.2 Methodology and Data Analysis

This study utilizes bibliometric visualization and analysis techniques. As a quantitative method, bibliometric analysis combines both evaluative and descriptive approaches to illustrate trends and characteristics within a collection of publications. In addition to examining the intellectual framework of a specific field, bibliometric analysis helps identify emerging trends in article and journal performance, collaboration patterns, and key research components (Donthu et al. 2021). For example, in the field of education, bibliometric analysis can provide valuable insights for practitioners, enabling them to expand research topics and offer more impactful recommendations (Feifei and Abdulla 2023), including providing important insights to support decision-making in the development of education policy and the management of research resources (Chaudhary and Singh 2024; McAllister and Torres 2023).

Furthermore, bibliometric analysis follows a structured process. According to Dewi et al. (2021), this methodology involves five key stages, which are illustrated in Fig. 3.2.

The sample for this study consists of 403 publications sourced from the Dimensions database, collected on September 2, 2024, focusing on publications from the last decade (2015–2024) and selected based on several criteria: (1) using "creative imagination in education" as a keyword in the title and abstract, (2) limiting the publication type to articles, (3) covering publications from 2015 to 2024, (4) focusing on open-access works, and (5) targeting the Sustainable Development Goal (SDG) category of *Quality Education* out of the 17 SDG goals.

The classification system in Dimensions emulates standard classification systems used globally, leveraging machine learning guided by expert topic input to ensure accurate classification representation. The selection of Dimensions is based on its broader coverage compared to other databases such as Web of Science (WoS) and Scopus, with the number of indexed journals being 82.22% higher than WoS and 48.17% higher than Scopus (Singh et al. 2021). Additionally, Dimensions is freely accessible, facilitating access for researchers with limited budgets (Ansorge 2023). This database also offers additional features such as tracking researcher profiles, awards, publications, and clinical trials, which can add value to comprehensive analyses of research trends in creative imagination in education (Becker et al. 2019).

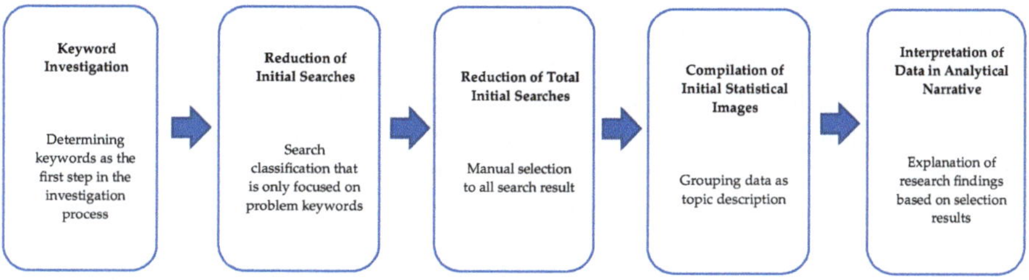

Fig. 3.2 Five stages of bibliometric analysis

The criteria for data searching from the Dimensions database also utilized open-access works, with the aim of assessing the extent of development of this topic within widely accessible open access. Although data obtained from open access cannot represent the entirety of publications developed on the research topic, in terms of data transparency and accountability, researchers can fully ensure that all analyzed data are accessible and verifiable by others, thereby enhancing the transparency and accountability of the research. Of course, this is done while strictly adhering to the Creative Commons license provisions, which regulate the use, sharing, and distribution of works in accordance with applicable rules (Mauludiah and Umam 2020). The data search in this study was conducted with consideration of the multidisciplinary nature of the topic "creative imagination in education," which encompasses fields such as education, psychology, humanities, technology, and others. This aligns with the findings of Spinaci et al. (2020), who stated that for broad and cross-disciplinary trend analysis, Dimensions is more appropriate to use compared to Crossref, WoS, or Scopus. However, for richer results, combining data from multiple databases is recommended, although this study used only Dimensions due to time and budget constraints.

The researchers also acknowledge that despite its advantages, Dimensions has limitations, particularly regarding incomplete metadata such as institutional affiliation data, which can affect the interpretation of analysis results (Guerrero-Bote et al. 2020; Robinson-Garcia et al. 2017). Therefore, this study also applied bibliometric analysis techniques involving quantitative data processing and visualization using software such as VOSviewer to systematically map research trends and relationships between publications (Gäal and Martins 2022; Gómez et al. 2025). The VOSviewer software is used to analyze, visualize, and evaluate the collected publication data, including bibliographic relationships among authors, countries, institutions, journals, and the co-occurrence of author keywords. As noted by Hewa Welege et al. (2021) and Van Eck and Waltman (2017), VOSviewer is widely used and validated through various guidelines, demonstrating its reliability and user-friendliness for bibliometric investigations. In addition to its capability to create detailed network visualizations, even for large-scale data, and to extract key terms from literature via text mining, VOSviewer also effectively analyzes the evolution of specific fields based on recurring terms (Huang et al. 2022). VOSviewer, developed as a free Java-based software, is recognized for these features.

It should be noted that all graphs and visualizations presented in this article are the results of analysis and production by the author using VOSviewer version 1.6.19, which is open-licensed software for bibliometric analysis. The data used were sourced from the Dimensions database, which provides open and licensed access for academic research purposes. The author assures that all data and visualizations used are original, have not been taken from third-party sources without permission, and comply fully with applicable copyright regulations. Every citation and data source used has been fully documented in accordance with academic standards and publication ethics.

3.3 Results and Discussion

Initially, a total of 8464 publications related to “Creative Imagination” were identified in the Dimensions database through a free-text search in the title and abstract fields. However, after narrowing the search criteria to focus on “Creative Imagination in Education,” restricting the publication period to the past ten years (2015–2024), and selecting only open-access articles, the dataset was refined to 403 publications. These results are summarized in Table 3.1:

The table above illustrates that the number of publications on creative imagination in education has shown an increasing trend over the years. During the initial period (2015–2019), the percentage of publications remained relatively stable at around 1%. However, starting in 2020, a significant surge occurred, with the percentage doubling compared to the previous year (2019). The number of publications continued to rise, reaching its peak in 2023, contributing 20% of the total or six times the amount recorded in 2015.

Overall, this trend indicates growing research interest in this topic. These findings can serve as a valuable reference for educational researchers and practitioners to further explore and deepen their understanding of the role of creativity and imagination in the educational context.

3.3.1 *Country Bibliography Pairs*

The country bibliographic pairs are visualized in Fig. 3.3. The researcher applied a threshold to identify significant countries contributing to the study. The threshold was set at a minimum of four publications per country. Out of 46 countries, 10 met this criterion. For countries exceeding the threshold, data on total link strength, citation counts, and number of publications were presented. The United Kingdom ranked first with a total link strength of 3, 40 citations, and 5 publications. Spain

Table 3.1 Number and percentage of creative imagination in education publications (2015–2024)

No.	Publication (year)	Number of publication	Percentage (%)
1.	2015	13	3
2.	2016	15	4
3.	2017	20	5
4.	2018	26	6
5.	2019	24	6
6.	2020	48	12
7.	2021	61	15
8.	2022	71	18
9.	2023	79	20
10.	2024	46	11
		403	100

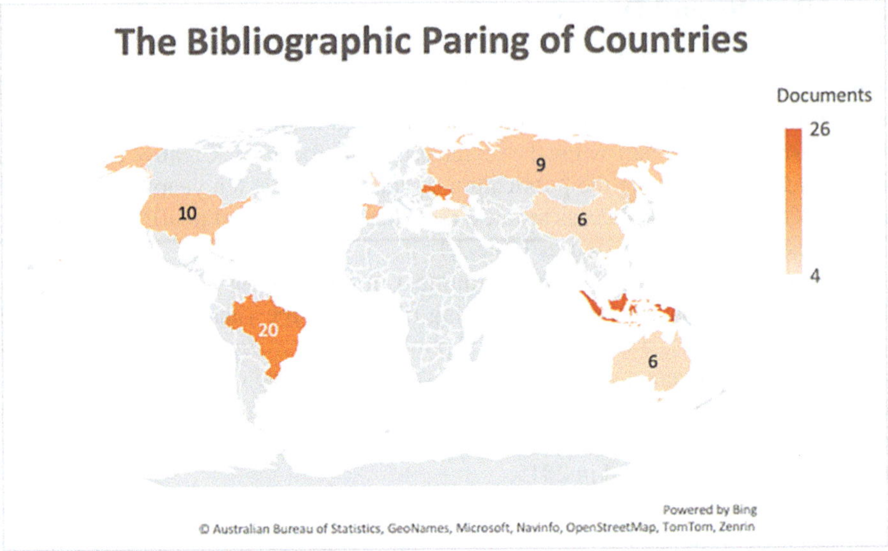

Fig. 3.3 Country bibliography pairs

came second with a total link strength of 2, 63 citations, and 11 publications. Following this was Ukraine, with a total link strength of 2, 39 citations, and 22 publications. The United States, with a total link strength of 1, achieved 162 citations and 10 publications.

The data reveal an intriguing dynamic between research quantity and impact. For instance, Ukraine's high total link strength and publication count suggest active research output and collaboration; however, its relatively lower citation count indicates that this research may have less visibility or influence within the broader academic community. This discrepancy could be attributed to factors such as language barriers, publication in less widely indexed journals, or limited international collaboration networks. Conversely, the United States, despite having a lower total link strength, commands the highest citation count, highlighting the significant impact and recognition of its research outputs. This observation suggests that citation metrics may better reflect research quality and influence than publication volume alone, consistent with bibliometric principles that emphasize both quantity and impact (Donthu et al. 2021). As Appio et al. (2014) have articulated, a fundamental technique in science mapping operates on the assumption that citations reflect intellectual linkages between publications formed when one publication cites another; thus, the impact of a publication is determined by the number of citations it receives.

Spain's position, characterized by moderate link strength but relatively high citation counts, indicates a balanced profile of both productive and influential research. Meanwhile, the United Kingdom's leading link strength but comparatively lower citation count relative to the USA suggests the presence of strong collaborative networks that may still need to translate into broader academic impact.

These findings underscore the importance of fostering not only research productivity but also international collaboration and visibility to enhance the impact of research on creative imagination in education. Countries with robust collaborative ties tend to produce research that gains wider recognition, thereby accelerating knowledge dissemination and practical application.

In terms of relationships between countries, the data visualization reveals clustering patterns, as shown in Fig. 3.4. Four major clusters were identified: the Ukraine group, the Spain group, the United Kingdom group, and the United States group.

The analysis of country bibliographic pairs identified four major country clusters in research on creative imagination in education: Ukraine, Spain, the United Kingdom, and the United States. Ukraine leads in total link strength and publication volume, reflecting strong internal collaboration and high productivity. However, as shown in Table 3.2, Ukraine's relatively lower citation count compared to the United Kingdom and even Indonesia (despite Indonesia not belonging to any cluster) indicates that factors beyond publication quantity and collaboration networks—such as research visibility, language barriers, and journal impact—play a significant role in determining international recognition.

The United States, although ranked second in total link strength, has the highest citation count, underscoring the greater impact and global acknowledgment of its research outputs. Spain's moderate link strength combined with relatively high citation counts reflects a balanced research profile in terms of productivity and influence. Meanwhile, the United Kingdom, with the highest link strength but lower citation counts compared to the United States, demonstrates strong collaborative networks that have yet to fully translate into broader academic impact.

The formation of these country clusters is likely influenced by regional proximity, shared language, and historical academic and institutional ties, as described in studies on the structure of scientific collaboration networks and the mechanisms driving the growth of international collaboration (Wagner and Leydesdorff 2005). Additionally, the relationship between research productivity and collaboration, along with personal and organizational factors affecting both, is an important aspect in understanding these dynamics (Abramo et al. 2017). Other studies highlight how the characteristics of international collaborations and intellectual property arrangements influence the quality of innovations produced (Fu et al. 2022). Nonetheless, disparities in citation impact within and across clusters indicate the need for

Fig. 3.4 Data visualization of the country clusters

Table 3.2 Number of publications on creative imagination in education

Country	Documents	Citation
Indonesia	26	47
Ukraine	22	39
Brazil	20	6
Spain	11	63
United States	10	162
Russia	9	6
Australia	6	19
China	6	18
United Kingdom	5	40
Turkey	4	11

strategic initiatives to improve research quality, accessibility, and international partnerships to strengthen countries' positions within the global research network.

Understanding these collaborative dynamics is crucial for policymakers and academic leaders seeking to foster an inclusive and impactful research ecosystem. Promoting cross-cluster collaboration and addressing barriers to international engagement can accelerate global progress in creative imagination research, which in turn contributes to achieving inclusive and quality education goals as emphasized in Sustainable Development Goal 4 (SDG 4).

It can be seen from the data found that although Indonesia has the highest number of publications related to creative imagination in education, the country has yet to join a significant international collaboration cluster. This indicates that Indonesia still faces limitations in its global academic network and citation recognition (Hamamah et al. 2023; Lakitan et al. 2012). Several studies have revealed that one of the main factors influencing this is language barriers, where the English proficiency of Indonesian academics remains a major challenge for publication in high-reputation international journals, given the dominance of English in the dissemination of scientific knowledge (Hamamah et al. 2023; Indrayadi 2023; Sugiharto 2021). Furthermore, the issue of predatory publishing is also a serious problem, with approximately 17% of articles from Indonesia published in journals suspected of predatory practices, which undermines the credibility and recognition of Indonesian research (Fahlevi et al. 2023; Kirtania 2023; Macháček and Srholec 2022).

The limited strategic international collaboration and the fact that funding predominantly comes from local institutions rather than international sponsors have also weakened Indonesia's research position on the global stage (Fahlevi et al. 2023; Lakitan et al. 2012; Pohl 2020). Therefore, there is a significant opportunity for Indonesia to strengthen international collaboration through policies that support multilingualism, language editing services, and access to online resources, as well as educational reforms emphasizing the improvement of English proficiency and continuous professional development for educators (Hamamah et al. 2023; Hartono 2021; Lauder 2020). With these strategic measures, Indonesia can enhance the quality of its research, expand its global academic network, and gain wider recognition internationally. It can thus be highlighted that Indonesia has the potential for further development and an important role in advancing education based on creative imagination. In this context, there is an opportunity to deepen research efforts, encourage cross-country collaboration, and implement best practices in education that focus on the development of creative imagination. This serves as a foundation to explore future possibilities and strengthen Indonesia's position as a leader in promoting education centered on creative imagination.

3.3.2 Institute Bibliographic Pairs

Institution bibliographic pairs are visualized in Fig. 3.5 using a Density Visualization. The yellow coloration represents the density of publications, with a more intense yellow indicating a higher concentration of publications from a specific institution. The analysis applied a threshold of at least two publications per institution, with each publication having been cited at least twice. Out of 274 institutions, only 11 met this threshold.

Based on citation counts, the top three institutions contributing to research on creative imagination in education are the University of Turku (30 citations), the University of South Australia (13 citations), and the University of Granada (12 citations). Other institutions meeting the minimum threshold of two publications include Sebelas Maret University, State University of Maringá, University of

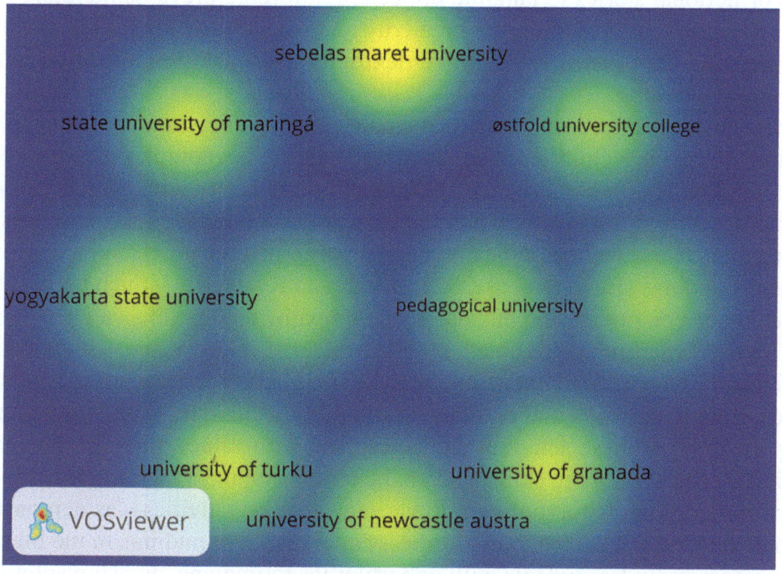

Fig. 3.5 Density visualization of publication documents by institution

Newcastle Australia, and Yogyakarta State University. Each of these institutions has contributed three publications.

These findings suggest significant potential for further research and publications related to creative imagination in education across various institutions. The implications of these results highlight opportunities for institutions to enhance their research efforts on this topic, foster cross-institutional collaborations for exchanging ideas and best practices, and promote diversified and relevant contributions to support creative imagination-based education.

The institution bibliographic pairs analysis indicates that certain institutions play a significant role in advancing research on creative imagination in education. One such influential institution is the University of Turku, which consistently contributes publications focusing on creativity in education, particularly in the context of educational psychology and child development. Renowned for its expertise in psychology and developmental sciences, research from the University of Turku often explores the psychological dimensions of creative imagination, such as its impact on children's cognitive development and social-emotional skills.

Three highly cited publications from the University of Turku exemplify the importance of creative imagination in education are Rönkkö and Aerila (2024), Heljakka (2023), and Nikkola et al. (2020). Rönkkö and Aerila (2024) offer a pedagogical innovation by integrating storytelling elements into practical crafting activities. Their research demonstrates how narrative components can stimulate students' creative expression, enabling them to design and create works freely, supported by the stories they construct themselves. This process not only encourages creative freedom but also develops students' narrative skills and multimodal thinking. This approach shows that imagination thrives when learners can connect personal experiences or fictional stories to tangible projects, reinforcing previous literature on the value of narrative as a medium for creativity and meaning-making. This model also highlights the importance of cognitive flexibility and intrinsic motivation in interdisciplinary learning.

These findings align with numerous studies highlighting the importance of story structure in helping children understand and internalize complex concepts (Egan 2004; Tambak 2016). Children tend to connect more easily with content presented in a narrative form, especially when the story draws on everyday experiences relevant to their lives, such as issues related to waste or climate change occurring in their surroundings (Burgess et al. 2022; Schatt and Ryan 2021). Such narratives not only enhance comprehension but also foster empathy and social awareness from an early age. Furthermore, as noted by Fleer (2012), Garcia and Mukhopadhyay (2019), and Shank (2016), the stimulation of imagination in education can be optimized through activities that combine storytelling techniques with simulation and role play. Each of these activities can be infused with values, thereby encouraging creativity, critical thinking skills, and character development.

The integration of storytelling into various creative activities—as demonstrated in Rönkkö and Aerila's research—further reinforces these findings in the literature. This process provides students with opportunities to imagine, envision solutions to challenges in their environment, and practice collaboration while facilitating the development of the twenty-first-century skills. Thus, storytelling serves not only as a pedagogical tool for academic learning but also as an effective means of instilling awareness, values, and social sensitivity that are relevant to the current era.

Furthermore, studies by Heljakka (2023) and Nikkola et al. (2020) both emphasize the central role of play in stimulating and nurturing children's creative imagination within educational settings. Both studies show that play-based learning environments not only offer space for the exploration of ideas, but also facilitate growth across various areas of children's development. In other words, play functions not only as a means of expression but also as a vehicle for learning that fosters courage, creativity, and collaborative problem-solving skills. Heljakka's (2023) study highlights that engaging in play is a strategy for building resilience, creativity, and innovation. Play environments are intentionally designed to support the development of children's resilience, willingness to take risks, and confidence to innovate, emphasizing active experiences and the courage to try as key factors in shaping adaptive learners. Slightly different, the study by Nikkola et al. (2020) places greater emphasis on the social interactions that occur within play and how these contribute to the development of imagination. The quality of relationships—both with peers and teachers—during play is shown to influence children's creative thinking skills. This focus on social dynamics and imaginative scenarios reveals that such interactions can collectively foster important aspects of child development, including collaboration, empathy, and communication.

By considering both of these studies, it can be concluded that learning through play is a fundamental foundation for fostering children's creative imagination. These findings are in line with various previous studies that emphasize the central role of play in stimulating children's imagination (Van Alphen 2011; Egan 1998, 2001, 2004; Judson 2016; Vygotsky 2004). Play not only provides children with opportunities to explore, but also serves as a medium to connect the real world with the world of imagination through a set of creatively designed rules and scenarios (Lima 2018; Schousboe and Winther-Lindqvist 2013; Sombrio et al. 2014).

Through active engagement in play activities, children also gain new experiences (Pelaprat and Cole 2011; Sombrio et al. 2014). As emphasized by Smolucha (in Tsai 2012), in childhood, imagination is essentially an internalization of play activities. Thus, the approach promoted by Heljakka (2023), which places play at the core of learning, not only develops creativity but also forms a strong and adaptive imaginative ability. This approach provides children with space to experiment, negotiate, and create solutions to challenges they encounter together—reflecting Vygotsky's (2004) view that children's imagination develops optimally through the integration of real-life experiences and the imaginative world realized in play activities. Therefore, the studies by Heljakka (2023) and Nikkola et al. (2020) further underscore the critical importance of providing learning environments rich in play opportunities as a key prerequisite for stimulating creative imagination and holistic personality development in children.

These three publications collectively promote a holistic approach to education, engaging students actively through activities that integrate emotional, social, and cognitive aspects. The studies emphasize experiential learning and student participation as key to enhancing engagement and creativity. Imaginative and supportive environments are crucial for facilitating creative learning, including providing spaces for experimentation, learning from mistakes, and exploring new ideas without pressure to achieve specific outcomes.

Creative imagination, as demonstrated in these contexts, not only enhances creativity but also supports the development of other skills such as collaboration, critical thinking, and resilience. These findings align with the work of some researchers (Ania et al. 2015; Brieger et al. 2020; Ofori-Attah 2021; Yusmaliana et al. 2020), who emphasize that imagination is a critical component of children's cognitive development. Through imagination, children can combine past experiences with new contexts, generating original creativity.

3.3.3 *Journal Bibliographic Pairs*

Journal bibliographic pairs are visualized in Fig. 3.6. At this stage, the researcher applied a threshold: a minimum of three publications per journal, with each cited at least three times. Out of 304 journals analyzed, only 11 met this threshold.

The journal rankings reveal that *Молодь і ринок/Youth and the Market* tops the list with 13 publications. Following this is the *Scientific Papers of Berdiansk State Pedagogical University* with 6 publications, and the *Journal of Physics: Conference Series* with 5 publications.

The results indicate that most journals have not yet made significant contributions to research on this topic. However, the presence of journals with publications

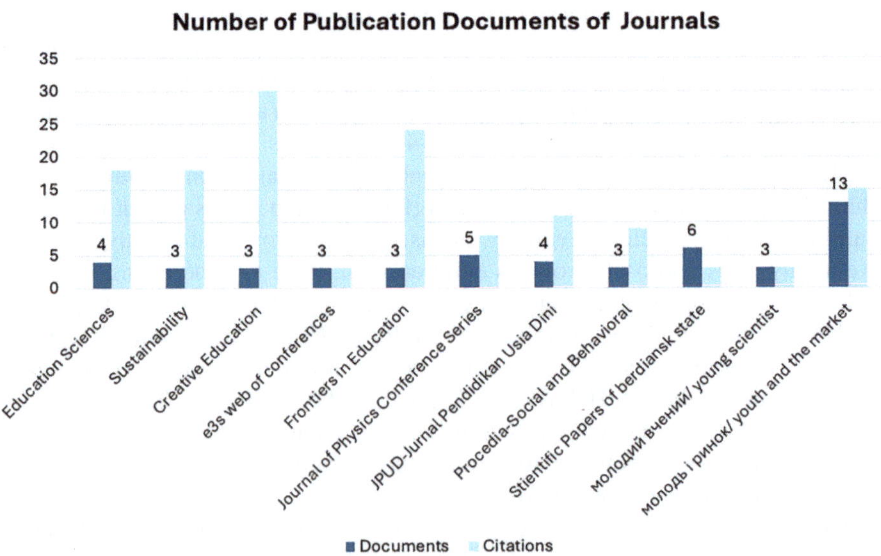

Fig. 3.6 Number of journal bibliography pairs

exceeding the threshold offers critical insights into the focus of research on creative imagination in education. Among these, *Молодь і ринок/Youth and the Market* stands out with 13 articles related to creative imagination in education. This journal primarily addresses topics such as general pedagogy, history of pedagogy, teaching theory and methods, correctional pedagogy, professional education, social pedagogy, education management, preschool pedagogy, and the integration of information and communication technologies in education and science.

Two overarching themes emerge from the articles published in *Молодь і ринок/Youth and the Market*:

1. The development of creative imagination in education
   (a) For students, fostering creative imagination involves integrating art, technology, and values for holistic learning. Stimulating imagination through art forms like music and poetry enriches students' thinking, emotions, and cultural appreciation (Dudnyk 2023; Novosiadla and Hutsol 2021).
   (b) For educators, developing creative activities for future teachers in vocational education includes fostering imagination, creative thinking, and fantasy, along with nurturing traits such as emotional intelligence, patience, and perseverance (Orshanskyi et al. 2022). In Ukraine, pedagogical heritage emphasizes imagination, intuition, and self-actualization through value-based learning (Martyniuk 2021)
2. Creative imagination as a foundation for creativity across educational levels
   (a) At the primary level, imagination can be stimulated through activities like storytelling to enhance values, vocabulary, and imagination (Perkhun and Klym 2021) and sandplay games to support spatial and motor skills (Potyuk 2022).
   (b) For special education, imagination plays a critical role in addressing challenges related to personality development (Matveieva and Bai 2024).
   (c) At the higher education level, imagination fosters adaptability, collaboration, and creativity in students (Khrapach 2022; Osadcha and Chemerys 2017; Senyk 2021; Vorozhbit-Horbatiuk 2021).

The interconnection across these educational levels underscores the critical role of creative imagination in addressing the challenges of the modern era.

In conclusion, the concept of imagination in education offers extensive opportunities for research within the learning context. Other significant contributors to this theme include the *Scientific Papers of Berdiansk State Pedagogical University, the Journal of Physics: Conference Series, and journals like Education Sciences* and *JPUD—Jurnal Pendidikan Usia Dini*, which also demonstrate notable contributions. These journals primarily focus on education, psychology, social sciences, and related fields, illustrating the ongoing significance of integrating creative imagination into education.

For journals that have yet to meet the threshold, there remains ample opportunity to expand their contributions to this field. These findings encourage reflection on the development of literature and the role of journals in supporting and advancing knowledge of creative imagination in education. Moreover, the data provides direction for researchers and educators in identifying relevant sources to enhance their academic work. While many journals still need to meet the threshold, the findings underscore the importance of collaborative efforts to expand the literature in this area and encourage more journals to engage in research on creative imagination in education.

Regarding citation records, a threshold of at least one document with a minimum of 10 citations was used. Among 304 sources, 19 journals met this threshold, as illustrated in Fig. 3.7.

The visualization in Fig. 3.7 highlights the top journals based on citation counts. *Art Education Journal* occupies the first position with 65 citations, followed by *Frontiers in Psychology* with 60 citations and *Policy Futures in Education Journal* with 52 citations.

These results indicate that the *Art Education Journal* has the most significant influence among journals in this research context, with citation numbers far exceeding the established threshold. This dominance suggests a strong focus on the development of arts education in relation to the research topic.

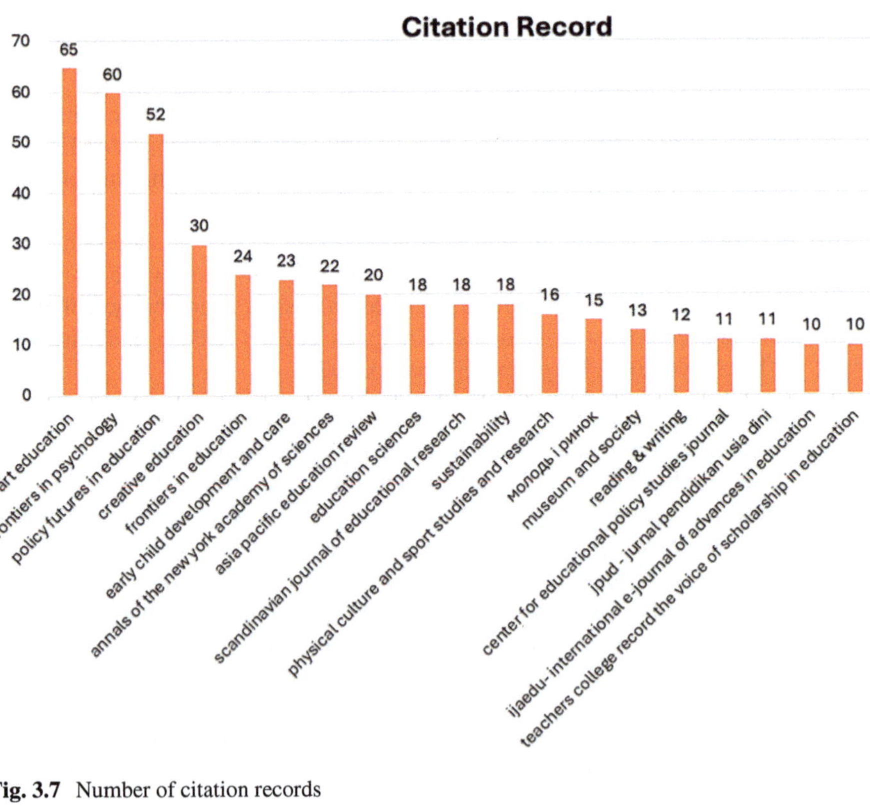

Fig. 3.7 Number of citation records

The second-ranked journal, *Frontiers in Psychology*, reflects the critical contribution of psychology to understanding creative processes and learning. This highlights the growing importance of interdisciplinary approaches, such as combining arts and psychology, in contemporary academic literature.

The *Policy Futures in Education Journal*, with its substantial citation count, underscores the relevance of educational policy issues within this research area. The high citation numbers suggest a need for more innovative policy strategies to support the integration of technology in creative learning practices.

Overall, these findings emphasize the importance of literature from various disciplines in supporting the understanding and implementation of creative imagination in education. Further analysis could identify patterns or trends within these citations, providing stronger arguments for the relevance and future directions of research in this field.

3.3.4 Bibliographic Pairs of Publications

The bibliographic pairs of publications are visualized in Fig. 3.8, presented using a network visualization. At this stage, the researcher applied a threshold: a minimum of two publications per author and at least two citations per publication. Out of 403 publications and 775 authors, only 13 publications met this threshold.

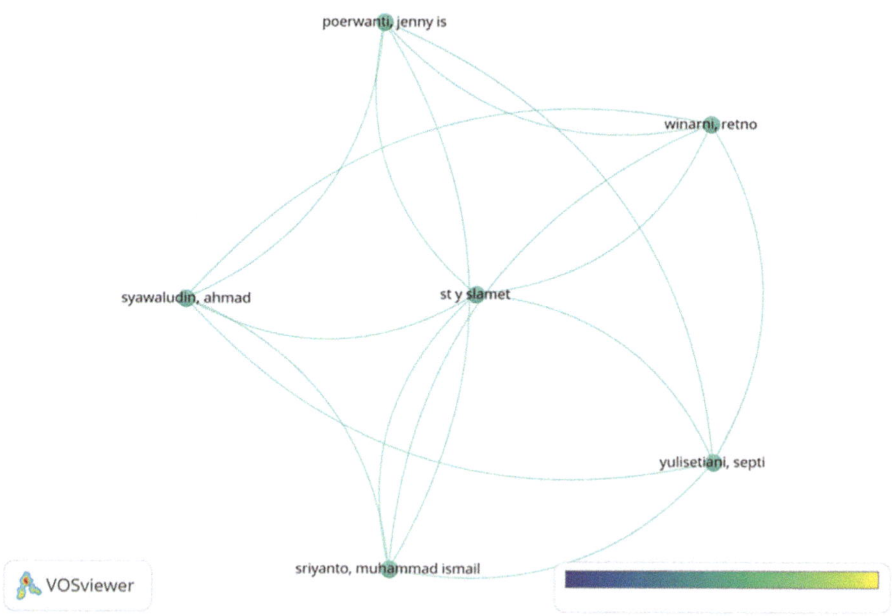

Fig. 3.8 Network visualization of publication bibliographic pairs

Under these criteria, six authors are interconnected through two publications (Winarni et al. 2021, 2022) and five citations within the bibliographic pairs. These authors are Jenny Poerwanti, Muhammad Ismail Sriyanto, Slamet, Ahmad Syawaludin, Retno Winarni, and Septi Yulisetiani.

The visualization in Fig. 3.8 indicates that many authors do not exhibit direct collaboration or follow each other in research on creative imagination in education.

To identify the most influential contributors, the researchers applied a threshold of a minimum of 10 citations per author. Out of 775 authors, 50 met this threshold. A detailed graph showing the authors, the number of citations, and the years of publication is presented in Fig. 3.9.

Figure 3.9 highlights the most influential articles related to creative imagination in education. The leading article is Rolling (2016), with 65 citations. This study emphasizes the importance of integrating educational imagination in STEAM education as a means of individual learning, fostering social responsibility, and promoting creative problem-solving. Rolling proposes an adaptable educational model that positions art and design practices as engines of innovation, facilitating more effective interdisciplinary integration within general education.

The second most cited article, Cole and Heinecke (2018), with 52 citations, explores how imagination can drive new ideas in higher education institutions. The authors discuss how, in the USA, student activities have utilized social imagination to challenge neoliberal influences in education. This work presents an alternative discourse to neoliberal ideals, such as entrepreneurial universities and competitive research institutions, by emphasizing an imagination-driven framework opposed to economic-centric thinking.

The third-ranked article, Gaunt et al. (2021), with 34 citations, focuses on the evolving role of higher music education (HME). The study proposes frameworks that balance artistic mastery, societal roles, and the cultivation of imagination amid

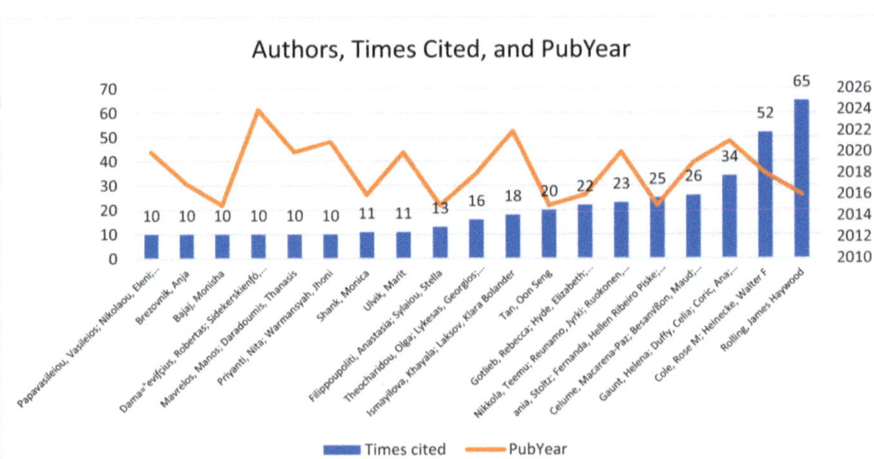

Fig. 3.9 Number of authors' publications and citations

global changes. Imagination, combined with values like artistic excellence, creativity, and social responsibility, is highlighted as essential for nurturing musicians who can deeply engage with both heritage and innovation. The authors introduce the concept of the "musician as a maker in society," emphasizing that artistic vision, societal engagement, and imaginative development collectively prepare musicians and institutions to contribute meaningfully to their communities.

The bibliographic and author collaboration analysis of research on creative imagination in education reveals critical patterns. These patterns indicate that integrating imagination and creativity into learning plays a vital role in fostering social responsibility and problem-solving skills. This approach supports the creation of a more holistic and adaptive education system, responsive to social changes and technological advancements (Kay et al. 2023; Przybylla and Romeike 2014; Rolling 2016).

While author collaborations play a significant role in advancing this research, focusing on high-impact works also provides valuable reflections for future research directions on this topic. Examining other highly cited studies further reinforces the importance of this emerging field in education.

3.3.5 Co-occurrence of Keywords from the Author

Figure 3.10 shows the network visualization of the co-occurrence of authors' keywords. From 13,125 keywords, a threshold of 50 occurrences per keyword was applied, resulting in 43 keywords being identified. A relevance score was calculated for each keyword, and the top 60% most relevant terms were selected, narrowing the list to 18 keywords. Among these, "imagination" ranks first in relevance (3.57) with 339 occurrences. Other prominent keywords include "education," "creative imagination," "students," "development," "study," "child," "learning," "ability," "activity," "creativity," "teacher," "training," "knowledge," "technology," "creative thinking," "educational process," and "art."

The visualization highlights two clusters, represented by distinct colors, which indicate groups of keywords frequently associated with one another. The largest cluster, marked in red, includes keywords such as "imagination," "education," "creative imagination," "student," "development," "study," "child," "creativity," "teacher," and "art."

The network visualization of authors' keywords provides valuable insights into trends and patterns in research on creative imagination in education. The dominance of keywords like "imagination" and "education" reflects the central focus and increasing awareness of the critical role of creative imagination in the educational context. The identified keyword clusters also reveal dimensions related to the development of creativity, such as "creative activity," "creative thinking," and "creative writing." These highlight a focus on fostering creativity in various forms—activities, thought processes, and written expression.

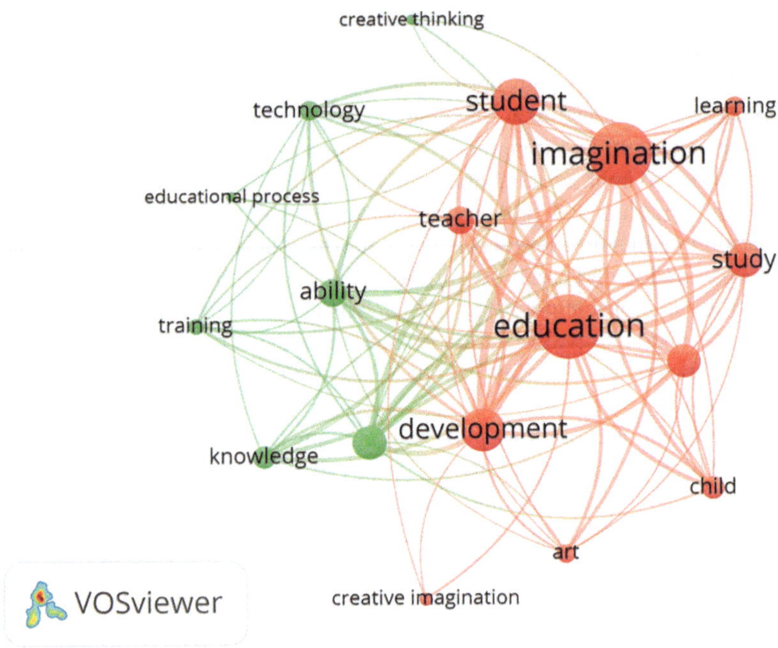

Fig. 3.10 Network visualization of co-occurrence of authors' keywords

The bibliographic analysis shows a strong thematic connection between imagination and education. Keywords like "imagination," "education," and "development" are prominent due to their frequent co-occurrence in the literature. The red cluster emphasizes the relationship between imagination, students, and learning, pointing to research focusing on how imagination impacts student development and overall education. This relationship aligns with the findings of research by Yusmaliana et al. (2020, 2022) and Yusmaliana and Suyadi. (2019), which emphasize the importance of creative imagination as a transformative tool in education. Their studies show that nurturing students' imagination not only enhances their cognitive and creative capacities but also equips them with vital problem-solving skills necessary to tackle real-world challenges.

Moreover, imagination is increasingly integrated with environmental education, reflecting its crucial role in addressing the global environmental crisis. Yusmaliana et al. (2023, 2024) highlight how imaginative approaches in education can foster a deeper understanding of sustainability and environmental management. By linking creative imagination with environmental education, this research underscores the importance of cultivating future generations who are not only creative thinkers but also responsible global citizens. This integration aligns with the broader educational goal of improving education quality by embedding global issues into the curriculum. It supports the notion that imagination is not merely a cognitive skill but a vehicle for addressing complex social and environmental challenges, thereby reaffirming its vital role in shaping a holistic and meaningful education system.

In contrast, the green cluster represents connections between technology, the educational process, and capabilities, illustrating the role of technology in supporting innovation in learning. Research suggests that digital technologies in education enrich learning experiences and offer students opportunities to explore creative ideas through virtual environments (Al-Shihri 2019; Jailani and Suyadi 2023; Mardati et al. 2024; Oleksiivna 2020; Vasenko 2021; Yarudin et al. 2022; Yemchuk 2022). These findings underscore the integration of imagination into education as central to creating responsive systems aligned with social and technological changes.

Keywords with smaller occurrences such as "art," "creative thinking," "educational process," and "creative imagination," although less frequent remain relevant due to their connections to dominant themes. These smaller keywords may indicate niche trends or emerging topics within the field. The color-coded clusters offer insights into the thematic structure where keywords with the same color share strong connections and form distinct sub-themes such as technological development or creative learning. Researchers may explore literature referencing these smaller keywords to understand their specific roles within broader research contexts. For instance connecting "creative imagination" with sustainability highlights the potential for creative imagination to contribute to education that supports sustainability and global responsiveness. Despite limited exploration these intersections have the potential to improve students' understanding of global sustainability-related challenges

The results of this bibliographic analysis have substantively shown that the trends and patterns of research on creative imagination in education are highly relevant and directly contribute to the achievement of SDG 4 (Quality Education). The emergence of dominant keywords such as "critical thinking," "creative learning," "educational process," and "technology" within the main research clusters indicates that the development of creative imagination is an essential part of the twenty-first-century learning, emphasizing critical thinking skills, technological adaptation, and innovation in the educational process. The mapping of co-occurring keywords clarifies research focus areas within creative imagination in education. As noted by Liu et al. (2022) creativity in education must adapt to contemporary developments and prepare students to face dynamic challenges, particularly in the context of Education for Sustainable Development (ESD). This requires fostering critical thinking stimulated by creative imagination. The increase in the number of publications and citations identified in this analysis can be interpreted as a measurable indicator of growing global attention to the importance of education that not only transfers knowledge but also nurtures students' imaginative and creative abilities. Thus, the bibliographic approach used in this study not only maps research trends descriptively but also provide quantitative evidence that research on creative imagination in education genuinely supports the enhancement of education quality, the development of relevant curricula, and students' readiness to face future challenges, all of which are central to SDG 4.

However, it should also be noted that although international literature widely highlights the importance of creative imagination in education, its implementation in various contexts still faces considerable challenges. Vygotsky (2004) emphasizes that imagination must be rooted in real experiences to retain its social relevance. For example, in Indonesia, the education system remains heavily exam- and memorization-oriented, which often limits the space for developing students' creative imagination (Cotrunnada et al. 2019; Hidayat et al. 2023). Furthermore, many teachers lack adequate training to apply creative teaching methods, compounded by heavy workloads and high demands, which also contribute to the low creativity and innovation indices in Indonesia (Kurniawati et al. 2022; Nelmira et al. 2022). Therefore, although creative imagination is believed to enhance innovation and readiness to face 21st-century challenges, its benefits can only be realized if supported by systemic changes in curriculum, teacher training, and school culture.

3.4 Conclusions

Overall, this study provides a comprehensive overview of the development of creative imagination within the context of education. By employing a meticulous bibliographic analysis, it reveals that the topic of Creative Imagination in Education encompasses a broad and deep scope. The number of publications on creative imagination in education has shown a significant upward trend since 2020, peaking in 2023 and continuing to grow in 2024, reflecting increasing academic interest in this topic.

Indonesia leads in terms of publication output, while the United States records the highest number of citations, demonstrating the global influence of its research. Country clustering reveals close collaboration among Ukraine, Spain, the United States, and the United Kingdom in advancing research on this topic. Institutions such as the University of Turku and leading journals like Art Education Journal and Frontiers in Psychology play pivotal roles in disseminating literature on creative imagination.

The primary themes of the research focus on integrating imagination into education through art, technology, and holistic approaches to foster creativity and critical thinking. Nonetheless, there are significant opportunities for further exploration in other relevant contexts, such as sustainability and cross-cultural education, which could broaden the contribution of this literature in the future. Similarly, findings related to less frequently mentioned keywords such as art, educational process, and technology suggest potential for deeper exploration in these areas.

3.5 Recommendations

The findings highlight the growing significance of creative imagination in education, offering numerous opportunities for advancing research and practice. Future studies should expand on emerging themes such as art, educational processes, and technology, which hold potential for fostering holistic and innovative learning environments. Moreover, integrating creative imagination with sustainability presents an untapped opportunity to prepare students for addressing global challenges through critical thinking and innovative solutions. Cross-cultural and interdisciplinary collaborations should be encouraged, building on the strong connections identified between countries like Ukraine, Spain, the United States, and the United Kingdom, while also involving nations like Indonesia, which leads in publication output.

To harness the transformative potential of creative imagination, educational institutions must leverage digital tools and virtual environments to enhance student engagement and creativity. Teachers play a pivotal role in this process; therefore, professional development programs should focus on equipping educators with strategies to integrate imaginative practices, such as storytelling and creative problem-solving, into their teaching. Additionally, there is a need to address underexplored areas, including special education and vocational training, to ensure the benefits of creative imagination extend across diverse educational contexts. By pursuing these directions, researchers, educators, and policymakers can shape an education system that is both adaptive to technological advancements and responsive to the dynamic needs of society.

3.6 Limitation

The bibliometric analysis used in this study has inherent limitations that need to be considered. This method only measures publication patterns, the number of documents, and citations related to the topic of creative imagination in education, without being able to capture the actual effectiveness or impact of applying creative imagination in learning contexts. Thus, the results of this analysis reflect academic popularity and research trends rather than the quality or success of the concept's implementation in educational practice.

Furthermore, bibliometrics cannot directly assess how creative imagination influences learning outcomes, skill development, or changes in student behavior. Therefore, the findings of this study should be regarded as an initial overview mapping the development of research rather than empirical evidence of the effectiveness of creative imagination-based education. To gain a more comprehensive understanding, further studies employing both qualitative and quantitative approaches are necessary to examine the real impact of creative imagination in the learning process.

Another limitation relates to the data coverage, which only includes open-access publications from the Dimensions database, potentially not representing the entirety of global research in this field. This may also affect the interpretation of the trends and publication patterns identified. Hence, the results of this study should be supplemented with broader and more in-depth reviews to provide a more accurate picture of the contribution and effectiveness of creative imagination in education.

References

Abramo, G., D'Angelo, A. C., & Murgia, G. (2017). The relationship among research productivity, research collaboration, and their determinants. *Journal of Informetrics*, *11*(4), 1016–1030. https://doi.org/10.1016/j.joi.2017.09.007

Ammar al-Talbi Al-Farabi (1993). *Prospects: The Quarterly Review of Comparative Education*, *XXIII*(1/2), 353–372. https://doi.org/10.1017/UPO9781844654642.005

Van Alphen, P. (2011). Imagination as a transformative tool in primary school education. *RoSE - Research on Steiner Education*, *2*(2), 16–34.

Al-Shihri, A. R. S. (2019). The effects of electronic games on the behaviors of students in elementary stages، their social states and their academic achievements. *Journal of Educational and Psychological Sciences*, *3*(13). https://doi.org/10.26389/ajsrp.a021218

Ania, S., Fernanda, H. R. P., Maria, F. Q. F., Marlene, S. D., & Járci, M. M. (2015). Creativity in Gifted Education: Contributions from Vygotsky and Piaget. *Creative Education*, *06*(01), 64–70. https://doi.org/10.4236/ce.2015.61005

Ansorge, L. (2023). Citation coverage by Dimensions and Scopus of articles published in European Science Editing. *European Science Editing*, *49*. https://doi.org/10.3897/ese.2023.e102691

Appio, F. P., Cesaroni, F., & Di Minin, A. (2014). Visualizing the structure and bridges of the intellectual property management and strategy literature: a document co-citation analysis. *Scientometrics*, *101*(1), 623–661. https://doi.org/10.1007/s11192-014-1329-0

Ariffin, A. (2018). *Reka Bentuk Model Aktiviti Penggunaan Lagu Kanak-Kanak Melayu Tradisional untuk Peningkatan Imaginasi-Kreatif Prasekolah*. Universiti Malaya Kuala Lumpur.

Becker, H., O'flinn, R., Scott, R. A., & Carlson, A. J. (2019). The new dimensions in scholcomm: How a global scholarly community collaboration created the world's largest linked research knowledge system. *Serials Librarian*, *76*(1–4), 72–78. https://doi.org/10.1080/0361526X.2019.1540259

Brieger, E., Arghode, V., & McLean, G. (2020). Connecting theory and practice: reviewing six learning theories to inform online instruction. *European Journal of Training and Development*, *44*(4–5), 321–339. https://doi.org/10.1108/EJTD-07-2019-0116

Burgess, S., Rogers, M., & Jefferies, D. (2022). Empathy and Narrative: A Discussion of Contemporary Stories from Childcare and Healthcare. *International Electronic Journal of Elementary Education*, *14*(5), 631–646. https://doi.org/10.26822/iejee.2022.268

Chaudhary, P., & Singh, R. K. (2024). Quality of teaching & learning in higher education: a bibliometric review & future research agenda. *On the Horizon*, *32*(4), 149–158. https://doi.org/10.1108/OTH-06-2023-0025

Cheng, L. (2023). Delving into the role of mindfulness on the relationship among creativity, anxiety, and boredom of young EFL learners. *Heliyon*, *9*(2), e13733. https://doi.org/10.1016/j.heliyon.2023.e13733

Cole, R. M., & Heinecke, W. F. (2018). Higher education after neoliberalism: Student activism as a guiding light. *Policy Futures in Education*, *18*(1), 90–116. https://doi.org/10.1177/1478210318767459

Cotrunnada, Z. C., Na'Im, M., & Sumardi. (2019). Comparison of creative and creative capabilities history learning results using the method problem solving and problem based learning. *IOP Conference Series: Earth and Environmental Science*, *243*(1), 0–10. https://doi.org/10.1088/1755-1315/243/1/012154

Dandarova-Robert, Z., Cocco, C., Astaneh, Z., & Brandt, P.-Y. (2023). Children's Imagination of the Divine: Creativity Across Religions. *Creativity Research Journal*, *35*(2), 254–275. https://doi.org/10.1080/10400419.2023.2177414

Dewi, P. S., Widodo, A., Rochintaniawati, D., & Prima, E. C. (2021). Web-Based Inquiry in Science Learning: Bibliometric Analysis. *Indonesian Journal of Science and Mathematics Education*, *4*(2), 191–203. https://doi.org/10.24042/ijsme.v4i2.9576

Dias, S. S. E. (2013). Imagination in Twenty-First Century Teaching and Learning Teachers as Creative-Adaptive Leaders in the Classroom. *ProQuest Dissertations and Theses*, 188. https://

search.proquest.com/docview/1433306385?accountid=10673%0Ahttp://openurl.ac.uk/redirect/athens:edu/?url_ver=Z39.88-2004&rft_val_fmt=info:ofi/fmt:kev:mtx:dissertation&genre=dissertations+%26+theses&sid=ProQ:Education+Database&atitle=&title=Imaginati

Dmytro Khrapach. (2022). Assessment Tools of The Formation of Technical and Technological Competence of Bachelors of Vocational Education in The "Transport" Major. *Молодь і Ринок*, *16*(1), 1–23.

Donthu, N., Kumar, S., Mukherjee, D., Pandey, N., & Lim, W. M. (2021). How to conduct a bibliometric analysis: An overview and guidelines. *Journal of Business Research*, *133*(April), 285–296. https://doi.org/10.1016/j.jbusres.2021.04.070

Dudnyk, Y. (2023). The Impact of Musical Art on The Emotional World of A Person and Society. *Молодь і Ринок*, 110–114.

Egan, K. (1998). *The Educated Mind: How Cognitive Tools Shape Our Understanding*. University of Chicago press.

Egan, K. (2001). The Cognitive Tools of Children's Imagination. *Annual European Conference on Quality in Early Childhood Education (11th, Alkmaar, Netherlands, August 29-September 1, 2001).*

Egan, K. (2004). Memory, Imagination, and Learning: Connected by the Story. *Phi Delta Kappan*, *70*(6), 455–459.

Fahlevi, H., Faradisa, F., & Dawood, R. (2023). Research grants, research collaboration, and publication in predatory journals: Evidence from publications by Indonesian social scientists. *Learned Publishing*, *36*(3), 439–452. https://doi.org/10.1002/leap.1561

Farklı, D., Gruplarındaki, Y., Yaratıcı, Ç., Güçleri, H., Etkisi, Ü., Gündoğan, A., Ari, M., & Gönen, M. (2013). The Effect of Drama on the Creative Imagination of Children in Different Age Groups*. *Journal of Education)*, *28*(2), 206–220.

Feifei, C., & Rohaya Abdulla. (2023). A 30-year Perspective of Bibliometric Analysis on Teacher Cognition : Toward Teacher Development. *International Journal of Instruction*, *16*(4), 209–230.

Fleer, M. (2012). Imagination, emotions and scientific thinking: What matters in the being and becoming of a teacher of elementary science? *Cultural Studies of Science Education*, *7*(1), 31–39. https://doi.org/10.1007/s11422-011-9365-z

Fu, X., Fu, X., Ghauri, P., & Hou, J. (2022). International collaboration and innovation: Evidence from a leading Chinese multinational enterprise. *Journal of World Business*, *57*(4), 101329. https://doi.org/10.1016/j.jwb.2022.101329

Gäal, L. P. M., & Martins, M. S. (2022). Open access in the context of Information Science research. *Transinformacao*, *34*. https://doi.org/10.1590/2318-0889202234e220016

Garcia, J. G., & Mukhopadhyay, T. P. (2019). The role and efficacy of creative imagination in concept formation: A study of variables for children in primary school. *Education Sciences*, *9*(3). https://doi.org/10.3390/educsci9030175

Gaunt, H., Duffy, C., Coric, A., Delgado, I. R. G., Messas, L., Pryimenko, O., & Sveidahl, H. (2021). Musicians as "Makers in Society": A Conceptual Foundation for Contemporary Professional Higher Music Education. *Frontiers in Psychology*, *12*. https://doi.org/10.3389/fpsyg.2021.713648

Gómez, M. C. S., García, J. L. C., Castro, S. V., & del Brio Alonso, I. (2025). Research Methods in the Educational Field: Bibliometric Analysis - A Comparative Study between Scopus and WoS. *Revista Espanola de Educacion Comparada*, *46*(46), 141–172. https://doi.org/10.5944/reec.46.2025.40201

Guerrero-Bote, V. P., Chinchilla-Rodríguez, Z., Mendoza, A., & de Moya-Anegón, F. (2020). Comparative Analysis of the Bibliographic Data Sources Dimensions and Scopus: An Approach at the Country and Institutional Levels. *Frontiers in Research Metrics and Analytics*, *5*(January), 1–12. https://doi.org/10.3389/frma.2020.593494

Hamamah, Emaliana, Hapsari, Y., Degeng, P. D. D., & Fadillah, A. C. (2023). Using Nominal Group Technique to Explore Publication Challenges and the Usefulness of AI-Based Writing Technologies: Insights From Indonesian Scholars. *Theory and Practice in Language Studies*, *13*(8), 2038–2047. https://doi.org/10.17507/tpls.1308.20

Hartono, H. (2021). A Study on the Communicative Competence of Indonesian Teachers at Bilingual Programs. *International Journal of Language Teaching and Education*, *4*(2), 109–116. https://doi.org/10.22437/ijolte.v4i2.12178

Heljakka, K. (2023). Building playful resilience in higher education: Learning by doing and doing by playing. *Frontiers in Education*, *8*, 1071552. https://doi.org/10.3389/feduc.2023.1071552

Hewa Welege, N. M., Pan, W., & Kumaraswamy, M. (2021). Social network analysis applications in sustainable construction and built environment management: a review. *Built Environment Project and Asset Management*, *11*(4), 511–528. https://doi.org/10.1108/BEPAM-03-2020-0047

Hidayat, D. N., Fitriah, Mahlil, & Mason, J. (2023). Factors Impacting English Teachers' Creativity in Teaching English as a Foreign Language in Indonesia. *Studies in English Language and Education*, *10*(1), 155–173. https://doi.org/10.24815/siele.v10i1.26145

Hidayat, T., & Syafe'i, M. (2018). Peran Guru dalam Mewujudkan Tujuan Pembelajaran Pendidikan Agama Islam di Sekolah. *Rayah Al-Islam*, *2*(01), 101–111. https://doi.org/10.37274/rais.v2i01.67

Huang, T., Zhong, W., Lu, C., Zhang, C., Deng, Z., Zhou, R., Zhao, Z., & Luo, X. (2022). Visualized Analysis of Global Studies on Cervical Spondylosis Surgery: A Bibliometric Study Based on Web of Science Database and VOSviewer. *Indian Journal of Orthopaedics*, *56*(6), 996–1010. https://doi.org/10.1007/s43465-021-00581-5

Indrayadi, T. (2023). Doctoral students' challenges in preparing and publishing research in reputable international journals. *Issues in Educational Research*, *33*(3), 1012–1029.

Jailani, M., & Suyadi, S. (2023). The Impact of Neuroscience-Based Modern Islamic Education on the Aspect of Creative Imagination and Development of Religious Moderation on Madura Island. *Islam Transformatif : Journal of Islamic Studies*, *6*(2), 157–172. https://doi.org/10.30983/it.v6i2.5793

Judson, G. (2016). *Two Key Points for Understanding Imagination in Education*. https://www.gettingsmart.com/2016/09/imagination-misunderstood/?utm_campaign=coschedule&utm_source=twitter&utm_medium=Getting_Smart&utm_content=Two

Karpyuk, & Davydenko. (2022). Spatial Thinking of Students when Studying Graphic Disciplines. *ВІСНИК СХІДНОУКРАЇНСЬКОГО НАЦІОНАЛЬНОГО УНІВЕРСИТЕТУ*, *2*(272), 23–27. https://medium.com/@arifwicaksanaa/pengertian-use-case-a7e576e1b6bf

Kasmar, I. F., Amnda, V., Mutathahirin, M., Maulida, A., Sari, W. W., Kaputra, S., Anwar, F., Taufan, M., & Engkizar, E. (2019). The Concepts of Mudarris, Mu'allim, Murabbi, Mursyid, Muaddib in Islamic Education. *Khalifa: Journal of Islamic Education*, *3*(2), 107. https://doi.org/10.24036/kjie.v3i2.26

Kay, L., Brandsen, S., Jacques, C., Stocco, F., & Zaffaroni, L. G. (2023). Children's Digital and Non-Digital Play Practices with Cozmo, the Toy Robot. *M/C Journal*, *26*(2). https://doi.org/10.5204/mcj.2943

Kirtania, D. K. (2023). The Top 100 Most Cited Papers in Bibliometrics: A Bibliometric Analysis. *Journal of Data Science, Informetrics, and Citation Studies*, *2*(2), 130–132. https://doi.org/10.5530/jcitation.2.2.17

Kurniawati, F., Saleh, A. Y., & Safitri, S. (2022). How to foster students' creativity? The effects of teacher subjective well-being mediation on the intellectual humility. *Cakrawala Pendidikan*, *41*(1), 31–42. https://doi.org/10.21831/cp.v41i1.40055

Lakitan, B., Hidayat, D., & Herlinda, S. (2012). Scientific productivity and the collaboration intensity of Indonesian universities and public R&D institutions: Are there dependencies on collaborative R&D with foreign institutions? *Technology in Society*, *34*(3), 227–238. https://doi.org/10.1016/j.techsoc.2012.06.001

Lauder, A. F. (2020). English in Indonesia. *The Handbook of Asian Englishes*, *2008*, 605–627. https://doi.org/10.1002/9781118791882.ch26

Lima, J. (2018). Learning Science through Art, Imagination, and Reflection. *Creative Academic Magazine*, *11A*, 33–36.

Liu, H., Sheng, J., & Zhao, L. (2022). Innovation of Teaching Tools during Robot Programming Learning to Promote Middle School Students' Critical Thinking. *Sustainability*, *14*(11), 6625. https://doi.org/10.3390/su14116625

Macháček, V., & Srholec, M. (2022). Predatory publishing in Scopus: Evidence on cross-country differences. *Quantitative Science Studies*, *3*(3), 859–887. https://doi.org/10.1162/qss_a_00213

Mardati, A., Annisa, L., Febrilia, Y., & Yusmaliana, D. (2024). The development of lectora inspires media in learning mathematics fractions material. *Alifmatika: Jurnal Pendidikan Dan Pembelajaran Matematika*, *6*(1), 24–41. https://doi.org/10.35316/alifmatika.2024.v6i1.24-41

Martyniuk, A. (2021). Formation of A Creative Personality's Worldview in The Conditions of The Educational Space of Ukraine. *Молодь і Ринок*, *31*(477), 17–20.

Matveieva, N., & Bai, I. (2024). Main Approaches to The Development of Creative Skills of Young School Students with Special Educational Needs. *Молодь і Ринок*, *15*(1), 37–48.

Mauludiah, E., & Umam, K. (2020). Potensi Doaj Sebagai Open Data Dalam Menumbuh kembangkan Open Educational Resources. *BIBLIOTIKA : Jurnal Kajian Perpustakaan Dan Informasi*, *4*(1), 1–9.

McAllister, J., & Torres, J. (2023). Discovering Open Access Trends in Engineering: A Bibliometric Analysis of Open Access Publications at a Large Research University. *Science & Technology Libraries*, *42*(1), 50–67. https://doi.org/10.1080/0194262X.2022.2032535

Nelmira, W., Efi, A., Elida, Adriani, & Sandra, Y. (2022). Efforts to Develop Creativity in Vocational Education through a Learning Model Based on Student Research Activities. *Educational Administration: Theory and Practice*, *28*(1), 01–09. https://doi.org/10.17762/kuey.v28i01.319

Nikkola, T., Reunamo, J., & Ruokonen, I. (2020). Children's creative thinking abilities and social orientations in Finnish early childhood education and care. *Early Child Development and Care*, *192*(6), 872–886. https://doi.org/10.1080/03004430.2020.1813122

Novosiadla, I., & Hutsol, T. (2021). The Practical Use of Integrative Learning Model of Schoolchildren on The Example of Studying Piano Works By Ukrainian Composers. *Молодь і Ринок*, *19*(2), 108–113.

OECD. (2018). The Future of Education and Skills: Education 2030. In *OECD Learning Framework 2030*. http://www.oecd.org/education/2030/E2030PositionPaper(05.04.2018).pdf

Ofori-Attah, K. (2021). Early childhood education and instructional ecology: A vygotskian approach in teaching early childhood social studies. *Athens Journal of Education*, *8*(2), 139–160. https://doi.org/10.30958/aje.8-2-2

Oktaviani, D. D. (2022). Kolaborasi Konsep Imajinasi Kreatif dan Intelektual dalam Adaptasi Pengembangan Media Film. *Rekam: Jurnal Fotografi, Televisi, Animasi*, *18*(2), 175–182.

Oleksiivna, P. T. (2020). A Role of Innovative Technologies in Forming of Artistic Abilities of Future Teacher of Musical Art. *Педагогічні Науки Journal*, *00534*(2), 142–146.

Orshanskyi, L., Nyshchak, I., & Pavlovskyy, Y. (2022). Pedagogical Conditions Formation of Creative Activity of Future Teachers of Labor Education in The Process of Professional Training. *Молодь і Ринок*, 12–18.

Osadcha, K., & Chemerys, H. (2017). Analysis Of The Mention Of The Concept "Graphic Competence" In The Preparation Of The Future Bachelor Of Computer Science. *Ukrainian Journal of Educational Studies and Information Technology*, *5*(3), 37–46. https://doi.org/10.32919/uesit.2017.03.04

Pelaprat, E., & Cole, M. (2011). "Minding the Gap": Imagination, Creativity and Human Cognition. *Integrative Psychological and Behavioral Science*, *45*(4), 397–418. https://doi.org/10.1007/s12124-011-9176-5

Perkhun, L., & Klym, M. (2021). Development of Preschool Children Creativity by Means of A Fairy Tale, Staged in A Game in The Environment of Preschool Educational Institutions. *Молодь і Ринок*, 100–103.

Pohl, H. (2020). Collaboration with countries with rapidly growing research: supporting proactive development of international research collaboration. *Scientometrics*, *122*(1), 287–307. https://doi.org/10.1007/s11192-019-03287-6

Potyuk, S. (2022). Sand Play as A Factor for The Formation of A Healthy Lifestyle. *Молодь і Ринок*, 166–169.

Przybylla, M., & Romeike, R. (2014). Physical computing and its scope - towards a constructionist computer science curriculum with physical computing. *Informatics in Education*, *13*(2), 241–254. https://doi.org/10.15388/infedu.2014.05

Rieckmann, M. (2017). Education for Sustainable Development Goals (SDGs): Learning Objectives. In *European Conference on Educational Research 2017*. United Nations Educational, Scientific and Cultural Organization.

Robinson-Garcia, N., Mongeon, P., Jeng, W., & Costas, R. (2017). DataCite as a novel bibliometric source: Coverage, strengths and limitations. *Journal of Informetrics*, *11*(3), 841–854. https://doi.org/10.1016/j.joi.2017.07.003

Rolling, J. H. (2016). Reinventing the STEAM Engine for Art + Design Education. *Art Education*, *69*(4), 4–7. https://doi.org/10.1080/00043125.2016.1176848

Rönkkö, M., & Aerila, J. (2024). *A Friend to Milli Mörriäinen The Storification Method Supporting Pupils ' Crafting Processes*. *31*(1), 1–13.

Schatt, D., & Ryan, P. (2021). Story Listening and Experience in Early Childhood. In *Story Listening and Experience in Early Childhood*. Springer Nature. https://doi.org/10.1007/978-3-030-65358-3

Schousboe, I., & Winther-Lindqvist, D. (2013). Children's Play and Development: Cultural-Historical Perspectives. In *Springer*. Springer Dordrecht Heidelberg New York London. https://doi.org/10.1007/978-94-007-6579-5

Senyk, A. (2021). Professional and Creative Self-Realization of Higher Education Applicants by Training Means. *Молодь і Ринок*, *4*(190), 87–91.

Shank, M. (2016). Imagination, Waldorf, and critical literacies: Possibilities for transformative education in mainstream schools. *Reading & Writing*, *7*(2), 1–9. https://doi.org/10.4102/rw.v7i2.99

Singh, V. K., Singh, P., Karmakar, M., Leta, J., & Mayr, P. (2021). The journal coverage of Web of Science, Scopus and Dimensions: A comparative analysis. *Scientometrics*, *126*(6), 5113–5142. https://doi.org/10.1007/s11192-021-03948-5

Sombrio, G. D. S., Ulbricht, V. R., & Haeming, W. K. (2014). Games and Gamification: A Proposal for a Creative Learning Process in Education. *Journal of Education and Human Development*, *3*(4), 117–129. https://doi.org/10.15640/jehd.v3n4a12

Spinaci, G., Colavizza, G., & Peroni, S. (2020). Preliminary Results on Mapping Digital Humanities Research. In *La svolta inevitabile: sfide e prospettive per l'Informatica Umanistica* (pp. 246–252). Associazione per l'Informatica Umanistica e la Cultura Digitale. https://doi.org/10.6092/unibo/amsacta/6316

Steiner, R. (2019). *Imagination : Enhancing the Powers of Thinking*. http://www.systems.wsu.edu/scripts/wsuall.pl?url=https://search.ebscohost.com/login.aspx?direct=true&db=nlebk&AN=2143333&site=ehost-live

Sugiharto, S. (2021). De-westernizing hegemonic knowledge in global academic publishing: toward a politics of locality. *Journal of Multicultural Discourses*, *16*(4), 321–333. https://doi.org/10.1080/17447143.2021.2017442

Tambak, S. (2016). Metode Bercerita dalam Pembelajaran Pendidikan Agama Islam. *Jurnal Al-Thariqah2*, *01*(01).

Tsai, K. C. (2012). Play, Imagination, and Creativity : A Brief Literature Review. *Journal of Education and Learning*, *1*(2), 15–20. https://doi.org/10.5539/jel.v1n2p15

Van Eck, N. J., & Waltman, L. (2017). Citation-based clustering of publications using CitNetExplorer and VOSviewer. *Scientometrics*, *111*(2), 1053–1070. https://doi.org/10.1007/s11192-017-2300-7

Vasenko, V. (2021). Technical Modeling in Design and Technological Activity of Schoolchildren. *East European Scientific Journal*, *3*(5(69)), 17–23. https://doi.org/10.31618/essa.2782-1994.2021.3.69.59

Vorozhbit-Horbatiuk, V. (2021). Competence of Pedagogical Partnership as A Basis of Quality of Higher Pedagogical Education. *Молодь і Ринок*, *10*(196), 63–68.

Vygotsky, L. S. (2004). Imagination and Creativity in Childhood. *Journal of Russian and East European Psychology*, *42*(1), 7–97. https://doi.org/10.2753/RPO1061-0405290173

Wagner, C. S., & Leydesdorff, L. (2005). Network structure, self-organization, and the growth of international collaboration in science. *Research Policy*, *34*(10), 1608–1618. https://doi.org/10.1016/j.respol.2005.08.002

Wang, Y., Highhouse, S., Lake, C. J., Petersen, N. L., & Rada, T. B. (2017). Meta-analytic Investigations of the Relation Between Intuition and Analysis. *Journal of Behavioral Decision Making*, *30*(1), 15–25. https://doi.org/10.1002/bdm.1903

Widiawati, N. (2019). A Study of the Epistemological Thought of al-Farabi. *Al-Afkar, Journal for Islamic Studies*, *2*(1), 48–63. https://doi.org/10.5281/zenodo.2546854

Winarni, R., Slamet, S. Y., Poerwanti, J. I. S., Sriyanto, M. I., Yulisetiani, S., & Syawaludin, A. (2022). Peningkatan Keterampilan Menulis Kreatif Bermuatan Pendidikan Budi Pekerti pada Guru-Guru Sekolah Dasar Melalui Hybrid Learning. *Widya Laksana*, *11*(1), 98–105. https://doi.org/10.23887/jwl.v11i1.37151

Winarni, R., Slamet, S. Y., Poerwanti, J. I., Sriyanto, M. I., Yulisetiani, S., & Syawaludin, A. (2021). An Analysis of Preservice Elementary Teacher Ability in Creative Writing Containing Character Education with Local Wisdom Perspective. *Jurnal Ilmiah Sekolah Dasar*, *5*(3), 498. https://doi.org/10.23887/jisd.v5i3.37069

Yarudin, Arqam, M. L. Wantini, Perawironegoro, D., & Yusmaliana, D. (2022). Development of Jurisprudence Learning Media for Android-based Islamic Economics at SMK Muhammadiyah Karangmojo. *Istawa : Jurnal Pendidikan Islam*, *7*(1), 27–44. https://doi.org/10.24269/ijpi.v6i1.3107

Yemchuk, T. (2022). Increasing The Digital Competence of The Teacher by Means of Information and Communication Technologies. *Academic Notes Series Pedagogical Science*, *1*(204), 114–119. https://doi.org/10.36550/2415-7988-2022-1-204-114-119

Yusmaliana, D., Fitriana, F., Zakaria, G. A. N., & Kurbiyanto, A. (2023). Ecological Education in Islamic Religious Learning Based on Creative Imagination. In A. Sharifi, D. Simangan, & S. Kaneko (Eds.), *Bridging Peace and Sustainability Amidst Global Transformations*. Springer, Singapore. https://doi.org/10.1007/978-981-99-7572-3_4

Yusmaliana, D., Kurbiyanto, A., Zakaria, G. A. N., & Pg Dr Hjh Nur-Ashikin binti Pg Hj Petra. (2024). Green Minds, Sacred Paths: Nurturing Environmental Affection Through Islamic Education and Creative Imagination. In J. IDEC Institute, Hiroshima University, Higashi-Hiroshima, Hiroshima & A. Sharifi (Eds.), *Sharifi, A. (eds) Environmental Sustainability and Resilience. World Sustainability Series*. Springer, Singapore. https://doi.org/10.1007/978-981-97-6639-0

Yusmaliana, D., & Suyadi. (2019). Development of Creative Imagination based on Neuroscience in Islamic Religious Learning. *Edukasia : Jurnal Penelitian Pendidikan Islam*, *14*(2), 267–296. https://doi.org/10.21043/edukasia.vi4i2.4213.

Yusmaliana, D., Suyadi, S., Widodo, H., & Suryadin, A. (2020). Creative Imagination Base on Neuroscience: A Development and Validation of Teacher's Module in Covid-19 Affected Schools. *Universal Journal of Educational Research*, *8*(11B), 5849–5858. https://doi.org/10.13189/ujer.2020.082218

Yusmaliana, D., Suyadi, Widodo, H., & Zakaria, G. A. N. (2022). Planning Framework dan Outline Model Imajinasi Kreatif Berbasis Neurosains (IKBN) Pada Pembelajaran Keagamaan Islam Tingkat Sekolah Dasar. *EDUKASI: Jurnal ...*, *20*(2), 233–251. https://doi.org/10.32729/edukasi.v20i2.764

Zuhaeriah, Z., Ali, M., & Yusra, Y. (2020). The Role of Islamic Education Teachers Competency in Improving the Quality of Education. *International Journal of Contemporary Islamic Education*, *2*(1), 108–130. https://doi.org/10.24239/ijcied.vol2.iss1.15

Zupic, I., & Čater, T. (2015). Bibliometric Methods in Management and Organization. *Organizational Research Methods*, *18*(3), 429–472. https://doi.org/10.1177/1094428114562629

Chapter 4
Artivism for Sustainability: The Pedagogical Potential of Artivism

Leni Garcia

4.1 Introduction: Education for Sustainable Development and Artivism

UNESCO promotes Education for Sustainable Development (ESD) because while Sustainable Development Goal (SDG) 4 on Quality Education is a goal on its own, it is also an essential key to achieving all the other SDGs. ESD aims to hone students in key sustainability competencies that will allow them to engage and positively contribute to global society (UNESCO 2017). ESD design, therefore, aims at translating learning into action. As such, it encourages non-traditional agentic learning.

Agentic learning affirms the learner's agency, their power to act and to take control of their learning process. Therefore, they are active participants—not passive recipients—directing themselves toward personal growth by setting their learning goals, finding learning strategies that suit their temperament, and being responsible for their advancement. They are led by their curiosity, desire to learn, and belief in their ability to achieve their goals. These characteristics support what has been called "four-life" learning (Huang 2022; Looi 2023), that is, lifelong, lifewide, lifedeep, and lifewise learning (Kwek et al. 2017).

These forms of learning, which are integrative of living and learning, are prioritized in the UN development agenda because they support the building of inclusive and sustainable societies (UNESCO n.d.). UNESCO still draws from the International Commission on Education (ICE) for the twenty-first century chair delors' report in 1996 that proposed the four pillars of education: *learning to know,*

L. Garcia (✉)
De La Salle University, Manila, Philippines
e-mail: elenita.garcia@dlsu.edu.ph

A. Sharifi et al. (eds.), *Frameworks and Practices Towards a Sustainable and Peaceful Future*, World Sustainability Series,
https://doi.org/10.1007/978-981-95-2310-8_4

learning to do, learning to be, and *learning to live together*—learnings that, in turn, can only be led by an agentic, lifelong learning.

This chapter discusses how artistic or creative activism, or simply artivism, can promote an agentic way of learning that can integrate the four pillars of learning. Primarily a work in applied aesthetics, it positions artivism within the cognitivist aesthetic framework to defend its educational character and, through its intentional element, provide the opportunity for learners to learn about the world (*learning to learn*), to express themselves creatively (*learning to be*), to encourage action for social change (*learning to do*), and to do all these communally while acknowledging and celebrating difference (*learning to live together*). As such, artivism can provide avenues for the four-life learning and help develop sustainability competencies through the three learning domains of "head," "heart," and "hands." It can contribute to the practice of Education for Sustainable Development (ESD). While studies show that activist learning helps promote agentic learning (Watson 2022; Hytten 2017; Bilon 2021; Hopwood 2024), little is written about the potential pedagogical benefits of creative activism, such as the ones engaged in by artivists.

One possible benefit of exploring the pedagogical potential of artivism for sustainability is that while it can contribute to honing sustainability skills, it may specifically address the problem of communication expressed by environmentalists when it comes to climate action advocacy. While scientific facts that explain the anthropogenic origins of the climate crisis abound, they are ineffective in prompting climate action (Stacey 2024; Garcia 2020; Marshall 2014). Climate action requires an experiential processing of information, which can trigger the emotions, create meaning, and move people into action (Marshall 2014, pp. 48–49). Radical environmentalism, often identified with violence and law-breaking to oppose practices that further the anthropogenic causes of climate change, is one such response. However, much is to be discussed about the moral questions such actions pose (Taylor 2010). Artivism, by its nature, has the potential to elicit emotional responses, create meaning, and prompt climate action without violence that creates further suffering. When developed within the educational pillars mentioned above, it can inspire peaceful self-transformation that can, in turn, positively contribute to sustainable futures.

Following is a brief description of the key sustainability competencies, the four-life learning, and the ICE's pillars of education. A case for creative engagements will be made through the cognitivist aesthetic framework to propose artivism as a potentially powerful pedagogical tool for sustainability competencies. The Copenhagen Experiment will then be described to show that creative activism can be assessed, rendering similar activities good candidates for ways of learning that promote learner agency, which is necessary in achieving sustainability skills. This chapter is, however, limited to discussing how artivism may be useful in promoting sustainability competencies. Its implementation in various curricula requires further study and is beyond its current scope.

4.2 Sustainability Competencies, Four-Life Learning, and the Pillars of Education

UNESCO's ESD program lists the learning objectives for achieving the SDGs as cross-cutting competencies (2020). These are (1) *systems thinking*, the ability to have a comprehensive view of things, analyze complex systems, and understand the interconnections among their elements; (2) *anticipatory or futures thinking*, the ability to anticipate scenarios, proactively plan for them, and envision possibilities and work to create them for a sustainable world; (3) *normative or values thinking*, the ability to see things from different perspectives, critically examine and reflect on them, maintain ethical principles, and reflect on personal and societal sustainability values; (4) *strategic competency*, the ability to develop with others and implement creative solutions or innovative actions to promote sustainability in different contexts and levels; (5) *collaboration or interpersonal competency*, the ability to connect with others, negotiate, resolve conflicts, and build relationships for sustainability through collaborative work; (6) *critical thinking*, the ability to think deeply about ideas, opinions, and events, and to question them, as well as critically reflect on one's values, ideas, and actions, and contribute to critical discourse on sustainability; (7) *self-awareness*, the ability to know oneself, learn one's role in different contexts while being in control of one's feelings and desires; and (8) *integrated problem-solving competency*, the ability to use different problem-solving frameworks in different contexts for inclusive and equitable solutions while applying the aforementioned competencies. These are identified within the *cognitive* (head), *socio-emotional* (heart), and *behavioral* (hands) domains (UNESCO 2020), which develop learner agency by empowering students to make informed decisions that will guide them in acting responsibly in their pursuit of economic, social, and environmental sustainability goals (UNESCO 2020). It is evident here that while cognitive faculties are involved in learning, a holistic development required for sustainability skills cannot be achieved without simultaneously activating the other faculties.

For agentic learning to happen within these domains, however, alternative pedagogies which are no longer confined to formal classroom settings are required. The UNESCO Institute for Lifelong Learning (UNESCO n.d.) encourages lifelong learning that integrates learning and living. *Lifelong* learning is the practice of learning, especially self-directed learning, that goes beyond educational institutions' spatial and temporal confines. It is founded on the belief that there is no end to learning as long as one has the curiosity to discover things. Learning is for everyone, young and old, can be done anytime, and continues for the rest of one's life (UNESCO n.d.). Scholars, however, have advocated for the expansion of lifelong learning to the four-life learning, which refers not only to lifelong, but lifewide, lifedeep, and lifewise learning as well.

Lifedeep learning refers to intentional disciplinal mastery that can be translated and applied to different contexts, opening it to meaningful interdisciplinary encounters and collaboration. It has also been identified with contemplative, reflective, and spiritual learning practices (Hoppers and Yekhlef 2013). Finally, *lifewise* learning is learning that focuses on values and character development, the development of wisdom that aids in establishing a work ethic and developing empathy that allows one to go beyond oneself and relate with others. Like lifedeep learning, lifewise learning is supported by spiritual and meditative learning activities (Chang and Kidman 2022; UNESCO n.d.; Kwek et al. 2017; Huang 2022; Looi 2023; Rahmian et al. 2024).

Furthermore, the UIL adopts the way the Delors Report of 1996 presents lifelong learning as resting on the four pillars of learning to learn, learning to do, learning to be, and learning to live together. *Learning to learn* is more than acquiring knowledge but acquiring tools for understanding and interpreting it. *Learning to do* means having practical expertise or knowledge that is translated into action, developing skills, and dealing with various situations, including the ability to work harmoniously with others. *Learning to be* is the commitment to personal development, self-awareness (physical, mental, and spiritual), belief in one‘s autonomy, capacity for judgment, and personal responsibility. Finally, *learning to live together* goes beyond the self, learning to connect with, understand, and effectively cooperate with others, celebrating diversity, and promoting intercultural understanding (Delors 1995).

4.3 Creativity, Learning Agency, and Artivism

Much has been written on the importance of creativity in learning. "Learning-in-creativity," where students are allowed to share their little or big discoveries and have them seriously considered by other participants in the learning activity, has been found to be effective in promoting the potential for students to direct and contribute not just to their learning but to the learning of others as well (Beghetto and Karwowski 2018). This means that creativity can contribute to the *learning to learn* (acquiring knowledge), *learning to be* (expression of discoveries), and *learning to live together* (offering diverse perspectives, sharing one's creativity in learning). It also contributes to *learning to do* insofar as creative learning allows one to apply discoveries to practical life. Creativity, therefore, is agentic (Zielinska et al. 2021). It enables the learner to intentionally transform existing knowledge into new knowledge, creating possibilities necessary to confront the world's complexities today and in the future (Zielinska et al. 2021; Lau 2011).

4.3.1 Artivism

While activist artists have been creating works since the latter half of the twentieth century, artivism has slowly coalesced into a familiar art movement only in recent times. "Artivism" is a compound of the words "art" and "activism," and is said to have been born from the association between the Chican@ artists of the US and the

Zapatistas in Mexico in 1997 (Williams 2022; Flores 2015; Komi n.d; Garcia 2023). The former were inspired by the latter's passion for upholding their identity, specifically those of the Mexican indigenous peoples. They started creating art in the streets—outside the mainstream channels of galleries and museums—as a way of peacefully invoking dialogue about the social ills that plague people in contemporary times, "in pursuit of more respectful ways of being" (Buckland 2011, Artivism section). Its primary concern is to create change through action, in order to create change in society (Hadjiosif 2021; Deal and Nicholson 2018; Williams 2022; Garcia 2023).

Given its origins in identity activism, artivism thrives on critical engagement with social issues. It calls attention to what is problematic, asks questions, and dares its spectators to question their ways of thinking. It sees "art" as presenting alternative ways of seeing and experiencing the world, often causing emotions to arise in the spectator. It is mainly responsible for *affect*. The "activism" part, on the other hand, aims at creating social change. Its success is measured in its *effect*. Artivism, the production of art with activist intentions, must therefore be both *affective* (eliciting emotions) and *effective* (creating change). Its affect must be effective, while its effect, affective, or what has come to be known as artivism's *æffect* (Duncombe 2016; Garcia 2023). Artivist works are, therefore, pieces that evoke a response that can be turned into action for change when reflected upon.

Some popular examples are Rand Jarallah's (2015) make-up artistry which she uses to call attention to the oppressive practices that affect women, Daniel Arzola's (n.d.) use of his posters to counter stereotypes and raise awareness and campaign for the LGBTQI+ community, and Banksy's murals, street graffiti, and public installations used as political commentaries on various social issues (Egnér 2025). Banksy's "I Don't Believe in Global Warming" on Regent's Canal in London in 2009, for instance, captured the climate change denial that blighted the UN Climate Summit in the same year. Frequent reference to this work, especially during the London Climate Action Week in 2023, shows its relevance to climate change discussion even in the current times (Egnér 2025).

Haring's (1989) pop-art which he created to express his anti-nuclear sentiments and to raise awareness about AIDS and promote compassion, and Barbara Kruger's collage work used to destroy hierarchical roles and prompt critical thought on various issues affecting women (The Broad 2025) are also clear examples of artivism. Their popularity and the conversations built around them point to the possibility of their having achieved *æffect*. Many of these advocacies are aligned with and helpful to one or several SDGs.

Artivism is collaborative in nature. Its goal is to get its audience to respond to their call to action, as social, economic, and environmental ills cannot be addressed by just one individual. They have to be addressed by all to create an impact. Because of this, many artists form collectives. For instance, DePART (n.d.), a transdisciplinary collective of artists and educators, partners with other organizations in different places and, through creative community work, questions dominant historical narratives and opens spaces for a more nuanced and inclusive practice of remembrance. The Detroit Narrative Agency (n.d.), on the other hand, while focusing on issues around the communities of Black, Indigenous, and People of Color (BIPOC) in Detroit, offers fellowship to young filmmakers who produce films that tell BIPOC stories. These films are publicly screened to create awareness, heal, and promote the use of art in the practice of resistance and resilience (Detroit Narrative Agency (NDA) n.d.). Posa Suto, an anti-racist, spiritual, and artistic collective, creates safe spaces for black, queer, trans, and other diverse people. Their name translates to "our house" in Palenque language, invoking the safety of spaces focused on collectivity, creativity, love, and the Afrocentric traditional spirituality and healing, to strengthen the existence of gender non-binary black people (Bertha Foundation 2025). The Javaad Alipoor Company (2017) is a theater company working for the marginalized Gypsy Roma and Traveller community in the UK. They tackle issues such as the gap between the rich and the poor, climate emergency, challenges to democracy, and the oppression of minorities. They ask questions on these issues through their internationalist art, that is, art engagements that are non-Euro-centric and post-colonial, using theater as their medium of expression.

These collectives underscore the importance of working with communities, of individuals working together for a common cause based on shared experiences, and global goals. Their art, like theater, film, and other collaborative creative forms, are used as means to bring forward their advocacy and change the way people view the world.

4.3.2 Aesthetic Frameworks and Artivism

As discussed in the foregoing, artivist works aim at *æffect*. However, æffect can only happen if there is authorial intent. It, therefore, has to be assumed that a piece of artivist work is created with a particular *intention* in mind that informs its meaning. This puts artivism on the intentionalist side of the debate in aesthetic interpretation, against the "Intentional Fallacy" which, according to aestheticists Wimsatt and Beardsley (1946), is committed when one assigns the author's intention—how the creator planned their work to be perceived—to be *the* meaning of the artwork. They insist that the meaning of the artwork must be derived from the work itself—its internal properties—and not from its origins. "The design or intention of the author is neither available nor desirable as a standard for judging the success of the work…" (Wimsatt and Beardsley 1946, p. 468).

Artivism embodies the intentionalist view, such as the one espoused by aestheticist E. D. Hirsch, who insists that authorial intent defines the parameters for the validity of interpretation (Hirsch 1967). It is difficult to deny the importance of the creator's intention in artivist works. For these works to induce social change and actually *do* something, they are crucial to the spectators' understanding (and appreciation) of their artworks. Authorial intent has to be conveyed, or the work fails in its mission. Therefore, committing to intentionalism seems necessary for artivism to work. In other words, its position has to be deliberately *intentionalist,* requiring the artists' engagement with the real world, much reflection on the ensuing experience of it, and deliberately manipulating their materials in order to deliver their message to their spectators. This element in artivism is essential to its pedagogical function. If it is to teach something about the world and develop agency in its participants, whether as creator or spectator, its intentional character has to be affirmed. In the same manner, for the intention behind an artivist work to be realized, it has to be *understood* by its audience.

Can art-making and artworks really be used as pedagogical tools? There have been objections to seeing art in this manner. First, if art is a product of the artist's imagination, then it is pure fantasy or merely fictional. As such, they cannot be about the world and, therefore, cannot teach anything about the real world, much less, effect change in it. Second, art does not argue like Philosophy or do experiments like the hard sciences. The information they provide cannot be tested. Unlike such disciplines, art also cannot be articulated differently since its content is limited by the form in which it comes. It cannot be reformulated or stated otherwise, as "normal" lessons can. Third, understanding deals with universals, that is, some form of generality, while art deals in particulars, meaning that art can only represent something in particular. If so, art cannot aid in understanding and cannot be used to teach anything about the world (Graham 2005).

Cognitivist aesthetics offers a defense of art as a learning tool. It is a form of normative aesthetics that claims that art, rather than just capturing the beautiful (art as contemplation on beauty), or a means of experiencing aesthetic pleasure (hedonist framework), or a way of expressing and evoking emotions (expressivist framework), can also be about making the world more comprehensible. While it does not deny that artworks can cause aesthetic pleasure, be expressive of emotions, or represent beautiful things and experiences, its focus is on the capacity of (some) art to *teach* something about the world. In other words, aesthetic cognitivism does not exclude from the value of art its capacity to cause aesthetic pleasure or art's inherent capacity to evoke emotions. It claims that apart from these, some artworks are also valuable because of their capacity to help us understand the world more and ourselves better. In this way, art is transformative. As it enables understanding of one's experience of the world, it helps effect change in oneself and, by extension, in the world (Gibson 2008; Graham 2005). Therefore, the core of cognitivist aesthetics aligns with the pedagogical pursuit. It places aesthetic experience in the realm of understanding and assigns art the task of meaning-making. The cognitivist view emphasizes art's reflective and thought-provoking role, but remains compatible with the other normative aesthetic frameworks.

On the other hand, not all works of art can be considered artistically valuable because of what they can teach their creators and their audience. Some artworks are best served by the aesthetic (beauty), hedonist (pleasure), or expressivist (emotion) frameworks, or a combination of them. However, certain pieces can be appreciated more, and the greatness of their value can be grasped better if interpreted in terms of what they contribute to our understanding of ourselves and the world. Artivist works are some of these. In fact, artivist works either circumvent or respond effectively to the objections against art as a learning tool.

Artivist works, by virtue of being works of art intended to do something—to affect its viewers to effect change in themselves and the world—cannot be said to be purely imaginational, fictional, or non-representational. Because of their nature, works of artivism are almost always direct representations of what is in the world. They are rarely purely fictional and tend to be almost documentary in their treatment of their subject matter. Even if they use fictional elements, they are *metaphors* and ultimately refer to elements in the actual world (Baumberger 2013). How could the call for change—which all artivist works are—be made without addressing real-world problems? This is the reason that artivist works are, more often than not, staged in public, everyday spaces. They require either actual contexts or a direct reflection of one. They require real people interacting with the artworks, realizing truths about themselves, the world, and how they could change themselves and act to change the world.

Consider, for example, Olafur Eliasson and Minik Rosing's "Ice Watch" (Eliasson and Rosing 2015), an installation of 12 huge chunks, totaling 80 tons of free-floating ice from a Greenland fjord, arranged in a clock-like formation in the Place du Pantheon in Paris. Spectators were given the opportunity to observe the ice melt from December 3 to 12, 2015 (Eliasson and Rosing 2015). The installation's context, crucial in understanding this work, is the 2015 Paris Summit on Climate Change. Aaron Staples' "Not Just Tuna" series, made for Greenpeace, depicts the harm commercial tuna fishing, using destructive methods and forced labor, has on the environment and the workers' lives (Greenpeace 2015). Banksy's "I Don't Believe in Global Warming" graffiti mentioned above literally shows how useless climate denial is in the midst of the climate crisis happening as one views the graffiti (Egnér 2025). Artivist works like these straightforwardly present to viewers matters that need to be attended to and conditions that urgently need to change, calling spectators to action. To respond, however, viewers must understand what is being called for and why. The *æffect* of artivism is therefore based on the effectiveness of their cognitive function (Garcia 2023).

While the unity of content and form prevents the reformulation of what is presented in an artwork, it does not require such reformulation in order to direct the mind to a different perspective. The cognitivist framework of art is not about the acquisition and addition of knowledge. It is about art's ability to produce an understanding of the world and the spectator's or the artist's self (Baumberger 2013). Art can "direct the mind" so that it sees and contemplates something other than what is ordinarily seen. It can present a different perspective so the mind can understand meanings and connections not available to everyday experiences (Graham 2005, p. 64). Works like the Guerrilla Girls' *Dearest Art Collector* in 1986 (National Museum of Women in the Arts 2025) or Peggy Diggs' *Beliefs* in 2019 (Funderburk 2021) take on the ordinariness of text, or words handwritten or printed. However, because they are presented in a particular form—mimicking a letter or a poster—they prompt questions, induce critical thought, and bring the spectator to a different mindset. They disturb the mind's peace and prod it to think further and reflect on everyday experiences of seeing artworks in museums, in the case of *Dearest Art*

Collector, and perhaps, check unexamined prejudices about race, in the case of *Beliefs.*

As to the last objection of understanding being about general things, while artworks present only particular things, artivist works call attention to social issues that are experienced by all. Seeing a painting in the series of Zaria Forman's *Greenland 2012: Chasing the Light paintings*, say "Greenland No. 54" (Forman n.d.), one does not get stuck with the particular iceberg presented on the canvas but contemplates on the meaning of the disappearing glaciers and what it means for us to live in a world where ice melts and causes environmental catastrophe. Aestheticist Graham (2005, p. 70) writes that the question when faced with such works should not be "is this how it really was?" (particular), but "does it open our eyes to new ways of seeing things of this sort?" (general/universalizable). Graham (2005, p. 73) writes further, "Great works of art enable us to understand what it is to be a human being not in the way that physiology, psychology, or anthropology do, but by providing images through which our experience may be illuminated."

4.4 Artivism as an Aesthetic Pedagogical Tool for ESD

The cognitivist aesthetic framework reveals artivism to have the capacity to teach something to both creators and their audience. As a pedagogical tool, engagement with it promotes agentic learning, encouraging self-directed learning and contributing to the learning of others. Artivism is goal-oriented, intending to initiate change in spectators who, in turn, hopefully initiate change in the world through concrete action. The infinite ways such an intention can be expressed in art point to the presence and affirmation of choice in order to create an impact on various social crises. Agency is also affirmed in its inherently collaborative nature, especially in the artists' engagement with their own experience, with the material they work with in the production of their art, with their audience, and the resulting courses of action that their work can inspire. The foregoing shows that artivism is not only compatible with the four-life learning, four pillars of education, the sustainability competencies, and the learning objectives of "head, heart, and hands" for targeting these competencies, it also has the potential to be a powerful pedagogical means that can be strategically employed to fulfill them.

It has to be emphasized that while the cognitivist framework is focused on artivist works' ability to foster understanding of the self and the world, it does not nullify art's emotive import. Artivism's æffect allows understanding and emotion to intertwine, creating the ground for more informed and holistic decision-making from which effective action ensues. It is this character of artivism that makes it a good candidate for motivating action in sustainability campaigns that require an emotional response to the urgent call for change, such as those involving the climate crisis.

4.4.1 Artivism and the Sustainability Competencies

The creative works discussed above show that artivist processes are conducive to developing sustainability competencies. Artivism, by its nature, is agentic, infusing work with authorial intent, demanding critical engagement with real-world issues, and requiring collaborative work. These characteristics make it conducive to honing the eight competencies, applied together in an integrated way.

To begin with, an artivist project cannot commence without an experience and understanding of current issues, their causes, and how they can be addressed. This requires *critical thinking,* which accompanies every stage of creation or production. Creators do research, reflect, and decide on what issue or set of issues they need to tackle. The process of conceptualization, deciding on what materials to use, and how to set them up to convey the intended message in the most effective way requires knowledge of systems, how elements work together, and how best to connect them to deliver one's intended purpose. This requires a holistic view of systems and skill in analyzing them, which makes up the *systems thinking* competency.

The call to action signaled by an artivist work, often urgent, is done in clear anticipation of how things could get worse if nothing is done. This hones the learner's *anticipatory thinking*, keeping an eye on the future, which informs what can be done at present. Related to this is the sharpening of *strategic competency,* as it is part of the creative process in artivism to choose the form through which the artwork could be disseminated. The site and time of its exhibition, the materials used, and how they are used are all part of the strategy in creating the greatest possible impact of its intended message on the spectators.

As mentioned above, artivism, by default, works on the values of justice, compassion, peace, and mutual understanding. They aim at community harmony and the betterment of the whole world and cannot be engaged in without activating the learner's *normative thinking,* which includes the monitoring of personal and communal values and shared experiences in order to persuade spectators of the work's goal. The same requires the learner to practice working collaboratively with others. In mounting an exhibition in a public place, for instance, creators require the help of co-workers or other artists in order to pull it through. Even post-installation, artivist works are never private or exclusive. They are shown in public spaces where encounters with others are immediate and inevitable. Even if artivist works can be created solo, the aim is always collaborative action to address the sustainability issue in question and promote well-being for all. Engagement with the public in the form of conversations, debates, inspiring new or derivative works that challenge existing problematic social structures and practices, which usually continues long after an impactful staging of a work, requires the practice of *interpersonal competency*.

Furthermore, because artivism aims at prompting people to know more, to think critically, reflect, and to act together based on the conclusion of those processes to effect change and solve problems, it can hone the learner's *integrated problem-solving* competency. Finally, activist artists cannot effectively do what they do without being self-aware individuals who hope to develop the same *self-awareness* in their audience. Their works, being triggers for critical thought and reflection, also promote self-knowledge and personal development, especially when spectators become engaged participants in the artivist agenda. This is how they learn and contribute to others' learning as well.

4.4.2 Artivism and the Four-Life Learning

Artivist engagements can be used as a non-conventional form of learning. Artivists often take to the streets to reach more people and take the opportunity to make each moment a learning moment, which is what *lifelong* learning is about. Artivism can also be done anywhere, anytime, using materials that are readily available, like paint and an existing wall, like in Banksy's works, or those obtained at great effort and expense, like the chunks of ice in Eliasson and Rosing's installation in 2015. It can be fleeting or temporary, as in performance artivism or installation art; or long-lasting or permanent, as in fixed visuals, recorded sounds, and printed texts. These activities provide a *lifewide* learning that thrives on various contexts in different modes while keeping the learner connected to the real world.

Opportunities for *lifedeep* learning are also presented by artivism as it anchors itself in observed realities and facts while enabling investigation and understanding of the connections among them, their implications, and the required action to address the challenges involved. Finally, by its being a creative form of activism, artivism promotes *lifewise* learning as its works are spurred by the principles of justice and compassion, aimed at promoting values such as peace, harmony, and well-being, while seeking to stop oppressive structures and practices with the aim of liberating beings from suffering.

4.4.3 Artivism and the Four Pillars of Education

In order to encourage *learning to learn,* artivism can be employed to promote the self-directed pursuit of knowledge and understanding of social issues that can lead to one's desire to do something about them. This feeds *learning to do* as the activist component of artivism assumes the will to implement the plan of action, to act, and make a difference. *Learning to be* is inevitable, as one cannot engage in artivism without self-awareness and the continuing project of self-knowing. It acknowledges one's responsibility as a member of the local and global community and one's autonomy in crafting a plan of action to create a better world. Finally, artivism cannot but promote *learning to live together* because its communal nature makes it very conducive to the promotion of inclusion, cooperation, diversity, and mutual understanding based on shared experiences in the world. The many artivist collectives established all over the world, including the ones mentioned in the previous sections, are proof of this.

4.4.4 Artivism and the Three Learning Domains

Artivism's capacity to engage collaborators and spectators intellectually through critical thinking, reflection, and consciousness-raising about global issues addresses the cognitive or "head" domain. Its capacity to engage them emotionally, producing its trademark "affect" addresses the socio-emotional or "heart" domain. Its capacity to prompt them into action, on the other hand, produces the "effect" it aims for and addresses the behavioral or "hands" domain. Artivist engagements, therefore, can offer a holistic educational experience that can transform the learner's whole person—whether as the creator or the spectator—to become a sustainability change agent in the future.

4.4.5 Artivism in Action: Example and Challenges

Duncombe and Harrebye (2022) conducted an experiment in 2018 on the comparative efficacy of creative and conventional forms of activism. They set up on Queen Louise's Bridge in Copenhagen to campaign against cattle-raising because of its negative environmental impact. They also petitioned for green taxes on meat production. They performed these two forms of anti-cattle-raising campaigns for 3 days.

For the creative version, they wore cow costumes, used machines that produced "farting" noises, and set up installations of actual cow dung while rap-singing, and distributing creatively designed and fun information flyers to passersby. For the conventional version, they delivered speeches, stopped passersby to collect signatures, and gave out information through formally formatted flyers. Observers in lab gowns noted the response of the passersby, those who signed the petition and those who refused. They conducted follow-up interviews with those who signed the petition.

The result of the experiment showed that creative activism was more effective in getting the passersby to do what they wanted them to do, even if the information they received was not as clear as those given in the conventional version. With the creative and "fun" format, people were more interested, easily signed the petition letters, and stayed to converse with the observers. Follow-up surveys also showed that there was better recall among those who responded to the creative format than those who did the conventional one (Duncombe and Harrebye 2022; Garcia 2023).

While more studies like this with different iterations are needed to get more definite findings about the impact of artivism, the Copenhagen Experiment, at least, demonstrated that metrics can be set up to measure the effect of creative activism (Duncombe and Harrebye 2022). Rubrics based on the targeted competencies can, therefore, be constructed for course assessment in an SDG-embedded curriculum. However, many things have to be considered before implementation can be done.

The types of artivist engagements, for instance, will have to be listed and defined for different year levels, based on the sustainability competencies targeted by academic programs. Budget requirements, access to materials, and safety issues will have to be covered. Networking with local artivists whose advocacy the educational institution can support and who can lead the younger students by example may have to be involved in the planning. A variety of safe public spaces where a peaceful exhibition can be mounted will have to be identified, including spaces within the campus where engagements can happen among diverse members of the institution. Since this will deal with physical spaces, environmental considerations will have to be part of the planning.

Since artivist engagements are participative, ethical issues, including intellectual property, data privacy, and mental health issues, will have to be considered, and safety measures incorporated in the activity plans. These, however, may be part of the students' task, as planning for these also hones their anticipatory, strategic, and critical thinking, as well as their self-awareness, normative, and collaborative skills. Responding to issues that may arise will also require them to practice systems thinking and integrated problem-solving skills. The implementation may require much work, but the potential of artivism for sustainability competencies and their accompanying forms of learning may be worth it. One activity, if done right, may be able to target all competencies. It may not be any more difficult than preparing for other forms of authentic assessments that will bring the students out of the classroom and deal with real-world issues. However, the artistic element in artivism may provide a more holistic development for the students due to its use of the learners' creative faculties.

4.5 Conclusion and Recommendation

This paper shows that artivism, through a cognitivist aesthetic framework, can foster critical thought about various social issues that are aligned with the SDGs, promote creativity in problem-solving, instill a sense of social responsibility in its spectators while affirming and strengthening it in its creators, and empower individuals and communities to take action. As such, it models agentic learning, which is essential in developing the four-life learning forms, achieving the pillars of education and the sustainability competencies through the cognitive, socio-emotional, and behavioral learning domains. It, therefore, can be explored as a potential creative pedagogical tool in SDG-embedded courses and ESD programs.

While sustainability competencies are identified to address the three pillars of sustainability—social, economic, and environmental—artivism's unique ability to produce æffect may directly address the difficulty in motivating environmental action. It is, therefore, deemed well-suited for environmental advocacy, where studies have shown that messaging has to get through not just the rational brain processing but, more importantly, the emotional as well, to prompt action. As far as directing minds to see things differently and triggering critical thought that can encourage people to decide better for sustainable futures, artivism can cover a wide range of advocacy that can contribute to the achievement of the SDGs.

Artivist engagements can be included as part of the curriculum in teaching sustainability concepts and in student-led learning activities. Students can be encouraged to use artivism as a campaign strategy in advocacy, as part of the solution in a problem-based project, or they can organize artivist engagements in the school community or join established artists in their work as inter- and transdisciplinary learning practices. Educators can encourage creativity and lend support to artivist projects, or lead artivist engagements if they are themselves activist creatives, and afterwards create safe spaces for meaningful reflection and dialogue. Studies on the implementation process and their impact on learning can be designed and conducted regularly to see how they contribute to the development of sustainability competencies. Results can enrich existing literature on the subject matter, while design and implementation can be made available for adoption by educational institutions.

Acknowledgments This paper is an offshoot of the study presented at the NERPS 2025 on Peace and Sustainability. Participation in the said conference was supported by the DLSU Office of the Vice-President for External Relations and Internationalization and subsidized by the Faculty Development Program under the DLSU Office of the Provost.

References

Arzola, D. (n.d.). *Featured projects.* Daniel Arzola: Artivist. https://www.danielarzola.com/s-projects-side-by-side

Baumberger, C. (2013). Art and understanding: In defence of aesthetic cognitivism. In M. Greenlee, et al. (Eds.), *Bilder sehen: Perspektiven der Bildwissenschaft* (pp. 41-67). Schnell & Steiner.

Beghetto, R., & Karwowski, M. (2018). Educational consequences of creativity: A creative learning perspective. *Creativity, 5*(2), 146-154.

Bertha Foundation. (2025). *Bertha artivism awards: Bertha artivists 2024.* Bertha Foundation. https://berthafoundation.org/story/bertha-artivism-awards-2024-cohort/

Bilon, A. D. (2021). Toward an agentic approach to activist learning: The interplay between agency and learning in the narrative of polish activists. *Adult Education Quarterly, 71*(20), 166-183.

Buckland, K. (2011). *Theory: Artivism.* Beautiful Trouble. https://beautifultrouble.org/toolbox/tool/artivism/

Chang, C., & Kidman, G. (2022). Life-long and life-wide education for our sustainable future. *International Research in Geographical and Environmental Education, 31*(2), 85-88. https://doi.org/10.1080/10382046.2022.2057895

Deal, G., & Nicholson, K. (2018). *On art & activism: A day of art action, January 20, 2018.* T/LT west. https://www.tiltwest.org/item/art-activism/

Delors, J. (1995). Learning: The treasure within. *Report to UNESCO of the international commission on education for the twenty-first century.* UNESCO Publishing.

DePART. (n.d.). *DePART: Decentralized practice of active remembrance and theatricalization.* https://www.depart.community/about-1

Detroit Narrative Agency (NDA). (n.d.). *What is Detroit narrative agency?* DNA. https://www.detroitnarrativeagency.org/

Duncombe, S. (2016). Does it work: The effect of activist art. *Social Research, 83(*1), 115-134.

Duncombe, S., & Harrebye, S. (2022). The Copenhagen Experiment: Testing the effectiveness of the creative vs conventional forms of activism. *Social Movement Studies, 21(*6), 741-765.

Egnér, V. (2025, January 1). *I don't believe in global warming by Banksy*. Street Art Utopia. https://streetartutopia.com/2025/01/01/banksy-climate-change/

Eliasson, O., & Rosing, M. (2015). *Ice watch: An artwork by Olafure Eliasson and Minik Rosing on the occasion of COP 21.* United Nations Conference on Climate Change. https://icewatchparis.com/

Flores, R. G. (2015). *Chicano artists and zapatistas walk together asking, listening, learning: The role of transnational informal learning networks in the creation of a better world.* In Motion Magazine. https://inmotionmagazine.com/auto/rf_informal1.html

Forman, Z. (n.d.). *Exhibitions.* Zaria Forman. https://www.zariaforman.com/exhibitions

Funderburk, A. (2021). *Using art as a means for social change, artivists can change the world.* Art & object. https://www.artandobject.com/articles/artivism-making-difference-through-art

Garcia, L. (2020). Filipino indigenous ecology: Stories to support an environment-friendly life. *13th DLSU Arts Congress Proceedings*, *4*(3), 105-114. https://www.dlsu.edu.ph/wp-content/uploads/pdf/conferences/arts-congress-proceedings/2020/13th-dlsu-arts-congress-proceedings.pdf

Garcia, L. (2023). Art and activism: Some philosophical issues in artivism. *16th DLSU Arts Congress Proceedings,* *7*(1), 1-12. https://16thdlsuartscongress.com/16th-dlsu-arts-congress-proceedings-volume-7/

Gibson, J. (2008). Cognitivism and the arts. *Philosophy Compass, 3(*4), 573-589. https://doi.org/10.1111/j.1747-9991.2008.00144.x

Graham, G. (2005). *The philosophy of the arts.* Routledge/Blackwell.

Greenpeace. (2015). *Not just tuna: The truth behind chicken of the sea's label.* Greenpeace. https://www.greenpeace.org/usa/not-just-tuna-the-truth-behind-chicken-of-the-seas-label/

Hadjiosif, S. (2021). *The rise of activism: Art for change.* Terra Movement: An Activist Hub. https://www.terramovement.com/the-rise-of-artivism

Haring, K. (1989). *Ignorance=fear.* The Keith Haring Foundation. https://www.haring.com/!/art-work/253

Hirsch, E. D. (1967). *Validity in interpretation.* Yale University Press.

Hoppers, W., & Yekhlef, A. (2013). *Common core skills for lifelong learning and sustainable development in Africa, Tunisia.* Association for the Development of Education in Africa.

Hopwood, N. (2024). Conceptualising praxis, agency and learning: A postabyssal exploration to strengthen the struggle over alternative futures. *Educational Philosophy and Theory,* 56(10), 956-966. https://doi.org/10.1080/00131857.2024.2333851

Huang, D. (2022). *A life-long process of purposeful learning.* SingTeach. https://singteach.nie.edu.sg/2022/04/19/a-life-long-process-of-purposeful-learning/

Hytten, K (2017). Teaching as and for activism: Challenges and possibilities. In Moses, M.S (Eds.), *Philosophy of education 2014* (pp. 385-394). Illinois: Philosophy of Education Society.

Jarallah, R. (2015). *Exhibitions.* Randistic. https://randistic.com/exhibitions

Javaad Alipoor Company. (2017). *Principles: Our manifesto.* Javaad Alipoor Company. https://www.javaadalipoor.co.uk/our-manifesto

Komi, K. (n.d.). *Artivism.* MigraMedia. https://www.uni-hildesheim.de/migramedia/home/unsere-medien/keywords/artivism/

Kwek, D., Hung, D., Koh, T. S., & Tan, J. (2017). *OER-CRPP innovations for pedagogical change: 5 Lessons.* Singapore: National Institute of Education. https://repository.nie.edu.sg/server/api/core/bitstreams/e78a6b92-0111-4ad7-9b89-d5b3e4d9c2c8/content

Lau, J. Y. F. (2011). *An introduction to critical thinking and creativity: Think more, think better.* John Wiley & Sons, Inc.

Looi, C. K. (2023). Towards a comprehensive perspective to making learning a part of life: A response to Fischer et al.'s contribution. *The International Journal of Information and Learning Technology, 40(*1), 31-39.

Marshall. G. (2014). *Don't even think about it: Why our brains are wired to ignore climate change.* Bloomsbury Publishing Plc.

National Museum of Women in the Arts. (2025). *Guerilla Girls.* National Museum of Women in the Arts. https://nmwa.org/art/artists/guerrilla-girls/

Rahmian, L., Hod, Y., Yuan, G., & Zhang, J. (2024). Supporting students' inquiry through doubly authentic learning designs: four configurations of interests, domain, and identity. *European Journal of Psychology of Education, 39,* 3381–3402. https://doi.org/10.1007/s10212-024-00818-5

Stacey, T. (2024). Religious repertoires of sustainability: Why religion is central to sustainability transitions, whatever you believe. *Environmental Innovation and Societal Transitions, 50,* 1-14.

Taylor, B. (2010). *Dark green religion: Nature spirituality and the planetary future*. University of California Press.

The Broad. (2025). *By Barbara Kruger in the collection.* The Broad. https://www.thebroad.org/art/barbara-kruger/untitled-your-body-battleground

UNESCO. (n.d.). *Institute for Lifelong Learning Technical Note*. Lifelong Learning. https://uil.unesco.org/fileadmin/keydocuments/LifelongLearning/en/UNESCOTechNotesLLL.pdf

UNESCO. (2017). *Education for sustainable development goals: Learning objectives*. UNESCO. https://unesdoc.unesco.org/ark:/48223/pf0000247444

UNESCO. (2020). *Education for sustainable development: a roadmap.* UNESCO. https://unesdoc.unesco.org/ark:/48223/pf0000374802.locale=fr

Watson, S. (2022). *Student activism: Learning through doing.* New Zealand Council for Educational Research.

Williams, G. T. (2022). *Artivism: The artist and artwork fighting for social change*. Sound of Life. https://www.soundoflife.com/blogs/experiences/artivism-artists-artwork-social-change

Wimsatt, W. K., & Beardsley, M. C. (1946). The Intentional Fallacy. *The Sewanee Review, 54*(3), 468-488.

Zielinska, A., Lebuda, I., Janskowska, D. M., & Karwowski, M. (2021). Self-regulation in creative learning. *Creativity, 8*(1), 52-71

Part II
Community-Led Pathways

Chapter 5
Youth at the Roots of Peace: Education, Environment, and Gender as an Approach to Peacebuilding

Kaila Jade Batuigas, Le'an Carlo Khrystoffer Ege, Kris Bernard Mutia, and Primitivo III Ragandang

5.1 Introduction

Globally, youth, numbering approximately 2.4 billion, hold a key role in peacebuilding endeavors (UNOY, 2021). This fact is acknowledged by numerous scholarly works and international resolutions (Berents and McEvoy-Levy 2015; Gillard 2001; McEvoy-Levy 2021; Podder and Ragandang 2022). The United Nations emphasizes the importance of youth in peacebuilding through Resolutions 2250, 2419, and 2535. Scholarly literature illustrates the active participation of young people in peacebuilding processes, while traditional practices exemplified by the role of youth further contribute to peacebuilding initiatives. Furthermore, recognizing the involvement of young people in peacebuilding as crucial for sustainable peace and local ownership has been a recurring theme in literature (Podder and Ragandang 2022).

However, previous studies show that young people were marginalized or excluded from the peacebuilding space. The field of peacebuilding has traditionally been adult-dominated and focused on addressing the immediate aftermath of war or conflict situations, aiming to prevent their recurrence. This focus often involved narratives centered on war and conflict resolution. For example, Barnett et al. (2007) describes peacebuilding as external interventions intended to reduce the risk that a state will erupt into or return to war. Even in the twenty-first century, studies around peace and peacebuilding continue this long-standing conflict-centric perception. However, there is a significant spatial gap in research addressing peacebuilding beyond conflict. An emerging perspective suggests that peacebuilding extends far beyond the cessation of hostilities into more interpersonal concepts (Redekop

K. J. Batuigas (✉) · L. C. K. Ege · K. B. Mutia · P. Ragandang
Mindanao State University-Iligan Institute of Technology, Iligan, Philippines
e-mail: kailajade.batuigas@g.msuiit.edu.ph

A. Sharifi et al. (eds.), *Frameworks and Practices Towards a Sustainable and Peaceful Future*, World Sustainability Series,
https://doi.org/10.1007/978-981-95-2310-8_5

2014). Additionally, an emerging trend considers youth-led organizations and their role in reshaping the narrative of peacebuilding.

Despite challenges such as limited resources, institutional barriers, and socio-political constraints, many youth-led groups have shown their capacity to actively address root causes of discord, amplifying the voices of marginalized communities, and advocating for social movements promoting justice, human rights, and economic development (Petcu 2020). By examining the challenges, contributions, and strategies of youth-led organizations, this research aims to explore how their activities and initiatives contribute to sustainable peacebuilding. By understanding the role of youth-led organizations in fostering peace within communities, this chapter offers insights that can add to the existing body of knowledge on peacebuilding practices, drawing examples from youth-led organizations in Iligan City.

This chapter explores youth-led initiatives that represent what we refer to as "other forms" of peacebuilding. Unlike conventional peacebuilding studies, this chapter focuses on alternative forms of peacebuilding in the fields of education, environment, and gender from the perspectives of youth-led organizations. These initiatives, grounded in the "Petals of Peace" concept by Toh and Cawagas, constitute vital forms of peacebuilding to foster a culture of peace led by the youth. By examining these alternative forms, this thesis hopes to contribute to our understanding of youth peacebuilding agencies.

This delves into the role of youth-led organizations in fostering sustainable peacebuilding in Iligan City, particularly through intrapersonal and interpersonal dimensions like gender, environment, and education, while the theoretical framework draws significant inspiration from Toh and Cawagas' (2017) "Petals of Peace Model" (PPM), and Berents and McEvoy-Levy's (2015) theorizing youth and everyday peacebuilding. These two frameworks offer complementary perspectives on how peace can be built both formally and informally within communities.

The PPM is considered an appropriate theoretical framework for the current study due to its comprehensive approach to addressing various dimensions of peacebuilding. In contrast to most theories of peacebuilding, which confine its definition toward conflict-centric scenarios, the PPM uses six key themes for peace represented by six flower petals, namely (1) cultivating inner peace, (2) living in harmony with the earth, (3) promotion of human rights and responsibilities, (4) building intercultural respect, reconciliation, and solidarity, (5) living with justice and compassion, and most importantly, (6) dismantling the culture of war (Toh and Cawagas 2017).

The study of Berents and McEvoy-Levy's entitled theorizing youth and everyday peacebuilding Berents and McEvoy-Levy (2015) highlights actions contributing at a grassroots level, focusing on small and everyday actions that build peace outside formal politics or institutions. It acknowledges that peace is not just the absence of violence but also the presence of positive relationships, mutual understanding, and the capacity to resolve tensions in everyday interactions (Berents and McEvoy-Levy 2015).

For youth-led organizations in Iligan City, the combination of the PPM and Everyday Peacebuilding is crucial. It allows for an inclusive, context-specific approach to peace that connects personal growth, community involvement, and collective action. These organizations, focusing on areas such as gender equality, environmental sustainability, and education, contribute to peace by addressing local issues in tangible, grassroots ways. This framework also offers a more comprehensive understanding of how young people, through their everyday activities, can shape peaceful communities by confronting personal and social challenges, whether related to gender, the environment, or educational access.

Existing literature often confine peacebuilding within the parameters of traditional, Western perspectives, predominantly focusing on conflict resolution and post-war reconciliation (Barnett et al. 2007). By adding to this Western-centric viewpoint, this chapter aims to present a more comprehensive understanding of peacebuilding that reflects the diverse revolution of peace, hence peacebuilding.

Moreover, this chapter explores the role and significance of youth-led organizations in advancing peacebuilding initiatives within the local context of Iligan City. Despite recognizing youth as agents of change in peacebuilding efforts, there needs to be more understanding of the specific contributions and challenges these organizations face. Through an in-depth analysis of youth-led initiatives and their impact on fostering peace and harmony, this chapter provides valuable insights into the efficacy of grassroots efforts in promoting sustainable peace and development.

Hence, this chapter attempts to answer several key research questions. It explores the nature of youth-led peacebuilding organizations and examines how their advocacies have evolved over time. It also investigates how these organizations in Iligan City contribute to local peacebuilding efforts and the ways in which their initiatives reflect the diverse needs and aspirations of the communities they serve. Furthermore, this chapter seeks to identify the challenges these organizations face in carrying out peace work and the coping mechanisms they employ to sustain their initiatives.

5.2 The Framework of the Study

5.2.1 Guiding Theories of Peacebuilding

This chapter is grounded by Toh and Cawagas' (2017) PPM and Berents and McEvoy-Levy's (2015) everyday peacebuilding. These two frameworks offer complementary perspectives on how peace can be built both formally and informally within communities.

The PPM presents peace as a multidimensional concept emphasizing six interconnected dimensions: inner peace, environmental harmony, human rights, intercultural respect, justice, and the dismantling of the culture of war. This model calls for a holistic approach to peacebuilding that extends beyond conflict resolution and includes the cultivation of values and actions that sustain peace at individual, community, national, and global levels.

Conversely, Everyday Peacebuilding highlights informal, grassroots efforts where peace is built through daily interactions, relationships, and addressing social tensions outside formal political institutions. This approach emphasizes that peace is not merely the absence of violence but the presence of mutual understanding, inclusivity, and community resilience.

The integration of these frameworks enables a context-specific approach where youth-led organizations address key areas such as gender equality, environmental sustainability, and education to promote peace. Gender equality correlates with overall societal peace (Hudson et al. 2009), environmental protection influences human security and conflict dynamics, and education provides stability while fostering awareness of conflict origins. By combining structured peace programs with everyday actions of justice and solidarity, these frameworks provide a comprehensive lens for understanding how youth contribute to long-term, sustainable peacebuilding in Iligan City.

To bridge the theoretical grounding of this study with its emergent themes, this research from Redekop (2014) calls for a shift from traditional conflict resolution to more holistic, systemic approaches to peacebuilding. While Redekop does not explicitly outline education, environment, and gender as the core dimensions of peace, his work opens space for frameworks that move beyond militarized and state-centric models and instead embrace interconnected and multidimensional pathways to peace. Building on this, the study identifies education, environmental sustainability, and gender as key thematic dimensions through which the six petals of PPM can be reorganized and synthesized. For instance, inner peace, justice, and human rights are examined in relation to education; environmental harmony and dismantling the culture of war align with environmental sustainability; and intercultural respect and justice are explored through the lens of gender equality. This thematic reorganization allows the study to honor the comprehensive scope of PPM while also aligning the framework with the actual peacebuilding practices and focus areas of youth-led organizations in Iligan City.

5.2.2 Conceptual Framework

Taking inspiration mainly from Toh and Cawagas' (2017) "petals of peace model" for peacebuilding, the conceptual framework of this study (Fig. 5.1) shows the basic features of what makes sustainable peacebuilding, and also highlights the significance of both youth-led organizations (YLOs) and the challenges which both harms and strengthens the peace process.

Three selected YLOs—AKINA, City Wide 4-H Club Iligan, and Pinas Safe Spaces (PSS), serve as peacebuilding conduits in Mindanao, particularly in the gender, environment, and education sectors. Represented as a watering pot, these organizations sustain peace through various initiatives, including youth gatherings and cleanup drives.

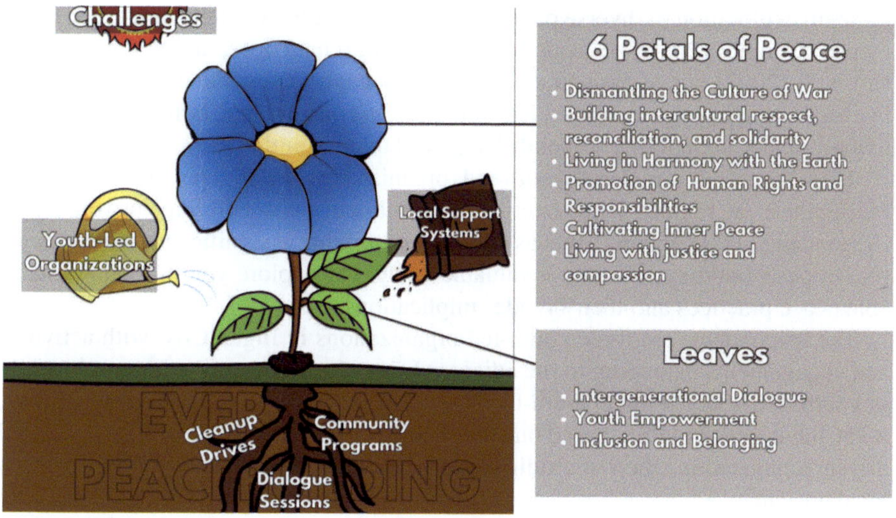

Fig. 5.1 Schematic diagram of the conceptual framework

The framework integrates Berents and McEvoy-Levy's (2015) Everyday Peacebuilding, depicted as the roots of a flower, signifying how community-level efforts (e.g., dialogues, cleanups, and local programs) form the foundation of peacebuilding. Additionally, local support systems, symbolized as fertilizer, represent the external resources and partnerships essential for YLOs to thrive.

The leaves of the flower, inclusion and belonging, intergenerational dialogue, and youth empowerment, highlight crucial aspects that foster participation, bridge generational gaps, and equip young leaders.

Despite their contributions, YLOs face challenges, symbolized by the sun, such as limited resources, political instability, and societal stereotypes. However, when addressed effectively, these challenges can strengthen both the organizations and the peacebuilding process.

Together, this conceptual framework offers a grounded and theory-informed lens to analyze how YLOs contribute to peacebuilding by addressing gender, education, and environmental concerns through everyday actions. It links abstract peacebuilding theories to the tangible, lived realities of youth in conflict-affected communities.

5.3 Methodology and Study Context: Iligan's Youth Peacebuilders

This chapter employs a qualitative case study design to examine youth-led peacebuilding beyond conflict. Following Baxter and Jack (2008), this approach provides a multifaceted understanding by analyzing phenomena within context. It enables an in-depth exploration of diverse perspectives, using interviews, field notes, and documents (Coombs 2022) to capture motivations, challenges, and outcomes. As Creswell (2007) notes, case studies effectively analyze specific cases or compare instances, making this method ideal for assessing factors that influence the effectiveness and sustainability of youth-led peacebuilding initiatives.

This chapter focused on youth-led organizations in Iligan City, Mindanao, Philippines, a region with over 150 youth-led peacebuilding organizations (Ragandang 2020). Iligan City was chosen for its cultural significance and strategic role in peacebuilding, offering a valuable context to explore youth-driven sustainable peace practices and their broader implications.

The respondents are three youth-led organizations in Iligan City, with activities and engagements extending to the adjacent Lanao Provinces and composed mainly of youth. We selected three youth-led organizations in Iligan City, namely, AKINA, 4-H Club Iligan, and PSS, based on a predefined criteria including being a youth-led driven organization, their multidimensional peacebuilding contributions, and its local presence.

This chapter used a qualitative approach to examine youth-led organizations' contributions to peacebuilding in Iligan City. Primary data was gathered through Key Informant Interviews (KIIs) with five founding members from three organizations. Semi-structured interviews allowed for an in-depth exploration of participants' experiences, providing valuable insights from key informants. Interviews, the main data source, offered perspectives on actions and events (Broadbent et al. 1998) and helped in understanding complex issues (Gerson and Horowitz 2002).

In addition to KIIs, this study utilized secondary data, including reports, documents, and media articles from the three youth-led organizations. Following the principles of triangulation (Denzin and Lincoln 2018), these sources validated and enriched the primary data. Organizational reports and media coverage provided additional context and depth, enhancing the credibility and comprehensiveness of the findings.

Ethical principles were central to this study, ensuring participants' rights and well-being (Resnick 2015; World Health Organization 2024). Informed consent was obtained, and participants were fully briefed on the study's purpose, risks, and benefits. Their identities and sensitive information were kept confidential, and they were reminded of their right to withdraw. Permission was requested to record interviews, ensuring accurate data collection. Key informants, being public figures, consented to having their names and roles disclosed, which helped clarify their involvement. The study maintained transparency while respecting participants' positions.

Interview responses were recorded, transcribed, and analyzed to identify emerging themes and perspectives. Microsoft Word was used for transcription, and inductive coding allowed for the continual addition of codes during analysis, ensuring a comprehensive evaluation (Urquhart 2013; Charmaz 2006). Responses were categorized into meaningful segments, preventing biases and promoting balanced analysis (Stake 2010).

Thematic analysis was used in this chapter, which involved the identification of themes through careful reading and re-reading of the transcribed data (King 2004; Rice and Ezzy 1999).

In addition to this, this chapter incorporated the six petals of peace model (Toh and Cawagas 2017) and the concept of everyday peacebuilding (Berents and McEvoy-Levy 2015) to interpret data. This framework provided insights into participants' experiences and their contributions to peacebuilding. The model offered a structured framework for interpreting interview responses and understanding peacebuilding dimensions (Toh and Cawagas 2017). Additionally, Berents and McEvoy-Levy (2015) highlight that grassroots actions in daily life can foster peace and address conflict. Lastly, thematic analysis was applied, involving careful reading and identification of recurring themes (King 2004; Rice and Ezzy 1999).

5.4 Origins and Motivations of Youth Organizations

AKINA was formally established in 2022 as part of USAID's "Opportunity 2.0" initiative, which seeks to provide second-chance education and workforce development for out-of-school youth (OSY) across various cities in the Philippines. In Iligan City, AKINA emerged as the local youth network formed under the initiative, implemented in partnership with the Education Development Center (EDC).

According to AKINA President Kelly John Robillos, the organization began as a collaborative effort between USAID and EDC to respond to the growing need for inclusive youth development structures in the city. Its formation was particularly centered on the mobilization and empowerment of OSY, enabling them to reintegrate into education and employment pathways while also fostering civic engagement.

> These NGOs (USAID and EDC) both had programs for Out of School Youth (OOSY). They looked for places all over the Philippines where they could implement these programs, and a part of these programs was to implement a local youth network or youth network.

AKINA's institutional identity is closely linked to its role as a mechanism for second-chance opportunities. Its programming reflects a commitment to peacebuilding through transformative education, emphasizing both capacity-building and inclusion. In alignment with Toh and Cawagas' (2017) Petals of Peace Model, AKINA's work resonates with the petals of inner peace, justice, and human rights, as it seeks to address structural inequalities through educational access and social reintegration.

Furthermore, AKINA exemplifies the principles of Everyday Peacebuilding (Berents and McEvoy-Levy 2015), particularly in how it supports youth in building meaningful connections within their communities. For instance, Tagorda (2024) examines how two Bangsamoro youth-led civil society organizations engage in combatting child marriage, highlighting youth participation as a central force in advancing social change. Using an institutional approach, the study explores the conditions that enable young people to act as empowered agents within their communities, positioning youth not just as beneficiaries but as critical drivers of advocacy and reform. By targeting marginalized groups such as OSY, AKINA in Iligan City contributes to sustainable peace through empowerment, participation, and local engagement.

4-H Club Iligan traces its roots to the broader international 4-H movement, which originated in the United States with a mission to promote youth involvement in agriculture, environmental stewardship, and community development. According to Cleo Bulos, a founding member and advisor of the Iligan chapter, the organization emphasizes the importance of agriculture and encourages young people to engage in farming and environmental initiatives as a means of fostering sustainable communities.

While initial efforts to establish a 4-H presence in Iligan occurred in previous years, Bulos notes that these attempts remained inactive and lacked sustained engagement. It was only in 2017 that the organization began to take shape informally, through the efforts of Bulos, Vera Bado (now serving as the club's president), and a group of environmentally committed youth from Iligan City. Their initiative aimed to strengthen environmental consciousness and promote agricultural values, particularly in barangays located in the city's hinterlands, where such efforts were most needed.

> Even though we (the officers) weren't farmers ourselves, we were very passionate about our cause in reaching out to and supporting the wider youth who share our love for agriculture and the environment. That is what drives us to organize ideas, lead activities, and initiate programs in-line with agriculture and promote activity.

The identity of 4-H Club Iligan is rooted in the principles of environmental responsibility, sustainability, and community engagement. Its programs reflect a commitment to raising awareness around ecological issues and supporting food security through localized agricultural practices. In line with Toh and Cawagas' (2017) Petals of Peace Model, the organization's efforts align with the petals of environmental harmony, justice, and dismantling the culture of war, particularly by addressing the structural neglect of rural and upland communities.

Moreover, the club's grassroots orientation reflects the principles of Everyday Peacebuilding (Berents and McEvoy-Levy 2015), as it facilitates community-based education and intergenerational dialogue. In Magno and Tagorda (2023) study, it explores the role of Bangsamoro civil society organizations (BMCSOs) in local peacebuilding, highlighting how their grounded knowledge of conflict dynamics enables them to foster long-term peace through advocacy, socialization, and dialogue facilitation. Through environmental campaigns, barangay outreach, and youth mobilization, 4-H Club Iligan contributes to building peace by fostering sustainable practices and empowering youth as stewards of their ecological and social environments.

The PSS emerged in Iligan City as a grassroots initiative dedicated to fostering inclusion, dialogue, and solidarity within the LGBTQ+ community. According to Co-Founder Ian Embradura, the group initially began as an informal gathering of

LGBTQ+ individuals seeking to celebrate Pride Month. These early meetups created a rare space for expression and visibility, providing participants with a platform to share their experiences in a safe and affirming environment.

Over time, what began as a single celebration evolved into a recurring community initiative. The group continued to hold annual Pride Month sessions, and by the third gathering, the founding members began to envision a more structured and sustainable form of organization. These sessions incorporated games, reflective lessons, and small group discussions, all designed to create a welcoming space for LGBTQ+ individuals—particularly youth—to express themselves and tell their stories. The increasing momentum of these efforts led to the formal registration of PSS in December 2023.

> We created this organization as an opportunity for the LGBTQ+ youth or folks to just tell their stories… It's one thing to acknowledge that you're gay or bi, but it's another thing to have an entire supportive community here in Iligan City who can actually understand you for being queer or non-heteronormative. We wanted to build such a community.

The identity of PSS is closely tied to themes of belonging, healing, and voice, particularly within the context of a fragmented and often marginalized LGBTQ+ community. Through storytelling and inclusive programming, the organization seeks to build interpersonal solidarity and reclaim cultural space within Iligan City's broader peacebuilding efforts. Within the framework of Toh and Cawagas' (2017) Petals of Peace Model, PSS's work resonates most strongly with the petals of inner peace, human rights, intercultural respect, and justice.

PSS also exemplifies the essence of everyday peacebuilding (Berents and McEvoy-Levy 2015), particularly in its informal, relational approach to creating change. By facilitating spaces for conversation, community-building, and storytelling, the organization nurtures the conditions necessary for interpersonal understanding and emotional healing. Its approach affirms the role of identity, representation, and inclusion as vital components of youth-led peacebuilding.

Table 5.1 provides an overview of the founding dates, motivations, and advocacy behind the establishment of the three prominent youth-led organizations: AKINA, 4-H Club Iligan, and PSS. Each of these organizations was created to address distinct social challenges within the local community of Iligan City.

5.5 Peace-in-Action: Gender, Environment, and Education as Practice

The pursuit of peace takes many forms and these three youth-led organizations each mobilizes their communities through distinct, context-driven strategies. According to the president of AKINA,

> My key function is to strategize the right direction for our organization, with guidance from the USAID Youth Coordinator. I also oversee our daily operations and engagements with

Table 5.1 The date and motivations behind the establishment of the AKINA, PSS, and 4-H Club Iligan organizations

Organization	Date of origin	Motivation	Advocacy
Asenso Kabatan-onang Iliganon Network Alliance (AKINA)	January 2021 (through USAID opportunity 2.0)	Established as a part of USAID's "Opportunity 2.0" initiative, which aims to provide second-chance education to 180,000 out-of-school youth across 15 Philippine cities, leading to better employment and livelihood opportunities	"Improving second-chance opportunities for out-of-school youth" Peace through education and opportunities
City wide 4-H club (4-H club Iligan)	2017 (informal organization) 2018 (official creation)	Established as the Iligan branch of the International 4-H Organization, focused on addressing community health inequities and advocating for equity and inclusion through civil discourse	"To make the best better" Peace through environmental awareness and self-sustenance
Pinas safe spaces (PSS)	July 2022 (informal organization) December 2023 (SEC registered)	Established to foster unity and togetherness within the fragmented LGBTQ+ community of Iligan City	"There is healing through storytelling" Peace through inclusion and storytelling

Note: Compiled by the authors based on key informant interviews, survey responses, and secondary data from the three organizations (2025)

partners and collaborators. In addition, I lead and initiate our grassroots activities within the community.

For the City Wide 4-H Club Coordinator, she said that,

I am responsible for organizing youth in our area. Last year, we were able to participate in the Regional Youth Camp. After that, we began conducting activities organized by the barangay, since the Department of Agriculture offers many youth-focused programs. As Coordinator, I make sure that we reach at least 44 barangays, although it is challenging because the 4-H Club has become politicized. This year, we're starting again, focusing on organizing by barangay, because the 4-H Club offers many opportunities, especially with support from the Department of Agriculture and the Agricultural Training Institute, which provide various forms of assistance.

Lastly, the executive director of the Pinas Safe Spaces stated that,

Setting the plan and direction for the PSS was mainly focused on expansion and networking. This time, the approach is more like a ripple effect. If we empower individuals to become Safe Space Advocates for others, I believe it will create a greater and more lasting impact.

With these strategies in mind, this part analyzes youth-led peacebuilding practices of the youth-led organizations that goes beyond institutional approaches through the lenses of education, environment, and gender using the framework, Petals of Peace Model by Toh and Cawagas (2017) as well as Everyday Peacebuilding of Berents and McEvoy-Levy (2015).

For education empowerment, AKINA is dedicated in their vision to have a community of empowered and self-reliant OSY contributing to positive youth development through bridging opportunities on education, employment, entrepreneurship, and youth engagements. This shared purpose creates a strong bond among members, fostering a sense of belonging and pride in their collective efforts to make education and opportunities accessible to vulnerable youth.

As Kelly Robillos mentioned,

> We reached around 800 Out-of-School youth who volunteered in the organization and we were able to enroll them through Technical Education and Skills Development Authority (TESDA). With this we were able to break the stigma slowly in our communities.

Through AKINA, members begin to see themselves as agents of change, not only within their organization but also in the broader context of societal transformation.

As former OSY and now Digital Officer Felmer Bonane states,

> I never thought I would gain anything. I just wanted to volunteer and help others like me who were OSY… Before, I felt I had no purpose, I was just wandering aimlessly, but now I can contribute as a person.

The organization's structure encourages active participation, offering roles in key areas such as education, youth volunteerism, and employment. For AKINA members, the work goes beyond helping others; it is about reclaiming and reshaping their own identities. By participating in initiatives that uplift the OSY community, members of AKINA realize their power to change the narrative around education and youth development. They see themselves reflected in the young people they help, and in doing so, strengthen their own sense of purpose and belonging within the organization.

Next, 4-H Club Iligan not only highlights the importance of agriculture and farming but also emphasizes the need for education through scholarships. Cleo Keven Bulos states that, "numerous students have availed the scholarships most especially those individuals who are children of the farmers. In Iligan, we also have interns who are exchange students who are currently finished in Taiwan, Japan. Numerous activities including immersions are even present in the organization." Lastly, Pinas Safe Spaces speaks for their mantra, there is "healing in storytelling." According to Ian Embradura, "our mission is clear for the LGBTQ+ Community, to create safe spaces through learning sessions and workshops to establish their feelings as part of the community."

For the environment, 4-H Club Iligan fosters a sense of identity among its members by grounding them in a shared commitment to agriculture and environmental stewardship. The organization creates a space where young people can engage in meaningful work and develop a strong personal connection to the causes they champion. Through their involvement, members come to see themselves as advocates for sustainability and leaders within their communities, reinforcing a collective identity that transcends the typical barriers faced by youth in agriculture. The 4-H Club's mission closely aligns with the personal and professional growth of its members, many of whom come from rural areas where agriculture plays a significant role in daily life.

However, as Bado noted, farming is often associated with negative societal perceptions, where "even if you're a farmer, people say you're at the margins of society." This stigma surrounding agriculture can make it difficult for young people to take pride in their work. The 4-H Club directly confronts this challenge by emphasizing the value and dignity of agricultural work. Through educational programs, advocacy, and community outreach, the organization helps members reframe their relationship with agriculture—not as a low-status occupation but as a vital and respected profession that contributes to both environmental sustainability and community well-being. By engaging with the 4-H Club, coordinator Ruth B. Panerio states that "it opens the minds of the youth to the fact that they can help the environment, and also themselves."

While not directly focused on environmental issues, they contribute to a harmonious society by promoting social inclusion and justice, which can indirectly impact environmental health by fostering community cohesion. AKINA contributes to environmental sustainability through education, entrepreneurship, and producing eco-friendly products from plastics and biodegradable materials while the PSS advocates for a society free from discrimination and inequality.

Lastly, for gender, AKINA, 4-H Club, and PSS highlight the role of inclusivity as a catalyst for peace, creating inclusive spaces for marginalized individuals which are central to their peacebuilding missions. Each organization provides a platform for underrepresented groups, such as LGBTQIA+ individuals, out-of-school youth (OSY), and rural youth, enabling them to find belonging, empowerment, and a voice in shaping their communities. For instance, PSS addresses internal fragmentation within the LGBTQIA+ community by offering a safe space where different gender identities can come together and share their experiences. As one co-founder of PSS explained,

> There is this tendency where trans people tend to gather only with other trans people, and the same happens with gay people, so there's fragmentation because they don't fully understand each other. We want to bridge that gap by providing a safe space where these different sub-sectors or gender identities can come together and learn from each other as well.

Similarly, AKINA focuses on empowering OSY, who often face barriers due to limited access to education and opportunities. Through vocational training, education, and leadership development, AKINA helps OSY reintegrate into society and take on active roles in peacebuilding. The organization's president noted,

> We reached out to around 800 out-of-school youth who volunteered and were enrolled in TESDA programs. I think that is one of our significant achievements because we were able to provide opportunities similar to those we received through our organization and from our founder, and we were able to involve them back in the community.

By addressing the stigma associated with OSY, AKINA promotes social inclusion and empowers marginalized youth to become agents of change in their communities.

4-H Club Iligan, on the other hand, uses agriculture as a tool for both economic and social inclusion. By engaging rural youth in sustainable farming practices, 4-H fosters food security and economic independence, helping young people gain practical skills while contributing to the community's long-term development. One 4-H member reflected on the organization's mission,

> The mission of 4-H is to develop young farmers who will become our food producers in the future. The average age of our farmers is around 59–60 years old, and human lifespan is decreasing. Without our 4-H Club providing early education on farming, especially focusing on basic methods rather than synthetic ones, this would be the core mission of the organization. For us, it is essential that this does not disappear from the youth sector.

The integration of the PPM Framework in grassroots initiatives demonstrates how these organizations align with the PPM. The PPM symbolizes peacebuilding as a flower with interconnected petals, each representing dimensions such as personal peace, social justice, environmental care, and nonviolence. In Iligan City, youth-led organizations like AKINA, 4-H Club Iligan, and PSS integrate various aspects of this model into their peacebuilding efforts.

For PSS, the focus is on cultivating inner peace and promoting human rights by supporting LGBTQIA+ individuals. The organization's co-founder explained that PSS does not dictate specific actions for its members but instead encourages them to share their stories, fostering emotional healing and resilience.

It is not really about giving specific instructions on what should be done or what should be learned, but more about asking, "What's your story?" When we learn about these stories and see how diverse we are, yet share the same pain and agony, that's where it has really helped me.

This approach not only supports personal peace but also addresses the marginalization of LGBTQIA+ individuals, aligning with the PPM's focus on human rights. 4-H Club Iligan, likewise, embodies the environmental care dimension of the PPM. Through sustainable agricultural practices, the organization promotes harmony with the earth, ensuring food security and community stability. One member noted that involving youth in sustainable farming is not just about agriculture, it is about ensuring the long-term stability of our community and preventing conflicts over resources. This highlights the role of environmental stewardship in maintaining peace.

AKINA's efforts align with the PPM's emphasis on dismantling the culture of war and promoting social justice. By providing education and vocational training to OSY, AKINA addresses social inequalities and helps marginalized youth reintegrate into society. As AKINA officers highlighted, the organization has successfully broken-down barriers for OSY, empowering them to contribute meaningfully to their communities and fostering a sense of social justice.

Here, we can see the significant role that youth-led organizations play in fostering peace through inclusivity, collective action, and the integration of a comprehensive peacebuilding framework. By addressing the unique challenges faced by marginalized groups such as LGBTQIA+ individuals and OSY, these organizations contribute to sustainable peace in their communities. Their efforts, grounded in the PPM, offer a holistic approach to peacebuilding that addresses personal, social, and environmental dimensions, demonstrating the powerful impact of youth-led initiatives in promoting long-term peace and development.

Table 5.2 highlights the diverse initiatives of three youth-led organizations, each addressing key community issues in line with their core values. Their activities contribute to peacebuilding and sustainable development, emphasizing the vital role of youth in creating positive change. As reflected above, the table also highlights how these organizations have integrated the Petals of Peace Model into their grassroots initiatives and programs.

5.6 Frictions, Fears, and Everyday Resistance

Youth-led peacebuilding in Iligan City is powerful, built from the initiatives of these organizations. However, these organizations face frictions, in cultural, political, and institutional contexts. Hence, this examines how young peacebuilders in the three organizations navigate challenges, assert their agency, and overcome structural discrimination while asserting their rights. This includes examining their challenges within the organization, potential conflicts between organizational work and personal life, and issues relating to budget and funding.

> '*Youth participation, mao ang pinaka main challenge*' (youth participation is our main challenge), said a key officer in one of our interviews. As peace researchers, we had the privilege of interviewing three organizations: AKINA, 4-H Club Iligan, and Pinas Safe Spaces. These organizations actively contribute to peacebuilding and creating a more just community, but they also face significant challenges. Beyond the issue of youth participation, they contend with officer commitment, funding shortages, political obstacles, and societal perceptions of the out-of-school youth, the LGBTQIA+ community, and agriculture.

AKINA has struggled with political interference, which has hindered their ability to mobilize their causes. As a key officer mentioned, "we are having difficulty implementing projects in barangays aligned with different political parties." He added, "AKINA does not support any party; it is neutral and a non-partisan organization."

4-H Club Iligan has faced a different challenge: the lack of commitment from officers and members. A key participant noted, "if the youth join organizations where they can't earn money, they tend to be less active. However, for us, if we aren't active as an organization, we can't access grants." Likewise, Pinas Safe Spaces has encountered financial constraints due to being a volunteer-driven organization reliant on donations. A key officer stated, "sometimes, we use our own money to hold sessions, as the materials, food, and drinks are affordable and necessary for the workshops."

Despite these challenges, these organizations continue to assert their agency through their actions and choices, drawing on Berents & McEvoy-Levy's everyday peacebuilding. To sustain their efforts, they remain driven by their advocacies. Pinas Safe Spaces expands its reach by establishing branches and continuing to offer safe space sessions, believing in the impact of their message that "there is healing in storytelling." They also emphasize the importance of diversity to ensure inclusivity and effectiveness. AKINA, on the other hand, actively seeks partnerships with

Table 5.2 The activities of the organizations and their messages

Organization	Activities in 2023–2024	Messages	Integration of PPM
Asenso Kabatan-onang Iliganon Network Alliance (AKINA)	Mangrove mission	Community awareness, education, and service	Nurturing inner peace: champions the rights of the out-of-school youth Living in harmony with the earth: contributes to environmental sustainability through education, entrepreneurship, and producing eco-friendly products from plastics and biodegradable materials Promoting human rights and responsibility: continues to champion the rights of out-of-school youth and fight against discrimination Building intercultural respect, reconciliation, and solidarity: encourages collective action among OSY, fostering solidarity through shared experiences and collaborative programs, such as the Savings and Internal Lending Community (SILC) Living with compassion and justice: Promotes justice through education and vocational training, allowing OSY to advocate for themselves and contribute positively to society Dismantling the culture of war: Addresses social inequalities faced by OSY and empowers them through education and vocational training, thus reducing the factors that lead to conflict
	Project "Bring Back Ganda"		
	Kabataan Kontra Basura	Sustainability and innovation	
	Operation Lingkod Kasiyahan Project 2.0		
	Project GreenNovation	Significance of youth in leadership	
	Youth leadership circle		
	E-PAFLO: YOUth informed seminar	Promotion of literacy	
	Operation Lingkod Kasiyahan		
	Book donation drive	Youth empowerment through education	
	Kabataang OSY Kabilang Ka Dito: BSKE 2023 cum opportunities session		

(continued)

Table 5.2 (continued)

Organization	Activities in 2023–2024	Messages	Integration of PPM
Pinas safe spaces (PSS)	Sari-sari safe space	Promotion of inclusivity and everyday Safe Spaces	Nurturing inner peace: advocates for gender equality Living in harmony with the earth: advocates for a society free from discrimination and inequality Promoting human rights and responsibility: Lobbying for local policies that protect youth rights and ensure their voices are heard Building intercultural respect, reconciliation, and solidarity: Addresses internal fragmentation within the LGBTQIA+ community, promoting understanding and respect among diverse gender identities Living with compassion and justice: empowers individuals by creating a safe and empathetic space for sharing experiences and healing. It advocates for human rights and social justice, addressing stigma and marginalization while promoting inclusion and equality for all Dismantling the culture of war: Fosters a culture of peace by supporting marginalized individuals and promoting understanding, which can prevent the escalation of conflicts within communities
	Pride march and celebration		
	PSS goes to Etivac	Honoring LGBTQ+ stories and experiences	
	Once upon a queer		
	Building systems of care and empathy	Exploring and challenging traditional concepts of gender	
	Late Night Chikahan and Kulitan		
	CAMP HEY: Bardagulan and quizmis night	Sparking dialogues on identity, desire, and expression	
	Gendered bodies, queering bodies		
	Gusto Mo o Gusto Ka?	Celebration of LGBTQ+ identity	
	New Year, Same Queer		

(continued)

Table 5.2 (continued)

Organization	Activities in 2023–2024	Messages	Integration of PPM
City wide 4-H club Iligan (4-H CLUB)	Nagsimula sa Pagbasa ang Bawat Batang Bida	Literacy as a foundation for agricultural knowledge	Nurturing inner peace: focuses on agriculture and environmental issues Living in harmony with the earth: raises awareness about agricultural practices and has worked on local ordinances concerning fertilizer use, while also exploring business opportunities in agriculture Promoting human rights and responsibility: Encourages youth participation in farming and advocates for farmers' rights Building intercultural respect, reconciliation, and solidarity: through community activities, such as coastal cleanups, members learn the value of teamwork and respect for diverse community members, promoting intercultural understanding Living with compassion and justice: instills compassion in youth by encouraging them to help their communities through sustainable agricultural practices and environmental initiatives Dismantling the culture of war: engages youth in activities that promote community resilience and stability, reducing tensions that can arise from resource scarcity
	Regional 4-H youth fest		
	Regional youth camp 2024	Significance of local youth engagement in the environment	
	E-PAFLO: YOUth informed		
	City-Wide youth camp	Environmental conservation and climate resilience	
	Regional 4-H youth camp 2023		
	Farm immersion: exploring agricultural learning sites	Recognizing the vital role of agriculture in community sustainability	
	Mangrove tree planting 2023		

Note: Compiled by the authors based on key informant interviews, survey responses, and secondary data from the three organizations (2025) as well as integrating the Petals of Peace Model by Toh and Cawagas (2017)

public agencies and private sectors, such as TESDA, CSWD, and the Local Government Unit of Iligan City, and they coordinate regular meetings and events for out-of-school youth. To generate income, AKINA has started producing tote bags and other products made from recyclable materials. Lastly, 4-H Club Iligan is committed to empowering youth through various opportunities. They demonstrate their agency by planning quarterly activities and fostering partnerships with grassroots communities to implement their projects.

While these three youth-led organizations are driven by their advocacies, they have faced discrimination. A key participant from AKINA highlighted this challenge, stating, "when people hear the term 'Out-of-School youth,' they often associate it with being idle or even label us as 'addicts.'" Similarly, a participant from 4-H Club Iligan pointed out, "we want to change the mindset of the youth, especially those in the hinterlands, that they are not just farmers, but are vital pillars of our community." Lastly, a key officer from Pinas Safe Spaces recalled the early days of the organization, saying, "when Safe Space started, we often overheard gossip, with people calling us an 'orgy group' simply because most of our leaders were gay men."

Despite these discriminatory attitudes, these organizations continue to fight for their causes, including gender rights, shifting the public's perception of out-of-school youth, and reshaping views on agriculture and environmental stewardship. Their commitment to their missions, particularly in peacebuilding, remains strong. Through their programs, projects, and activities, they have shown resilience and determination, actively working to combat discrimination and make meaningful contributions to their communities.

As Berents and McEvoy-Levy (2015) defined, youth peacebuilding is the proactive participation of young people in initiatives aimed at fostering peace, settling disputes, and addressing the underlying causes of violence. This engagement goes beyond traditional approaches, including everyday actions and initiatives undertaken by youth to promote peaceful coexistence (Berents and McEvoy-Levy 2015). Thus, the authors challenged traditional views by emphasizing the diverse nature of youth involvement in peacebuilding, highlighting the importance of their actions within the complexities of daily life.

Based on the main findings of the study, there is a consistent pattern that the three youth-led organizations have encountered. These organizations are advocacy driven entities in the fields of education, environment, and gender, which address social issues in the community that contributes toward peacebuilding initiatives. Despite having challenges along the way, they were able to provide measures or solutions in order to sustain the organization through their advocacy. Furthermore, the youth organizations encountered inconsistent youth involvement in their programs. They seek to motivate members to actively participate, even in the absence of organizational incentives.

As noted by a participant from 4-H Club Iligan, "the mobilization of the youth (is difficult), because sometimes, their activeness can only be sensed from the start. Sometimes, the youth are impulsive in a sense that they are committed from the beginning and then suddenly become cold." Another pattern that is observed in the findings of the study revealed that the youth-led initiatives coming from three different youth-led organizations have been integrated in the PPM framework by Toh and Cawagas. The PPM constitutes six factors of peacebuilding that make up a flower, signifying the interconnectedness with one another and these six petals were incorporated.

Contrary to expectations, this chapter revealed that the three youth-led organizations maintain a high degree of collaboration and partnerships with diverse stakeholders, encompassing local government agencies. This is one of their strategies in order to sustain their organizations. The AKINA has partnered with the TESDA, CSWD, and the USAID as their official partner. 4-H Club Iligan is under the City Agriculture's Office of Iligan City and has made collaborations little by little with the forty-four (44) barangays to disseminate their projects. Lastly, the Pinas Safe Spaces has already had events being partnered by the Local Government Unit of Iligan City as well as having partners in an academic institution. Moreover, each of the youth organizations possessed the essence of passion in serving the community driven by their pursuit of peace. This suggests that youth-led peacebuilding is increasingly recognized and supported by formal institutions.

Considering the driving elements that boost the three youth-led organizations as they contribute to the youth peacebuilding agency in Iligan City, there are certain underlying factors that block them in fostering a culture of peace in their advocacies. The role of politics has prevented these organizations from mobilizing their causes. A key participant from 4-H Club Iligan has expressed concerns about being drawn into partisan politics. He stated that "being in an organization since we are non-partisan, we should stand up and not be carried by the politics of the city. We should not hesitate to any offices or people, at least we have shared our organization and that is it." AKINA has also experienced the same case because their name is being tagged to a certain party even though they are a non-partisan organization. This contradiction highlights the challenge as the youth face in sustaining their initiatives and suggests that while their contributions are significant, they still encounter barriers to broaden the importance and its impact.

5.7 Conclusion

Compared to previous models that portray youth as passive individuals seeking validation within existing social or political structures, youth-led organizations in Iligan City actively establish inclusive spaces. These groups are not passively waiting for recognition or opportunities; instead, they are establishing new platforms directly tailored to the needs of marginalized communities. This self-generated inclusive space changes the conventional perception of youth in peacebuilding, presenting them as independent agents shaping inclusive and fair environments. Thus, the main contribution of our research delves into an extensive detail that youth-led peacebuilding is an active platform for inclusion.

In contrast to the prevailing conflict-focused paradigms often present in peacebuilding literature, this chapter provides a more interpersonal and preventive viewpoint. Through organizations like AKINA, 4-H Club Iligan, and PSS, we illustrate that peacebuilding goes beyond post-conflict recovery and encompasses initiatives addressing gender equality, environmental conservation, and educational empowerment. By aligning these initiatives with Toh and Cawagas (2017), we emphasize how these organizations embody a comprehensive approach to peacebuilding, aiming to foster long-term social cohesion rather than just preventing conflicts.

Another significant contribution of this chapter is that youth-led organizations act as agents of structural change. These organizations do not solely rely on existing institutional structures; instead, they engage in grassroots empowerment, working directly with marginalized groups such as LGBTQIA+ communities and out-of-school youth. By addressing the root causes of structural violence, as conceptualized by Galtung (1969), these organizations advocate for equity and justice, emphasizing that peacebuilding is not just about resolving conflicts but also about dismantling systems that perpetuate inequality.

This chapter also highlights the localized aspect of peacebuilding. By focusing on Iligan City, we provide a unique perspective to the broader field of peacebuilding, demonstrating how local conditions—such as socioeconomic challenges, cultural factors, and environmental issues—influence peacebuilding efforts in specific contexts. This contextual approach offers a more nuanced understanding of how peacebuilding unfolds in smaller urban settings, adding depth to the global discourse, which often overlooks localized grassroots initiatives.

Lastly, it emphasizes the role of youth in driving peacebuilding initiatives. Unlike traditional research that often positions youth as peripheral figures in peacebuilding efforts led by adults or external entities, this chapter emphasizes the independent and central role of youth as leaders in propelling peacebuilding processes. By framing youth as primary agents of change, we present a fresh perspective on youth peacebuilding, showcasing their ability to initiate transformative efforts to a more inclusive and peaceful society.

References

Barnett, M., et al., (2007). Peacebuilding: what is in a name? Global Governance, 13(1), 35–58. http://www.jstor.org/stable/27800641

Baxter, P. & Jack, S. (2008). Qualitative case study methodology: Study design and implementation for novice researchers. The qualitative report, 13(4), 544-559

Berents, H. & McEvoy-Levy, S. (2015). Theorising youth and everyday peace(building), Peacebuilding, 3:2, 115-125, https://doi.org/10.1080/21647259.2015.1052627

Broadbent, et al. (1998). Successfully completing case study research: Combining rigor, relevance and pragmatism. Information Systems Journal,8(4), 273–289.

Charmaz, K. (2006). Constructing grounded theory. A practical guide through qualitative analysis. Thousand Oaks, CA: Sage

Coombs, H. (2022). Case study research: single or multiple [White paper]. Southern Utah University. https://doi.org/10.5281/zenodo.7604301

Creswell, J. W. (2007). Qualitative inquiry and research design: choosing among five approaches (2nd ed.). New Delhi, India: Sage.

Denzin, N. K., & Lincoln, Y. S. (2018). The sage handbook of qualitative research. SAGE Publications.

Galtung, J. (1969). Violence, Peace, and Peace Research. Journal of Peace Research, 6(3), 167–191. https://doi.org/10.1177/002234336900600301

Gerson, K. & Horowitz, R. (2002). Observation and interviewing: Options and choices in qualitative research. In T. May (Ed.), Qualitative research in action (pp. 199–224).London: Sage Publications.

Gillard, S. (2001). Winning the peace: Youth, identity and peacebuilding in Bosnia and Herzegovina. International Peacekeeping, 8(1), 77–98. https://doi.org/10.1080/13533310108413880
Hudson, V. M., et al., (2009). The Heart of the Matter: The Security of Women and the Security of States. International Security, 33(3), 7–45. https://doi.org/10.1162/isec.2009.33.3.7
King, Nigel. (2004). Using Templates in the Thematic Analysis of Text. https://doi.org/10.4135/9781446280119.n21.
Magno, F. A. & Tagorda, Y. A, (2023). Unveiling the local-turn in peacebuilding: exploring the roles and contributions of Bangsamoro civil society organization in Mindanao peacebuilding. In *APISA Official Conference Proceedings* (pp. 19-31).
McEvoy-Levy, S. (2021). Youth as social and political agents: issues in post-settlement peace building. Notre Dame: Kroc Institute
Petcu, C. (2020). Recognizing youth as peacebuilders. in disarmament, demobilization and reintegration programming with Colombia's youth (pp. 7–8). Stimson Center. http://www.jstor.org/stable/resrep25152.9
Podder, S. & Ragandang, P. (2022). Youth inclusion in peace processes: the case of the Bangsamoro transition authority in Mindanao, Philippines. Conflict, Security & Development. 22. 1-20. https://doi.org/10.1080/14678802.2022.2151201.
Ragandang, P. (2020). Philippines: youth as conflict managers. peacebuilding of two youth-led non-profit organizations in Mindanao. Conflict Studies Quarterly. 87-106. 10.24193/csq.30.5
Redekop, P. (2014). Inner peace and conflict transformation. peace research, 46(2), 31–49. http://www.jstor.org/stable/24896064
Resnick, D. (2015). What is ethics in research and why is it Important? National Institute of Health.
Rice, P. L., & Ezzy, D. (1999). Qualitative Research Methods: A Health Focus. South Melbourne (Australia): Oxford University Press.
Stake, R. (2010). Qualitative research: Studying how things work. New York, NY: The Guilford Press
Tagorda, Y. A. (2024). Combatting child marriage: a study of youth-led civil society initiatives in the Bangsamoro Autonomous Region in Muslim Mindanao, Philippines. *Journal of Contemporary Sociological Issues*, *4*(1), 21-41.
Toh, S. & Cawagas, V. (2017) Building a culture of peace through global citizenship education: An enriched approach to peace education, childhood education, 93:6, 533-537, https://doi.org/10.1080/00094056.2017.1398570
UNOY. (2021). Youth Shaping the Peacebuilding Landscape. Global Network Strengthening Sustainable Youthdriven Peacebuilding. https://unoy.org/strategy2021-25/context/
Urquhart, C. (2013). Grounded theory for qualitative research. London: Sage Publications.
World Health Organization (2024). Ensuring ethical standards and procedures for research with human beings.

[illegible]

Chapter 6
From "Sirmata" to Reality: A Climate-Smart Agriculture Adoption Success in Benito Soliven, Philippines

Beata Maria F. De Ocampo

6.1 Introduction

In the heart of Isabela Province, Philippines, the municipality of Benito Soliven has achieved remarkable success in widespread climate-smart agriculture (CSA) adoption. The transformation moved from a limited government initiative into a model of community-wide engagement. This success is crucial given the Philippines' extreme climate vulnerability. The country ranks first in the 2022–2023 World Risk Index (Hilft 2022, 2023). Agriculture plays a dual role as both contributor to and victim of climate change (Intergovernmental Panel on Climate Change 2023).

The Adaptation and Mitigation Initiative in Agriculture (AMIA) Village program was launched in 2014. This climate-smart agriculture initiative was developed by the Philippines' Department of Agriculture. The program operates in partnership with the Consultative Group for International Agricultural Research, a global research partnership with 15 research centers addressing food security and sustainable agriculture in developing countries (Consultative Group for International Agricultural Research 2022; Vidallo et al. 2021).

AMIA has demonstrated remarkable growth. The program expanded from 21 villages in 2019 to 226 communities across 71 provinces by 2025. The program implements comprehensive strategies that combine climate-resilient technologies, community mobilization, and capacity building (Koerner et al. 2019; Vidallo et al. 2019).

Despite these strides, many smallholder farmers still face significant challenges. These challenges hinder the adoption of climate-smart agriculture practices. This chapter revisits the situation of smallholder farmers in the Philippines. It proposes a

B. M. F. De Ocampo (✉)
De La Salle University, Manila, Philippines
e-mail: beata.deocampo@dlsu.edu.ph

A. Sharifi et al. (eds.), *Frameworks and Practices Towards a Sustainable and Peaceful Future*, World Sustainability Series,
https://doi.org/10.1007/978-981-95-2310-8_6

trust-building framework that provides a lens through which to understand the success of the government strategy. This framework helps explain how the strategy led to community-wide adoption.

6.1.1 The Challenges of Smallholder Farmers and Climate-Smart Agriculture

Filipino farmers battle severe climate impacts. These include unpredictable seasons, intensified droughts, and devastating floods threatening their livelihoods (Mendez et al. 2021b; Peria et al. 2016). Despite these challenges, climate-smart agriculture (CSA) adoption typically remains low nationwide. Implementation programs attempt to leverage traditional Filipino "bayanihan" (communal unity). However, cultural values like "pakikisama" (social conformity) no longer sufficiently motivate participation (Alcantara et al. 2022).

Farmers face significant barriers to CSA adoption. These include financial constraints, increased labor requirements, technical complexity, and knowledge gaps. Many farmers revert to traditional practices despite training efforts (Mendez et al. 2021a, b, c).

The importance of irrigation for smallholder farmers cannot be overstated, particularly for rice production. Irrigated systems allow double to triple cropping annually. This compares to rainfed farms that are limited to single cropping (Global Yield Gap Atlas n.d.; World Bank 2022). Despite the critical need for irrigation infrastructure, only 65% of irrigable land in the Philippines has operational facilities (Department of Agriculture Communications Group 2022). This leaves upland and water-scarce areas marginalized.

Filipino smallholder farmers often become trapped in predatory lending arrangements. These arrangements involve traders and informal financiers charging exorbitant interest rates. Sometimes these rates reach 10–12% monthly, translating to over 200% annually (Catelo et al. 2017; Ecojesuit 2019). This financial exploitation results in farmers receiving only about 5% of revenue while intermediaries capture nearly half.

Climate-Smart Agriculture tackles climate challenges while boosting farmers' economic resilience. It operates through three strategic pillars: increased productivity, enhanced climate adaptation, and reduced greenhouse gas emissions (Food and Agriculture Organization 2013). CSA programs combine weather-resilient farming practices with diversification strategies. These strategies protect farmers during climate disasters (Food and Agriculture Organization 2013; Labios et al. 2021; Zheng et al. 2024).

Beyond teaching efficient planting techniques, these approaches diversify income streams to empower farming communities. Many diversification strategies serve dual purposes. They generate income while advancing climate goals (Manilay et al. 2021a, b, c; Rosales et al. 2020). Native pig raising reduces emissions by decreasing reliance on commercial feeds. Duck raising provides ecological pest control for snails in paddy fields (Rosales et al. 2020; Mendez et al. 2021b).

However, the success of CSA programs ultimately depends on individual farmer decisions. Farmers must choose to adopt and sustain these practices. Farmer engagement with CSA practices is influenced by three psychological factors. First, capacity significantly affects adoption likelihood. Farmers must believe they can successfully implement new practices (Triste et al. 2018). Second, relatedness shows mixed influence. Interpersonal relationships and emotional connections through "special trust" are crucial for farmer decision-making (Zhou et al. 2023). Third, autonomy emerges as critical for sustained engagement. Farmers highly value control over farming processes (Charatsari et al. 2016).

Building on this understanding of farmer psychology, trust emerges as a fundamental component of successful CSA adoption. The next section explores how trust develops between farmers and implementing institutions.

6.1.2 Trust-Building Framework

Trust emerges as fundamental to CSA adoption. Trust is conceptualized as the willingness to accept vulnerability based on positive expectations of others' actions. This occurs despite lacking monitoring or control capabilities (Mayer et al. 1995). This study employs Lumineau and Schilke's (2018) trust-building framework. It integrates this with Mayer et al.'s (1995) model of organizational trust. The combined framework examines how institutional relationships with farmers develop in CSA programs.

Lumineau and Schilke's (2018) trust-building framework examines how organizations develop trust through three distinct mechanisms. When applied to smallholder farmers, these mechanisms align with the trust antecedents identified in Mayer et al.'s (1995) integrative model. The three antecedents are benevolence, competence, and integrity. This alignment is particularly relevant in Climate-Smart Agriculture (CSA) initiatives. In these initiatives, trust depends on upholding promises of increased productivity and climate adaptation. Organizations must demonstrate both the capability and ethical commitment to deliver these outcomes.

First, motivational mechanisms foster trust through structured information sharing and incentive systems. In CSA programs, smallholder farmers' trust develops through clear communication about agricultural benefits and available incentives. This approach demonstrates organizational benevolence by prioritizing farmer interests and welfare.

Second, attentional mechanisms focus on operational excellence and problem-solving capabilities. Trust in CSA programs grows when implementing organizations demonstrate competence in technology, provision of extension services, and resolution of issues.

Third, interpretive mechanisms shape how information and decisions are perceived through organizational structures. Trust emerges when CSA implementing institutions demonstrate integrity through equitable policies, consistent implementation, and transparent operations.

6.2 Literature Review: Factors Influencing CSA Adoption

Climate-smart agriculture adoption among smallholder farmers is influenced by a complex interplay of institutional, social, and economic factors that can either enable or constrain sustainable farming transitions. Research demonstrates that effective approaches move beyond traditional top-down extension models toward collaborative, farmer-centered methodologies that respect local knowledge and decision-making processes. However, the path to widespread adoption remains filled with interconnected challenges that collectively create formidable obstacles for farming communities worldwide. Review of literature from local and global case studies generates five general categories that influence farmers to adopt climate-smart agriculture practices: provision of training and technical services, farmer involvement in decision-making, financial support, end-to-end value chain assistance, and institutional integrity.

6.2.1 Training, Technical Support, and Knowledge Systems

Technical complexity often exceeds farmers' existing capabilities, creating knowledge gaps that prevent successful implementation even when farmers are interested in adopting sustainable practices. Nguyen et al. (2025) provide concrete examples of this challenge, observing that Pakistani farmers lack technical knowledge for sugarcane pest management while Indonesian farmers have limited land management understanding. Similarly, Selvan et al. (2023) identify particular difficulties in assessing the fertilizing effectiveness of organic materials and weed control methods, while Rivero et al. (2017) report that some farmers did not implement vermiculture and fascine techniques simply because they lacked implementation knowledge (Chandra et al. 2017). In this regard, accessible training emerges as a critical enabler. Multiple studies confirm that free comprehensive training significantly improves adoption rates, including Climate-Smart Village approaches providing free training on climate systems, technologies, and practices (Labios et al. 2021), and Climate-resiliency Field Schools offering weekly sessions over 4–5 months (Chandra et al. 2017). Zheng et al. (2024) directly correlate training access with adoption, while Nguyen et al. (2025) emphasize that extension workers, field researchers, and government specialists fulfill essential functions in delivering timely training to agricultural communities.

Demonstration sites emerge as another critical adoption factor across multiple studies. Research confirms that providing farmers opportunities to observe and test new technologies significantly increases adoption willingness (Mendez et al. 2021a; Chandra et al. 2017). Sebastian and Bernardo (2019) describe climate-smart villages where demonstration sites act as field laboratories allowing firsthand implementation observation. This experiential learning addresses farmers' natural hesitancy toward unfamiliar practices by providing tangible effectiveness evidence before requiring time and resource investments (Labios et al. 2021; Wang et al. 2020).

Extension services are also very important as farmers are provided essential technical support in implementing CSA practices. Ma and Rahut (2024) recognize these services as critical for minimizing information gaps and enhancing farmers' understanding of sustainable practice advantages, costs, and risks. Municipal-level agricultural technicians offer guidance for farmers experimenting with new technologies (Mendez et al. 2021a), while specialists collaborate closely to implement climate-smart strategies (Labios et al. 2021). Taylor and Van Grieken (2015) highlight trust's significance in extension relationships, noting that local networks establish credibility and deliver continuous transition support.

However, individual farmer characteristics create varying levels of adoption readiness that must be carefully considered in intervention design. Age presents a particularly complex dynamic, as Li et al. (2024) and Ma and Rahut (2024) observe that older farmers may possess valuable experience that facilitates adoption but might simultaneously show greater resistance to new approaches. In the Philippines, research shows that 57-year-old average farmers tend to avoid training opportunities and show less inclination toward innovation (Dikitanan et al. 2017). Educational backgrounds and awareness levels compound these challenges, as farmers with limited educational backgrounds may need specialized interventions to understand the value of sustainable practices (Ma and Rahut 2024). Moreover, insufficient climate awareness remains a fundamental barrier, with only 37% of farmers aware of climate change impacts in some contexts (Alegado et al. 2019), highlighting the need for extension services to reduce information gaps and increase understanding of sustainable practice benefits, costs, and risks.

6.2.2 Collaborative Approaches and Farmer Autonomy

Central to CSA promotion is treating farmers as active collaborators rather than passive information recipients. Studies by Barbon et al. (2021) and Sebastian and Bernardo (2019) highlight how farmers learn effectively through participatory methods, while Chandra and McNamara (2018) explain that community-led approaches promote fair agricultural transformation. This collaborative approach creates ideal learning conditions through two-way conversations between farmers and researchers (Selvan et al. 2023), addressing limitations of conventional methods that ignore local conditions, traditional knowledge, and farmer-specific challenges (Mendez et al. 2021a).

The success of farmer-led initiatives stems from respecting farmer autonomy in decision-making and technology selection. Effective programs facilitate farmers' involvement in creating action plans and selecting suitable technologies, leading to higher sustained adoption rates because practices better fit local conditions and preferences (Mendez et al. 2021a; Labios et al. 2021; Chandra et al. 2017). This autonomy proves particularly important given that farmers typically make decisions independently or in consultation with trusted community members (Ruzol et al. 2021).

Beyond immediate climate responses, farmer-led adaptation enables comprehensive economic and social transformation. The participatory nature of successful farmer-led initiatives prioritizes inclusive participation and community ownership. Given that smallholder farmers manage 80% of the world's farms and produce food for approximately 2.5 billion people globally (Jack et al. 2022), these collaborative approaches represent a scalable pathway for building resilient agricultural systems that respond effectively to climate variability while supporting broader community development across diverse contexts.

6.2.3 *Financial Support and Economic Constraints*

Financial support mechanisms play important roles in encouraging CSA adoption. Basic provisions include additional income during calamities (Food and Agriculture Organization 2013; Food and Agriculture Organization, and United Nations Development Program 2018). Subsidies and affordable loans significantly increase adoption rates (Dinesh et al. 2017) while microfinance programs provide dividends and emergency fund access (Manilay et al. 2021a). Credit access represents a consistently identified facilitator for supporting CSA programs. Innovative mechanisms through cooperatives and agricultural credit policy councils assist farmers with pre- and post-disaster financing (Dikitanan et al. 2017; Go and Brummer 2023; Ma and Rahut 2024).

However, the research reveals a paradoxical situation where the very factors that could facilitate CSA adoption—such as credit access and market access—can simultaneously become barriers when poorly designed. Financial constraints emerge as the most pervasive obstacle, encompassing not only the immediate costs of implementing new technologies that many Filipino farmers cannot afford (Nguyen et al. 2025) but also the temporal mismatch between upfront investments and delayed returns that characterizes many sustainable practices. Farmers become trapped in debt cycles with high interest rates and short payment periods (De Ocampo 2024; Jack et al. 2022; Mwangi and Kariuki 2015) while limited customer reach leads some farmer cooperatives to become inactive or bankrupt due to market linkage challenges (Nguyen et al. 2025; Miano et al. 2024).

This temporal challenge is particularly acute, as Walsh et al. (2024) observe that yield improvements typically occur after several years with potential short-term yield reductions, while some practices like fruit tree cultivation can take up to 8 years before becoming productive, creating implementation–benefit gaps that challenge farmers needing immediate returns to maintain their livelihoods (Manilay et al. 2021b).

These financial challenges are compounded by significant labor requirements that represent a substantial hurdle, particularly as many CSA practices demand more intensive management than conventional approaches. For example, techniques like Alternate Wetting and Drying (AWD) and organic farming require more weeding work than conventional systems, demanding considerable labor and increasing operational costs (Walsh et al. 2024; Selvan et al. 2023). The situation is further complicated as rural communities simultaneously face labor shortages due to migration and off-farm employment opportunities, with farmers reporting that practices like contour farming are more labor-demanding and challenging to establish than traditional methods (Mizik 2021; Rivero et al. 2017).

6.2.4 Market Access and Value Chain Integration

Market access proves crucial, with strong evidence that farmers more readily implement CSA when they can reliably sell products at fair prices (Mwangi and Kariuki 2015; Labios et al. 2021). Successful approaches include farmer organizations helping access lucrative markets (Kangogo et al. 2020) and non-government organizations serving as intermediaries connecting producers with buyers (Go and Brummer 2023). Walsh et al. (2024) emphasize supply chain integration benefits, suggesting regional production focus can consolidate transportation and storage requirements within comprehensive value chain support systems (Mataia et al. 2020).

Comprehensive resource provision greatly improves adoption willingness. Research shows that minimizing initial farmer investments leads to higher adoption rates, demonstrated by programs distributing free fruit tree seedlings (Manilay et al. 2021b) or providing community seed banks, irrigation infrastructure, and support systems (Vidallo et al. 2019). Successful programs provide transportation, irrigation, infrastructure, safety measures (Grigorieva et al. 2023), and integrated decision-support tools including advisory services, training, financial assistance, insurance, and market connections (Dikitanan et al. 2017).

6.2.5 Systemic and Policy Challenges

The challenges faced by smallholder farmers are exacerbated by systemic policy failures that undermine long-term adoption efforts. Despite major commitments from various institutions and governments, many sustainable farming initiatives receive funding only for brief periods, with programs operating merely as pilots without scaling potential (Nguyen et al. 2025). This short-term approach conflicts fundamentally with the long-term nature of sustainable transitions. Grigorieva et al.

(2023) stress that short-term adaptation approaches may fail without sustained investments toward institutional change, while complex institutional structures generate uncoordinated policy agendas that lack proper coordination for supporting integrated sustainable farming strategies (Chandra et al. 2017).

Governance and power dynamics also introduce additional layers of complexity that can undermine even well-intentioned CSA promotion efforts. Research reveals that beneficiary selection often depends on personal connections to local authorities rather than actual need, creating unfair access to sustainable farming opportunities (Alegado et al. 2019). Benefit distribution varies considerably despite supposed eligibility, with some farmers receiving substantial support while others receive minimal or no assistance (Dulin and Lalican 2015). These power imbalances extend beyond local contexts, as Wigboldus et al. (2017) observed hierarchical relationships between different ethnic groups affecting implementation processes, resulting in farmer distrust of government policies perceived as harmful to agricultural livelihoods, which could potentially undermine participation in government-led initiatives (Ruzol et al. 2021).

Understanding these multifaceted enablers and barriers provides crucial context for examining how successful programs like the AMIA Village initiative overcome these challenges through comprehensive, trust-based approaches that address farmers' holistic needs while respecting their autonomy and local knowledge systems.

6.3 Research Methodology

Understanding the factors influencing CSA adoption and how trust develops between farmers and implementing institutions in climate-smart agriculture programs requires a comprehensive methodological approach that captures both community dynamics and the nuanced nature of trust-building. This research employed an explanatory Type 2 single-case embedded design following Yin's (2018) five-point case study framework to investigate the central question: What are the aspects of the AMIA Village program that earned the trust of Filipino smallholder farmers toward community-wide engagement in climate-smart agriculture practices?

The embedded units of analysis were the barangays of Lucban (AMIA Village) and Yeban Sur (non-AMIA Village) within Benito Soliven Municipality. These communities were considered as a single case study operating under the same local government leadership, climate vulnerability context, and market transformation experience. This design allowed for comparison between a successful AMIA implementation and a traditional non-AMIA community while controlling for external variables that might influence farmer behavior and program outcomes.

The case selection focused on the exceptional nature of the Lucban implementation, which achieved complete community-wide adoption—a rare outcome in climate-smart agriculture programs globally. This was compared not only to non-AMIA Village practices in Yeban Sur but also to other AMIA Village implementations across the country. Evidence of the model's success lies in its replication: the

Lucban AMIA Village program has been extended to three additional barangays in Benito Soliven—Maluno Norte, Maluno Sur, and Yeban Norte—specifically to replicate the identified success factors.

The research was grounded in two primary propositions that emerged from literature. The first proposition posits that farmers engage in CSA programs because of their trust in both the program and the institution implementing it. The CSA program and implementing institution constitute the external environment of the farmer, becoming external factors that either build or destroy trust based on how the program operates. The second proposition suggests that farmers engage in CSA programs because they have the capability to engage in the practices. This capability represents internal factors—farmer attributes that enable their participation in CSA practices.

The study employed a multi-layered sampling approach to capture diverse perspectives. Purposeful sampling was used to select participants for focus group discussions, with the researcher organizing discussions involving 2–4 farmers from both Lucban and Yeban Sur communities.

A total of 26 farmers participated, encompassing both male and female participants representing either rainfed or irrigated agriculture practices. Participants were selected regardless of their additional income sources or prior involvement in climate-smart agriculture, ensuring a diverse range of perspectives on farmer experiences and attitudes toward the AMIA program. Theoretical saturation was achieved when additional interviews ceased to provide new insights, with consistent themes emerging and subsequent conversations reinforcing rather than contradicting previously identified trust-building patterns.

Critical sampling complemented the purposeful approach by targeting key informants who could provide deeper contextual understanding of community dynamics and program implementation. This involved local government officials, expert farmers functioning as quasi-agricultural extension workers, and climate-smart agriculture experts in in-depth one-on-one interviews. The extended interview duration allowed for comprehensive exploration of complex topics such as trust development, community relationships, and the evolution of farming practices. A second round of interviews was conducted with local government and climate-smart agriculture experts specifically for validation of preliminary findings, ensuring that interpretations accurately reflected both the technical aspects of program implementation and the lived experiences of participants.

The qualitative analysis employed a semi-structured interview guide designed to explore relationships between trust antecedents and farmer behavioral outcomes. Atlas.ti software was used to visualize and analyze complex relationships between trust factors and farmer attributes, allowing for systematic pattern recognition and thematic development.

Four distinct triangulation methods were implemented to validate results and ensure credibility. First, the researcher conducted a comprehensive review of published AMIA Village program case studies by climate-smart agriculture practitioners and researchers, providing comparative context for the Benito Soliven findings. Second, CSA experts were consulted to verify whether the identified factors of CSA adoption were realistic and consistent with broader experiences in the Philippine context. Third, current developments were verified through government websites and news agencies to ensure that policy changes or external factors had not influenced the observed outcomes. Fourth, findings were validated through stakeholder consultations at the "Anticipatory Climate Change Adaptation: ClimBeR Learning Exchange" conference, organized by the Consultative Group for International Agricultural Research (CGIAR) and the International Institute of Rural Reconstruction (IIRR). The conference, held on October 28–29, 2024, at the Yen Center, Silang, Cavite, Philippines, provided an opportunity to test interpretations against the expertise of regional climate adaptation specialists and to receive feedback from practitioners working in similar contexts.

This methodological approach ensured that the research captured both the depth of individual farmer experiences and the breadth of institutional and community factors that contributed to the AMIA Village program's success. The combination of comparative case analysis, multiple stakeholder perspectives, and rigorous validation processes provided a robust foundation for understanding how trust-building mechanisms translate into sustained community-wide adoption of climate-smart agriculture practices. The methodology also positioned the research to contribute meaningfully to broader discussions about scaling sustainable agricultural innovations while respecting local contexts and farmer autonomy.

6.4 Results and Discussion

Benito Soliven is a landlocked municipality in Isabela province (Cagayan Valley, Region II, Philippines). The area experiences strong typhoons periodically. However, its primary climate-related vulnerability is drought. Farmers reported that drought affects all agricultural areas indiscriminately. This results in total crop failure or stunted growth. In contrast, with heavy rains, some areas remain unaffected, allowing for partial harvests.

In response to this specific vulnerability, local government CSA initiatives primarily focused on water management technologies. The AMIA Village program emphasized installing irrigation water distribution facilities beyond traditional National Irrigation Authority systems. These included solar-powered water pumps and rainwater harvesting facilities.

The AMIA Village program in Benito Soliven transformed previous approaches from selective farmer engagement to a whole-community model, achieving community-wide climate-smart agriculture (CSA) adoption within the target community. Unlike the earlier Community-Based Participatory Action Research (CPAR) program that worked with selected farmers, this model engaged entire communities beginning with a participatory rural appraisal.

Community-Based Participatory Action Research (CPAR), the local implementation of farmer learning groups (FLG) mentioned by Labios et al. (2021) and Mendez et al. (2021a, b, c), represented the earlier approach to introducing climate-smart agriculture practices in the region as a pilot program. This program contracted approximately twenty farmers with adjoining plots, providing them with training, resources, and opportunities for additional income as covered by individual contracts between the implementing institution and the selected farmers. The underlying strategy involved empowering selected farmers with knowledge and resources who would then disseminate their learning to fellow farmers in the community. However, despite being participatory and allowing farmers to decide which technologies to use on their farms, CSA adoption remained limited to program participants. This outcome aligns with experiences from other regions where farmers showed low adoption rates after program completion (Mendez et al. 2021a, b, c). The community saw this program as benefiting only a handful of chosen farmers due to their community status.

In contrast, the AMIA Village program transformed this approach by involving the entire community rather than just select farmers. The success of this transformation relied on a coordinated multi-tiered government strategy that operated seamlessly across national, provincial, and municipal levels, creating a comprehensive approach that addressed every stage of the agricultural value chain from production inputs to final market sales.

At the national level, implementation encompassed multiple critical components including infrastructure development through farm-to-market roads that connected rural communities to trading centers. Risk mitigation via calamity insurance from the Philippine Crop Insurance Corporation (PCIC) that protected farmers from climate-related losses was an existing benefit from the government. The provision of essential farming inputs and machinery was facilitated by the Rice Competitiveness Enhancement Fund (RCEF). Direct financial support through the Rice Farmers Financial Assistance (RFFA) for farming households was also provided.

Building on this national foundation, the regional and provincial governments contributed by fundamentally transforming market access through the strategic elimination of exploitative arrangements with intermediaries. They established new market systems that provided transportation to each community, allowing farmers to bring their produce directly to buyers, while creating farmer cooperatives with linkages to the National Food Authority (NFA) and other organizations that facilitated produce processing and ensured access to fair pricing mechanisms. This market opening shifted the traditional power dynamic where buyers dictated farm gate prices into a seller's market where farmers gained genuine negotiating power.

At the local government level, this multi-tiered strategy represented a significant evolution from the previous Community-Based Participatory Action Research (CPAR) model that worked with select farmers. Implementation began with the *Sirmata* process, where farmers documented their aspirations and identified necessary resources for climate resilience. While initial skepticism emerged due to their previous CPAR experience—where programs had benefited only select individuals and where the "same faces" repeatedly received benefits—this resistance dissolved when promised support materialized according to their *Sirmata* vision. The AMIA whole-community model provided a comprehensive local framework encompassing farming inputs, multiple irrigation options, technical training, entrepreneurial opportunities, extension services, and responsive feedback systems that addressed farmer concerns in a timely manner. This transformation created a sustainable model where agricultural prosperity became achievable for entire smallholder farming communities rather than just select individuals.

6.4.1 Building Trust Through Comprehensive Support

The AMIA Village program succeeded through three interconnected trust-building mechanisms that addressed farmers' most fundamental concerns while demonstrating institutional reliability. The first mechanism, motivational trust-building, earned farmer confidence by addressing their most pressing concern—economic vulnerability—through unprecedented comprehensive support that freed farmers from predatory debt cycles. Enhanced financial capacity became the primary enabler for farmers to trust and engage in the local government unit's prescribed CSA practices.

Farmers were freed from debt cycles. As evidenced among the farmer respondents, the majority expressed low interest in loans and did not need crop insurance proceeds after the onslaught of calamities like drought or typhoon. As one farmer explained, they received various forms of support, but what they appreciated most was the regular provision of seeds and fertilizers, which resulted in a considerable reduction of expenses. The continuous provision proved more valuable than one-time payments, aligning with Mills et al. (2017), who found that financial aid may initiate behavioral change but fails to achieve long-term engagement goals. Farmers expressed little interest in the government's annual P5000 assistance (Relativo 2023), preferring sustained resource provision over one-time cash payments.

The most transformative strategy was the "Isabela Model," which provided market access through end-to-end value chain support at local and regional government levels. This converted traditional dependence on intermediaries for loans and produce disposal into a seller's market where farmers gained choice in where to sell their goods and at what price (Department of Agriculture-Agriculture and Fisheries Information Division 2019).

The market access mechanism operated through a network of farmer cooperatives, the National Food Authority (NFA), and private enterprises. The cooperative named *Nagkaisang Magsasaka ng* Isabela (translated as "United Farmers of Isabela") further strengthened market access by offering higher prices than traditional traders. Key enablers included building farm-to-market roads and providing trucks to facilitate transport of produce to market.

Internet accessibility allowed price canvassing through the community-based "marites" system (meaning "*Mare, anong* latest?" or "Friend, what's the latest?"), a term that colloquially encapsulates the Filipino cultural phenomenon of informal, rapid information sharing akin to electronic word-of-mouth more frequently shared through Facebook Messenger. With the opening of the market, traders now approached individual farmers who handled transport, making it easier for farmers to dispose of crops at fair prices.

Government support covering every stage of the value chain (Mataia et al. 2020) created dual benefits: reduced production expenses upstream and enhanced income through fair market prices downstream. This fostered high trust. One farmer stated confidently that there have been no promises from the government that went unfulfilled.

In line with CSA climate adaptation principles, farm diversification opportunities were provided to help communities during calamities. Beyond farming main crops, the AMIA Village program established facilities for livestock raising, mushroom spawn production, and *pandesal* baking—a popular Filipino breakfast bread—creating climate-resilient income streams (Food and Agriculture Organization 2013; World Bank 2022). This diversification approach is supported by Zheng et al.'s (2024) comprehensive review finding that CSA practices are mostly combinations of farming techniques and diversification strategies for empowered farming communities. Training sessions helped farmers develop income-generating activities through project proposal creation, enabling them to apply for funding opportunities.

6.4.2 Demonstrating Competence Through Responsive Governance

The second trust-building mechanism, attentional mechanisms, manifested through innovative extension service delivery that respected farmers' "to see is to believe" mentality while addressing practical implementation challenges. One particularly impressive strategy appreciated by farmers was holding annual or biannual general assemblies attended by unit heads to ensure quick resolution of farmers' concerns. This comprehensive support system addressed typical extension service challenges by providing timely technical assistance and resolving farmer concerns quickly.

The establishment of quasi-extension services complemented the limited number of formal Agricultural Extension Workers (AEW) serving the 29 barangays of Benito Soliven, providing a timely venue for addressing farmers' concerns. This addressed the globally identified challenge of insufficient extension services in CSA adoption (Wang et al. 2020) through a cascading mechanism of technical support where AEWs collaborated with expert farmers within the community, including Farmer-led Extensionists (FLE) for corn production and Local Farmer Technicians (LFT) for rice cultivation. Mobile and social media channels were utilized for communication toward immediate problem resolution.

A contradiction of literature accounts of the success of demonstration sites as a promoter of climate-smart practices exists in Benito Soliven. General skepticism existed regarding new technologies and their introduction through demonstration sites as proof of concept due to perceived controlled conditions. The Agricultural Extension Worker explained the farmers' stance of "to see is to believe" and therefore, would not believe any claims until they themselves have tried the technology.

Several farmer respondents in the study criticized demonstration sites for unrealistic controlled conditions of an ideal situation where testing is done on a small plot of land, which is provided with ample irrigation, and monitored and maintained daily by a dedicated technical person.

Addressing this skepticism, a commercial company employed a local farmer to test a new corn seed variety on his entire farmland. Initial results were promising enough to convince other farmers to invest in the seeds. However, the subsequent planting season revealed unexpected pest problems and other significant technical challenges. One farmer noted that the test results were excellent at the demonstration site. So, in the succeeding planting season, many were convinced and purchased the supposed pest-resistant seed variety but were frustrated when worms attacked their crops.

It is important to note that each growing cycle represents a significant investment of time and resources for cash-strapped farmers. Therefore, they must approach these technical decisions with utmost care. Their livelihoods depend on making correct technical choices. They cannot afford the three-month setback and additional costs associated with replanting. Therefore, intensive training and responsive extension services rather than just demonstration of products and technology are necessary.

Farmer respondents of the study willingly engaged in various trainings, even those spanning four months. From literature, free and accessible training emerges as a critical enabler of sustainable farming adoption, which significantly improves adoption rates (Labios et al. 2021; Chandra et al. 2017). Training is traditionally delivered through extension services (Dulin and Lalican 2015) and increasingly through field schools as demonstration sites (Asia-Pacific Association of Agricultural Research Institutions 2018; Chandra et al. 2017; Vidallo et al. 2019).

6.4.3 Establishing Integrity Through Inclusive Governance

The third mechanism, interpretive trust-building, projected institutional integrity by addressing elite capture concerns and honoring farmers' aspirations and wisdom through participatory processes. "Elite capture" is a phenomenon wherein benefit distribution reaches the most influential members of the community first (Bask and Bask 2015), with Dr. Julian Gonsalves, one of the study's subject matter experts, identifying an even deeper manifestation as "cumulative advantage," where the same individuals receive benefits repeatedly across time. This selective engagement reinforces existing inequalities, making power dynamics a critical trust factor in CSA adoption, as confirmed by literature on institutional trust development (Dulin and Lalican 2015; Wigboldus et al. 2017).

While elite capture remained a typical challenge in local and global program implementations, the AMIA Village program introduced a transformative approach that combined equitable benefit distribution with genuine community participation. Rather than simply implementing top-down solutions, the program engaged farming communities through a bottom-up strategy beginning with participatory rural appraisal, where farmers articulated their needs through a visualization exercise called *Sirmata* (vision or dream in Ilocano dialect). Faithful to the CSA framework's pillars of farmer productivity and climate resilience, Municipal Agriculture Officer Engr. Edmund Gauiran reported that all identified needs from the farmers' *Sirmata* were subsequently provided, including infrastructure, machinery, farming inputs, and livelihood opportunities, demonstrating genuine commitment to farmer-identified priorities.

The program's design strategically minimized elite capture through mandatory farmer association formation, requiring communities to organize themselves as legally accredited associations to receive benefits collectively rather than individually. This structure created a multiplier effect that increased farmer confidence in the farmer association and the government as the financial capacities of farmers grew individually and collectively.

The associations evolved far beyond compliance mechanisms, with communities winning awards, purchasing their own trucks, and managing equipment rental services, which essentially developed what Bourdieu (1987) termed as "symbolic capital"—accumulated wealth in the form of prestige and collective identity convertible into economic benefits. Members demonstrated remarkable collective responsibility through Sunday "obra kumon" (communal work) sessions, maintaining critical infrastructure like rainwater harvesting facilities. The farmers proudly described their unity using a powerful local metaphor, comparing themselves to the individual strands of a broom that gain strength when tightly bound together.

This farmer association structure proved instrumental in creating dual layers of trust that sustained climate-smart agriculture engagement far beyond previous individual contract models. The legal personality of associations established formal trust relationships with government that guaranteed equitable benefit distribution to all community members. Simultaneously, internal trust was fostered among members through shared responsibility and collective accountability. This dual trust mechanism proved far more effective than earlier farmer learning group models, where selected farmers signed individual contracts with government but often ceased CSA engagement once contracts ended (Mendez et al. 2021b).

Trust emerges as essential for sustained farmer engagement (Fielke et al. 2022). The formal association structure provided more than contractual frameworks. It provided a good vehicle for establishing sustainable models for imbibing a climate-smart mindset through an equitable distribution of benefits while empowering farmers as key stakeholders in their communities. Beyond mere compliance, the legality of the farmer association created binding mutual obligations that reinforced both vertical trust between farmers and government and horizontal trust among community members.

Indeed, establishment of integrity through inclusive governance fostered trust, which is a strong foundation for a sustainable CSA engagement. However, an unexpected paradox emerged in how farmers experienced autonomy within the supportive AMIA Village program framework.

6.4.4 The Contra-Autonomy Paradox

While these three trust-building mechanisms successfully resulted in community-wide CSA adoption, they also revealed an unexpected paradox that challenges conventional understanding of farmer autonomy. This study introduces the concept of "contra-autonomy," wherein farmers do not feel coerced to comply with prescribed sustainable farming practices but instead feel gratitude for their hyper-dependence on government support. They construe their compliance as autonomous even though prescribed farming practices did not originate from their own sustainable farming decisions. This phenomenon illustrates how farmers become not just willing but enthusiastic participants in CSA programs when comprehensive support leads to considerably increased financial capacities. Still, it raises the critical question: Will they continue these practices when external support is reduced or eliminated?

This paradox manifested clearly in the practice of Alternate Wetting and Drying (AWD), a water-saving technique using a tube with holes called an observation well to measure irrigation water levels (Allen and Sander 2019). The Agricultural Extension Worker related that while nearly all farmers initially adopted AWD when it was first introduced, subsequent random visits revealed that many observation wells had disappeared, indicating farmers had reverted to previous practices. The extension worker attributed this to occasional farmer complacency, noting that enthusiasm for the new technique gradually diminished without ongoing monitoring and support. This observation reveals that since the practice was not a conscious choice, farmers tended not to do so without ongoing supervision, as the practice was externally motivated rather than internalized.

However, one elderly woman farmer exemplified true autonomy in CSA adoption, demonstrating what genuine environmental stewardship looks like beyond external incentives. Despite limited formal education, she achieved remarkable crop yields through serious commitment to training and determination to overcome educational limitations, earning agricultural awards at provincial and regional levels. Her persistent organic fertilizer production—meticulously collecting dried leaves and animal waste despite the labor-intensive nature of the practice—stood in stark contrast to widespread chemical fertilizer use among fellow farmers. Her case demonstrates how true autonomy transcends common constraints of age, gender, and educational attainment. While literature indicates that additional labor requirements often serve as adoption barriers (Mendez et al. 2021b), her success shows that effective training combined with intrinsic motivation creates lasting impact beyond provision of incentives and external support systems.

This contrast between externally motivated compliance and internally driven commitment raises fundamental questions about the long-term sustainability of CSA adoption and the ultimate goal of environmental stewardship. Within the Philippine context, the government regulates farmgate prices to keep basic food commodities affordable for consumers, particularly rice, which serves as the primary staple for the population. This price regulation policy necessitates comprehensive input subsidization to ensure that farmers can maintain productivity levels.

Without such subsidies covering seeds, fertilizers, machinery, and other production inputs, farmers would face financial losses that could threaten both their livelihoods and the nation's food security. This creates a structural dependence where farmers rely on government support not due to inefficiency, but as a direct consequence of policies designed to balance affordable food access for consumers with viable farming operations for producers.

Such dependence represents structural reality rather than policy failure, reflecting the complex economic dynamics of agricultural systems where social welfare considerations override pure market mechanisms. However, this reality underscores the challenge of distinguishing between farmers who adopt CSA practices out of genuine environmental conviction versus those who comply merely as conditions for receiving benefits. The path forward lies in transforming these necessary support systems from mere productivity enhancement toward cultivating farmers as environmental stewards who recognize their critical role in protecting our common home. This transformation requires moving beyond prescribed adoption of CSA practices toward genuine farmer-led adaptation that honors both economic necessities and environmental imperatives. When effective training combines with years of farmer experience and indigenous knowledge, it creates the foundation for farmer-led innovation where farmers become adaptive leaders rather than passive implementers, drawing on both scientific knowledge and generational wisdom to develop locally appropriate solutions that ensure agricultural systems serve multiple purposes: feeding the nation, supporting farming livelihoods, and protecting the ecological foundation upon which future generations depend—regardless of external support levels.

6.5 Conclusion and Recommendations

This examination of Benito Soliven AMIA Village implementation reveals the transformative potential of comprehensive, multi-tiered government strategies when systematically implemented to build trust and achieve community-wide climate-smart agriculture adoption. The case study demonstrates that exceptional success emerges not from isolated interventions, but from coordinated approaches that operate seamlessly across national, regional, provincial, and local levels while fundamentally transforming traditional farmer–institution relationships through trust-building mechanisms.

The remarkable transformation from the selective Community-Based Participatory Action Research (CPAR) model to the whole-community AMIA Village approach illustrates how systematic trust-building can overcome historical farmer resistance to agricultural programs. Unlike previous initiatives that benefited only select individuals and reinforced elite capture patterns, the AMIA Village program achieved complete community-wide adoption through three interconnected trust-building mechanisms: demonstrating institutional benevolence by addressing farmers' economic vulnerabilities through comprehensive value chain support;

showing institutional competence through responsive extension services and innovative governance; and projecting institutional integrity by honoring farmer aspirations through the culturally sensitive "Sirmata" participatory process and the equitable distribution of benefits through farmer associations.

The multi-tiered coordination proved essential, with national-level support providing infrastructure, risk mitigation, farming inputs, and financial assistance; regional and provincial governments transforming market access by eliminating intermediaries; and local government implementing whole-community models through participatory rural appraisal and comprehensive support frameworks. This systematic approach created dual layers of trust—vertical trust between farmers and institutions and horizontal trust among community members—ensuring that climate-smart agriculture practices became embedded in the community social fabric.

However, this success revealed the "contra-autonomy paradox," wherein farmers experienced apparent autonomy while remaining dependent on government support. This highlights the critical challenge of transitioning from externally driven compliance to genuine farmer-led adaptation, particularly within structural realities where government price regulation necessitates input subsidization for both farmer productivity and national food security. The transformative success demonstrates that sustainable agricultural transformation must emerge from deep respect for farming communities' values, knowledge systems, and aspirations, with farmers seeing climate-smart agriculture as pathways to realizing their own dreams rather than fulfilling external demands. The following sections detail recommendations for scaling success.

6.5.1 Establish Coordinated Multi-Tiered Government Strategies

Replicate the Benito Soliven model's systematic approach across national, regional, and local levels. Policymakers must create frameworks enabling seamless collaboration between government agencies, ensuring each governance level contributes complementary components while avoiding duplication or gaps in service delivery. This requires institutionalizing coordination mechanisms where national levels provide infrastructure and financial support, regional levels focus on market transformation, and local levels implement equitable distribution of resources and provide timely extension services. The evidence demonstrates that when all levels work in concert, trust development accelerates and community-wide adoption becomes achievable.

6.5.2 *Prioritize Market System Transformation*

Replicate the "Isabela Model" approach that converts buyer's markets into seller's markets. While input subsidies remain necessary within price-regulated food systems, policy emphasis should shift toward creating farmer cooperatives with direct linkages to processing facilities and marketing networks that eliminate exploitative arrangements with intermediaries. This includes building farm-to-market roads, providing transportation support, and establishing pricing transparency mechanisms that enable farmers to make informed marketing decisions. The goal should be to create structural changes that enhance farmers' bargaining power and income stability rather than temporary assistance programs.

6.5.3 *Create Policy Frameworks for NGO–Government Partnerships*

Collaboration between non-governmental organizations and national government agencies in delivering comprehensive end-to-end value chain support for smallholder farmers must be institutionalized. While the AMIA Village program has now expanded to 226 communities, many farming communities across the country still have not been reached or may not be included among the target communities. Considering that the current model necessitates provision of all farming inputs, questions of sustainability arise regarding how farmers could be weaned from total dependence.

It is therefore important to engage non-government organizations (NGO), in alignment with UN SDG 17 (Partnerships for the Goals), especially to cover more farming communities toward poverty alleviation (UN SDG 1), ensuring food security (UN SDG 2), and promoting climate action (UN SDG 13). Policymakers should establish formal mechanisms that leverage NGO expertise in community mobilization, technical trainings, and entrepreneurship toward developing diverse sources of income while utilizing government resources for infrastructure development, market linkages, and financial support systems. Policy frameworks should incentivize NGO participation through co-financing mechanisms, technical assistance provision, and preferential access to government programs for NGO-supported farmer groups, creating coordinated partnerships that address farmers' holistic needs more effectively than either sector working independently.

6.5.4 Foster Farmer-Led Innovation on Climate-Smart Practices

Encourage innovative climate-smart practices led by farmers through culturally sensitive participatory processes that transition from externally driven compliance to autonomous environmental stewardship. Programs should explicitly design progression pathways that gradually build farmer capacity for independent decision-making and innovation while maintaining necessary support systems. This requires institutionalizing approaches that honor farmer aspirations and wisdom while creating genuine ownership of development goals. Extension services should evolve into collaborative learning platforms where local agricultural champions become primary voices for change, supported rather than replaced by external technical experts. The ultimate goal involves creating farmer-to-farmer innovation networks that combine effective training with farmer experience and indigenous knowledge to develop locally appropriate solutions. Programs should include clear timelines and milestones for increasing farmer autonomy while identifying which support elements require permanent institutional frameworks versus those that can be gradually transferred to farmer control.

The Benito Soliven experience illuminates that when comprehensive, coordinated support systems demonstrate institutional trustworthiness while respecting farmer dignity and aspirations, remarkable transformation becomes possible. The journey from *Sirmata* to reality proves that sustainable practices flourish when communities unite around shared economic goals while recognizing their farms as both livelihood and legacy. The path forward requires replicating this systematic approach to trust-building while fostering farmer-led innovation networks that ensure climate adaptation toward supporting farming livelihoods, and protecting the ecological foundation upon which future generations depend. When farmers see their land not just as production sites but as home for generations to come, climate adaptation evolves from external requirement to internal commitment, ensuring sustainable agriculture that endures beyond any external program.

Acknowledgments The researcher would like to thank the following for their invaluable support: (1) Municipal government of Benito Soliven, particularly Mayor Roberto T. Lungan, Municipal Agriculture Officer, Engr. Edmund Gauiran, and Agricultural Extension Worker, Ms. Fides Amor Reyes; (2) her dissertation adviser, Dr Patrick Aure; and the subject matter experts, Dr Julian Gonsalves, Ms. Apple Eda Suplido, Mr. Carlito Balingbing, and Dr. Cholo Javier.

References

Alcantara, A. J., Basmala, J., Sepillos, S. Y., & Pasquin, E. (2022). The influence of Filipino culture on the mechanization program in Nueva Ecija, Philippines. *International Journal of Social and Management Studies*, *3*(1), 208–220. https://doi.org/10.5555/ijosmas.v3i1.100

Alegado, J. L., Pabuayon, I., Cateloc, S., & Camacho Jr., J. (2019). Measuring the performance of communal irrigation systems in Bohol, Philippines. *Asian Journal of Agriculture and Development, 16*(2), 25–44. https://doi.org/10.22004/ag.econ.298423

Allen, J., & Sander, B. (2019). *The diverse benefits of alternate wetting and drying (AWD).* CGIAR. https://cgspace.cgiar.org/handle/10568/101399

Asia-Pacific Association of Agricultural Research Institutions. (2018). *Climate smart farmers' field school as extension modality for climate change adaptation in rice farming: Bicol, Philippines.* http://www.apaari.org/web/wp-content/uploads/downloads/2018/Climate%20Smart%20Farmer-Success%20Story_10-8-2018_Final.pdf

Barbon, W. J., Punzalan, B., Wassman, R., Bui, V.L., Vidallo, R., Villanueva, J., Talsma, T., Bayot, R., Gonsalves, J. (2021). *Scaling of climate-smart agriculture via climate-smart villages in Southeast Asia: Insights and lessons from Vietnam, Laos, Philippines, Cambodia and Myanmar.* (CGIAR Research Program on Climate Change, Agriculture and Food Security Working Paper no. 376). https://hdl.handle.net/10568/115833

Bask, M., & Bask, M. (2015). Cumulative (dis)advantage and the Matthew effect in life-course analysis. *PLoS ONE, 10*(11), Article e0142447. https://doi.org/10.1371/journal.pone.0142447

Bourdieu, P. (1987). The force of law: Toward a sociology of the juridical field. *The Hastings Law Journal, 38*, 805–853.

Bündnis Entwicklung Hilft. (2022). *World risk report 2022.* https://weltrisikobericht.de/wp-content/uploads/2022/09/WorldRiskReport-2022_Online.pdf

Bündnis Entwicklung Hilft. (2023). *World risk report 2023.* https://reliefweb.int/attachments/c095742d-f207-4cdd-a998-18e0cc6cac31/WRR_2023_english_online-1.pdf

Catelo, S. P., Catelo, M. A. O., & Mina, C. S. (2017). Rural financial markets and credit delivery system in the Philippines. In C. E. C. Gan & G. V. Nartea (Eds.), *Microfinance in Asia* (pp. 177–226). World Scientific.

Chandra, A., & McNamara, K. E. (2018). Climate-smart agriculture in Southeast Asia: Lessons from community-based adaptation programs in the Philippines and Timor-Leste. In *Resilience: The science of adaptation to climate change* (pp. 165–179). Elsevier. https://doi.org/10.1016/b978-0-12-811891-7.00013-x

Chandra, A., McNamara, K. E., & Dargusch, P. (2017). Climate-smart agriculture: Perspectives and framings. *Climate Policy, 18*(4), 526–541. https://doi.org/10.1080/14693062.2017.1316968

Charatsari, C., Lioutas, E. D., & Koutsouris, A. (2016). Farmers' motivational orientation toward participation in competence development projects: A self-determination theory perspective. *The Journal of Agricultural Education and Extension, 23*(2), 105–120. https://doi.org/10.1080/1389224x.2016.1261717

Consultative Group for International Agricultural Research. (2022, December 6). *Philippines ClimBeR inception workshop report.* https://cgspace.cgiar.org/bitstream/handle/10568/129052/ClimBeR%20Philippines%20Inception%20Workshop%20Report_06Dec22.pdf?sequence=1

De Ocampo, B. M. (2024). Models of smallholder farmer upliftment: A multiple case study on Philippine rice farming communities. *DLSU Business Notes and Briefings, 12*(1), 1–12. https://www.dlsu.edu.ph/models-of-smallholder-farmer-upliftment-a-multiple-case-study-on-philippine-rice-farming-communities/

Department of Agriculture-Agriculture and Fisheries Information Division. (2019, October 12). RCEF rollout in Isabela marks milestone for PH rice industry agriculture. *Department of Agriculture.* https://www.da.gov.ph/rcef-rollout-in-isabela-marks-milestone-for-ph-rice-industry-agriculture/

Department of Agriculture Communications Group. (2022, June 2). NIA: Projects expanded PHL's irrigated areas. *Department of Agriculture.* https://www.da.gov.ph/from-business-mirror-nia-projects-expanded-phls-irrigated-areas/

Dikitanan, R., Grosjean, G., Nowak, A., & Leyte, J. (2017). *Climate-resilient agriculture in Philippines. CSA Country Profiles for Asia Series.* https://amia.da.gov.ph/wp-content/uploads/2018/08/CRA_Profile_Philippines.pdf

Dinesh, D., Aggarwal, P., Khatri-Chhetri, A., Loboguerrero Rodríguez, A. M., Mungai, C., Radeny, M., Sebastian, L., & Zougmoré, R. (2017). *The rise in climate-smart agriculture strategies, policies, partnerships and investments across the globe* (Vol. 30). https://cgspace.cgiar.org/rest/bitstreams/b53e0bef-1a83-4c65-980a-02a3cdfd4227/retrieve

Dulin, J., & Lalican, N. (2015). Socio-economic impacts of climate change support-policies to farming systems in a village in Tuguegarao city, Cagayan, Philippines. *Journal of Environmental Science and Management, 18*(1), 61–70. https://doi.org/10.47125/jesam/2015_1/07

Ecojesuit. (2019, October 24). *Why the margins persist in a growing economy* [Video]. YouTube. https://www.youtube.com/watch?v=d9TY1uUe3q4

Fielke, S., Taylor, B. M., Coggan, A., Jakku, E., Davis, A. M., Thorburn, P. J., Webster, A. J., & Smart, J. C. R. (2022). Understanding power, social capital and trust alongside near real-time water quality monitoring and technological development collaboration. *Journal of Rural Studies, 92*, 120–131. https://doi.org/10.1016/j.jrurstud.2022.03.022

Food and Agriculture Organization. (2013). *Climate-smart agriculture sourcebook.* https://www.fao.org/3/i3325e/i3325e.pdf

Food and Agriculture Organization, & United Nations Development Program. (2018). Integrating agriculture in national adaptation plans programme: The Philippines case study. In *FAO*. https://openknowledge.fao.org/items/f1b56538-17d2-458f-85fd-b52e0233c73f

Global Yield Gap Atlas. (n.d.). *Philippines: Rice production in the five Southeast Asian countries, Vietnam, Thailand, Myanmar, Philippines, and Cambodia.* Retrieved May 3, 2023, from https://www.yieldgap.org/philippines

Go, K. C., & Brummer, M. (2023). Power, governance, and global value chains: Case studies on NGO agri-food innovation intermediaries in the Philippines. *Asian Journal of Technology Innovation*, 1–22. https://doi.org/10.1080/19761597.2023.2225777

Grigorieva, E., Livenets, A., & Stelmakh, E. (2023). Adaptation of agriculture to climate change: A scoping review. *Climate, 11*, 1–37. https://doi.org/10.3390/cli11100202

Intergovernmental Panel on Climate Change. (2023). *Climate change 2023: Synthesis report.* https://doi.org/10.59327/IPCC/AR6-9789291691647.001

Jack, G., Plahe, J., & Wright, S. (2022). Development as freedom? Insights from a farmer-led sustainable agriculture non-governmental organisation in the Philippines. *Human Relations,* 75(10), 1875-1902. https://doi.org/10.1177/00187267221090779

Kangogo, D., Dentoni, D., & Bijman, J. (2020). Determinants of farm resilience to climate change: The role of farmer entrepreneurship and value chain collaborations. *Sustainability, 12*(3), Article 868. https://doi.org/10.3390/su12030868

Koerner, J., Bayot, R. S., Rosimo, M., Vidallo, R., & Gonsalves, J. (2019). *Scaling the capacities to adapt to a changing climate experiences of the AMIA climate resilient villages, Philippines.* CGIAR. https://cgspace.cgiar.org/bitstream/handle/10568/105717/CCAFS%20Info%20Note_Scaling%20the%20capacities%20for%20innovation.pdf?sequence=3

Labios, R. V., Galang, E. I. N., Orencio, P. M., Malayang, D. B. N., Vallez, A. G. C., & Lapitan, R. B. (2021). *ASEAN regional guidelines for promoting climate smart agriculture (CSA) practices* (Vol. 3). ASEAN Climate Resilience Network. https://asean-crn.org/wp-content/uploads/2023/05/2023_ASEAN-Guidelines-on-the-Promotion-of-CSA-Practices-Vol-3_adopted.pdf

Li, J., Ma, W., & Zhu, H. (2024). A systematic literature review of factors influencing the adoption of climate-smart agricultural practices. *Mitigation and Adaptation Strategies for Global Change, 29*(1). https://doi.org/10.1007/s11027-023-10098-x

Lumineau, F., & Schilke, O. (2018). Trust development across levels of analysis: An embedded-agency perspective. *Journal of Trust Research, 8*(2), 238–248. https://doi.org/10.1080/21515581.2018.1531766

Ma, W., & Rahut, D. B. (2024). Climate-smart agriculture: Adoption, impacts, and implications for sustainable development. *Mitigation and Adaptation Strategies for Global Change, 29*(5), 1-23. https://doi.org/10.1007/s11027-024-10139-z

Manilay, A. A., Barbon W. J., Monville-Oro, E., Thy, O., Rondon, M., Cabriole, M. A., & Sokchea, M. (2021a). *Microcredit and savings associations for building rural household resilience: A case study of selected village development fund and savings groups in Koh Kong and Mondul Kiri, Cambodia* (Working Paper No. 373). CGIAR. https://hdl.handle.net/10568/115997

Manilay, A. A., Cabriole, M. A., Itliong, K., Jordan, R., Rosimo, M., Monville-Oro, E., Barbon, W. J., & Gonsalves, J. (2021b). *Coconut-based systems in the Philippines: Intensification and diversification with climate-smart agriculture* (Working Paper No. 367). CGIAR. https://hdl.handle.net/10568/115290

Manilay, A. A., Phen, B., Cabriole, M. A., Thy, O., Gonsalves, J., Monville-Oro, E., & Barbon, W. J. (2021c). *A financial analysis of homestead native chicken raising: A climate-smart agriculture option adopted in the province of Koh Kong, Cambodia* (Working Paper No. 362). CGIAR. https://hdl.handle.net/10568/114901

Mataia, A. B., Beltran, J. C., Manalili, R. G., Catudan, B. M., Francisco, N. M., & Flores, A. C. (2020). Rice value chain analysis in the Philippines: Value addition, constraints, and upgrading strategies. *Asian Journal of Agriculture and Development*, *17*(2), 19–42. https://doi.org/10.22004/AG.ECON.307976

Mayer, R. C., Davis, J. H., & Schoorman, F. D. (1995). An integrative model of organizational trust. *Academy of Management Review*, *20*(3), 709–734. https://doi.org/10.5465/amr.1995.9508080335

Mendez, K., Vidallo, R., Rosimo, M., Angeles, D., Bernardo, E., & Gonsalves, J. (2021a). *Learning groups: Refining technologies and social processes for climate resilient agriculture*. CGIAR. https://cgspace.cgiar.org/bitstream/handle/10568/114778/Learning%20Groups%20research%20report_2021.pdf

Mendez, K., Vidallo, R., Rosimo, M., Angeles, D., Rosales, B., Bernardo, E., & Gonsalves, J. (2021b). *Portfolio of climate resilient options for farming and fishing communities in Guinayangan, Quezon*. CGIAR. https://hdl.handle.net/10568/114770

Mendez, K., Vidallo, R., Rosimo, M., Servano Jr, G., Urdelas, F., Bernales, L., Bernardo, E., & Gonsalves, J. (2021c). *Portfolio of climate resilient options for farming and fishing communities in Ivisan, Capiz*. CGIAR. https://cgspace.cgiar.org/bitstream/handle/10568/114772/Ivisan%20Research%20Report_ver02%20(2).pdf?sequence=1

Miano, C., Cabrera, G. A., Ann Yao, C. M., Cayabat, C. V, & Laguador, A. A. (2024). Scoping the needs and challenges of farmers and fisherfolks in the AMIA (Adaptation and Mitigation Initiatives In Agriculture) Villages in San Francisco, Quezon towards developing a community-based enterprise. *Educational Administration:Theory and Practice*, *30*(5), 789–796. https://doi.org/10.53555/kuey.v30i4.2947

Mills, J., Gaskell, P., Ingram, J., Dwyer, J., Reed, M., & Short, C. (2017). Engaging farmers in environmental management through a better understanding of behaviour. *Agriculture and Human Values*, *34*(2), 283–299. https://doi.org/10.1007/s10460-016-9705-4

Mizik, T. (2021). Climate-smart agriculture on small-scale farms: A systematic literature review. *Agronomy*, *11*(6), 1-16. https://doi.org/10.3390/agronomy11061096

Mwangi, M., & Kariuki, S. (2015). Factors determining adoption of new agricultural technology by smallholder farmers in developing countries. *Issn*, *6*(5), 2222–1700. https://core.ac.uk/download/pdf/234646919.pdf

Nguyen, H. T. T., Le, T. Q. A., Tuyen, M. C., & Hung, P. X. (2025). A review of climate-smart agriculture in Asia: Critical achievements, key challenges, and potential prospects. *Journal of Agriculture and Rural Development in the Tropics and Subtropics*, *1*(126), 25–42. https://doi.org/10.17170/kobra-2025011410816

Peria, A., Pulhin, J., Tapia, M., Predo, C., Peras, R. J., Evangelista, R. J., Lasco, R., & Pulhin, F. (2016). Knowledge, risk attitudes and perceptions on extreme weather events of smallholder farmers in Ligao city, Albay, Bicol, Philippines. *Journal of Environmental Science and Management*, *19*(Special Issue), 31–41. https://doi.org/10.47125/jesam/2016_sp1/03

Relativo, J. (2023, September 30). Marcos approves P5,000 financial assistance for 2.3 million rice farmers. *PhilStar Global*. https://www.philstar.com/headlines/2023/09/30/2300189/marcos-approves-p5000-financial-assistance-23-million-rice-farmers

Rivero, J. M. C., Carandang, W. M., De Luna, C. C., Carandang, V. Q., Gevaña, D. T., Galapia, G. A., & Cruz, R. V. O. (2017). Socio-economic and environmental impacts of a conservation farming village (CFV) project on upland farmers in General Nakar, Quezon, Philippines. *Ecosystems & Development Journal*, *7*(2), 46–58. www.generalnakar.gov.ph/

Rosales, B., Palima, C., Hernandez, G., Gonsalves, J., Rosimo, M., Santiago, R., Vidallo, R., Locaba, R., & Yim, S. (2020). *Native pigs: A climate resilient business enterprise*. CGIAR. https://hdl.handle.net/10568/108081

Ruzol, C., Lomente, L. L., & Pulhin, J. (2021). Cultural consensus knowledge of rice farmers for climate risk management in the Philippines. *Climate Risk Management*, *32*, 1–13. https://doi.org/10.1016/j.crm.2021.100298

Sebastian, L., & Bernardo, E. B. (2019). Making the smallholder farmers in Southeast Asia climate smart - The CCAFS R4D thrust. In Y. Shirato & A. Hasebe (Eds.), Climate smart agriculture for the small-scale farmers in the Asian and Pacific Region (pp. 201–226). National Agriculture and Food Research Organization & Food and Fertilizer Technology Center for the Asian and Pacific Region. https://cgspace.cgiar.org/server/api/core/bitstreams/8d1e9e7d-ec4e-4e89-b88c-51fc6d3c7a59/content

Selvan, T., Panmei, L., Murasing, K. K., Guleria, V., Ramesh, K. R., Bhardwaj, D. R., Thakur, C. L., Kumar, D., Sharma, P., Digvijaysinh Umedsinh, R., Kayalvizhi, D., & Deshmukh, H. K. (2023). Circular economy in agriculture: Unleashing the potential of integrated organic farming for food security and sustainable development. *Frontiers in Sustainable Food Systems*, *7*, 1-17. https://doi.org/10.3389/fsufs.2023.1170380

Taylor, B. M., & Van Grieken, M. (2015). Local institutions and farmer participation in agri-environmental schemes. *Journal of Rural Studies*, *37*, 10–19. https://doi.org/10.1016/J.JRURSTUD.2014.11.011

Triste, L., Vandenabeele, J., van Winsen, F., Debruyne, L., Lauwers, L., & Marchand, F. (2018). Exploring participation in a sustainable farming initiative with self-determination theory. *International Journal of Agricultural Sustainability*, *16*(1), 106–123. https://doi.org/10.1080/14735903.2018.1424305

Vidallo, R., Bayot, R., Rosimo, M., Monville-Oro, E., Gonsalves, J., Ilaga, A., Sebastian, L., Manalo, U., & Baltazar, P. (2019). *Pathways to attaining a food secure Philippines through a competitive and climate-resilient agri-fisheries sector*. CGIAR. https://hdl.handle.net/10568/106789

Vidallo, R., Bayot, R., Rosimo, M., Monville-Oro, E., Gonsalves, J., Ilaga, A., Sebastian, L. , Manalo, U., & Baltazar, P. (2021). *The AMIA experience: Supporting local actions for climate resilient agriculture*. CGIAR. https://cgspace.cgiar.org/bitstream/handle/10568/106787/Synthesis%20Brief.pdf

Walsh, C., Renn, M., Klauser, D., de Pinto, A., Haggar, J., Abdur, R., Hopkins, R. J., & Zamil, F. (2024). Translating theory into practice: A flexible decision-making tool to support the design and implementation of climate-smart agriculture projects. *Agricultural Systems*, *219*, 1-15. https://doi.org/10.1016/j.agsy.2024.104060

Wang, G., Lu, Q., & Capared, S. C. (2020). Social network and extension service in farmers' agricultural technology adoption efficiency. *PLoS ONE*, *15*(7), 1-14. https://doi.org/10.1371/journal.pone.0235927

Wigboldus, S., Hammond, J., Xu, J., Yi, Z., He, J., Klerkx, L., & Leeuwis, C. (2017). Scaling green rubber cultivation in Southwest China—An integrative analysis of stakeholder perspectives. *Science of The Total Environment*, *580*, 1475–1482. https://doi.org/10.1016/j.scitotenv.2016.12.126

World Bank. (2022). *Philippines: Country climate and development report 2022*. https://thedocs.worldbank.org/en/doc/4ec3282919652f7545bc25c49c1811e4-0070012022/original/PHCCDR-FINAL-formatted.pdf

Yin, R. K. (2018). *Case study research and applications: Design and methods* (6th ed.). Sage Publications.

Zheng, H., Ma, W., & He, Q. (2024). Climate-smart agricultural practices for enhanced farm productivity, income, resilience, and greenhouse gas mitigation: A comprehensive review. *Mitigation and Adaptation Strategies for Global Change*, *29*(4), Article 42. https://doi.org/10.1007/s11027-024-10124-6

Zhou, W., He, J., Liu, S., & Xu, D. (2023). How does trust influence farmers' low-carbon agricultural technology adoption? Evidence from rural Southwest, China. *Land*, *12*(2), Article 466. https://doi.org/10.3390/land12020466

Chapter 7
Everyday Peace in Multicultural Neighborhoods: Immigrant Strategies from an Agonistic Peace Perspective

Büşra Kuplay

7.1 Introduction

In recent years, Turkish immigrants, particularly Turkish Kurds, in Kawaguchi, have become one of the main targets of escalating anti-immigrant hate speech, neighborhood conflict narratives, and acts of hate (Akasaka and Sunami 2025). In this context, strategies for managing everyday conflicts and countering these narratives have become increasingly essential for navigating life in a foreign country.

This chapter draws on the everyday peace framework and links it to discussions on agonistic peace by focusing on Turkish immigrants in Kawaguchi as a case study. It seeks to identify the triggers of conflicts articulated in the neighborhood conflict narratives, examine their impact on the daily lives of Turkish immigrants, and explore the role of immigrant-led individual and collective peacebuilding initiatives in promoting peace and mutual understanding. Additionally, it discusses critical factors that must be addressed to enhance the impact and sustainability of these activities. By addressing the gap in the academic literature on Turkish immigrants in Japan, this chapter contributes to post-liberal peacebuilding studies by offering empirical insights into how the frameworks of everyday peace and agonistic peace can be applied in diverse, migrant-rich societies in the context of rising hate speech and neighborhood conflict narratives.

B. Kuplay (✉)
Graduate School of Asia-Pacific Studies, Waseda University, Tokyo, Japan
e-mail: kuplay.busra@fuji.waseda.jp

A. Sharifi et al. (eds.), *Frameworks and Practices Towards a Sustainable and Peaceful Future*, World Sustainability Series,
https://doi.org/10.1007/978-981-95-2310-8_7

7.2 Literature Review

7.2.1 *Conceptual Frameworks: Everyday Peace and Agonistic Peace*

Starting with Richmond (2010, 2020) and MacGinty (2014, 2021), critical scholars shaping post-liberal peace studies raised questions about how liberal peace studies approach peacebuilding and how focusing on everyday settings and local agency can promote peace in conflict-afflicted societies. Everyday peace, among other concepts, focuses on everyday settings to develop a methodology for identifying practices and methods adopted for peacebuilding in divided societies (MacGinty 2014, p. 549). Due to peace and conflict studies usually being addressed under international studies, their connection with local agencies and everyday settings was limited until their post-liberal and post-colonial critics. Following the critical approaches, the idea of everyday peace is built on Scott's (1985, 2008, 2013) notion of everyday resistance and hidden transcripts. As everyday forms of resistance, Scott (1985) identifies practices ordinary people use to resist the oppression, without any open confrontation which would put them in danger, such as "foot-dragging, dissimulation, false compliance, pilfering, feigned ignorance, slander, arson, sabotage, and so forth" (p. 29). "Hidden transcripts," on the other hand, are direct expressions that disguise and allow the subordinate group to speak their critique of dominant groups out in the forms of "rumors, gossip, folktales, songs, and gestures" while taking advantage of anonymity and protecting themselves from direct negative retaliations (Scott 2008, p. xiii). Both everyday forms of resistance and hidden transcripts are strategies to negotiate in unequal societies and form resistance where oppressed groups have limited chances of changing their life conditions through direct confrontation (Scott 1985, pp. 41, 241–242, 246; Scott 2008, p. 119; Scott 2013, p. 71). Implementing Scott's idea of everyday resistance to peace studies and looking into how people shape or protect peace in their daily lives is a result of a need to look into the role of local agencies in peacebuilding for more sustainable societies. Richmond (2020) argues that new approaches to peacebuilding that build upon post-structural and post-colonial critiques of liberal peace are a result of the need to "transform biases that are extensions of historical power-relations through resistance" (pp. 210, 217). By looking into the transformation of biases and the role of everyday resistance practices in everyday settings, we can see the "basis for peace formation" (Richmond 2020, p. 213).

Post-liberal and post-colonial critics of peacebuilding not only change perspectives toward where to look for peace but also what peace itself means. Critical peace theories argue against any universal frameworks, including a universal idea of peace, which they consider a form of hegemonic relations (Richmond 2020, p. 192). The theory of agonistic peace, drawing upon this critical perspective, offers new frameworks on how one can perceive "peace." Agonistic peace, originating from Mouffe's (2000) idea of agonistic pluralism and democracy, is an understanding of peace that refers to "agonistic practices" that change the friend/enemy relations through mutual respect and recognition that can be developed in confrontation (Shinko 2008; Shinko 2022). It aims to transform antagonistic relations into agonistic ones, the enemies into adversaries, by recognizing opponents and their rights to defend their ideas (Strömbom and Bramsen 2022, p. 1240). With this framework, González Martín et al. (2022) studied the peacebuilding process in Colombia and how clown art transforms the antagonistic modes of conflict into agonistic ones by replacing narratives representing others as an "enemy" to be eliminated with violence (pp. 1277–1278). Clowns open a "symbolic space" that allows people to develop new narratives and identities in a safe environment while developing an understanding toward each other (González Martín et al. 2022, pp. 1278–1279). Similarly, Göker and Çelik (2022) explore how agonistic listening among women from diverse political, ethnic, and ideological backgrounds can lead to a mutual understanding without consensus. They argue that even though the listening process in a discussion is usually based on antagonistic inclinations to develop counter-arguments, agonistic listening and having open dialogue in a safe environment allow one to accept the "other," despite disagreeing with each other (Göker and Çelik 2022, p. 1253). Lehti and Romashov (2022) investigate the daily interactions between Armenians and Azerbaijanis in a border bazaar between Georgia and Armenia after the war in the Nagorno-Karabakh in 2020. They argue that the border bazaar provided a space for building trust through the confrontation of antagonistic narratives that simultaneously exist and for suspending antagonism through everyday peace activities such as avoiding political topics (Lehti and Romashov 2022, pp. 1297, 1300).

These studies provide examples of how agonistic peace, in this regard, lies in constant contestation within various groups and identities, rather than a consensus, and focus on the production of new narratives for transforming these relations without denying the other. However, due to this focus, agonistic peace is questioned for its potential to promote a sustainable peace. Mouffe's (2000) idea of agonistic pluralism, on which the concept of agonistic peace is built, is criticized for romanticizing conflict, not giving clear definitions of antagonism and agonism, and ignoring the possibility of further polarization under continuing conflict situations (August 2022). Similar to the agonistic peace, the concept of everyday peace is also questioned in terms of whether or how it can contribute to a sustainable form of peace. The everyday resistance, which is the idea of everyday peace mostly built on, is focused on "individual self-help" and does not aim to change the system (Scott 1985, p. 29). Due to that, everyday peace is also associated with negative peace. Because of the circumstances in conflict-ridden or divided societies, and the need to prioritize the protection of themselves, everyday peace might only give a limited form of peace as in the form of tolerance and co-existence (MacGinty 2021, p. 4).

I argue that instead of adopting everyday peace and agonistic peace frameworks separately, combining them provides us with a more comprehensive approach for studying peacebuilding in diverse societies and contributes to existing academic discussions. There are various ways to examine how everyday interactions transform conflicts; however, the agonistic peace perspective offers a valuable analytical framework for understanding this transformation through the reconstruction of identities and the disruption of conflict narratives. By focusing on everyday settings, Lehti and Romashov (2022) show that "avoidance of controversial topics" in divided societies can be seen as a practice promoting "agonistic respect" (p. 1299). Similarly, Aggestam et al. (2015) argue that everyday peace strategies disrupt the narratives of conflicts and transform antagonistic relations into agonistic ones by creating spaces where problems can be identified and solved. These studies show how everyday settings function as spaces for agonistic peace practices through which conflict narratives can be reconstructed. Yet, the number of studies implementing the theory of agonistic peace within case studies is very limited, and existing studies mostly focus on conflict or post-conflict settings with traumatic experiences of violent wars (Strömbom and Bramsen 2022, p. 1241).

7.2.2 Literature on Turkish Immigrants and Turkish Kurds in Japan

The community of Turkish immigrants in Kawaguchi dates to the 1990s when, according to Nakajima (2003), the migration of Turkish Kurds to Japan began. While Japan had no visa restrictions for Turkish citizens at that time, allowing them to enter the country, many of them could not obtain a residency visa as unskilled laborers due to Japan's contemporary immigration policies and became visa overstayers (Ahmadian 2024, p. 260). Currently, Kawaguchi City is assumed to host more than 2,000 Turkish immigrants, with the majority of Turkish Kurds. As of February 4, 2025, 1513 Turkish citizens are registered at Kawaguchi Municipality, excluding individuals who are on provisional release,[1] in the supervisory measures system,[2] and those with other residency statuses (Kawaguchi City Official Site 2025). According to official data from December 2024, this group represents nearly one-third of the total Turkish immigrant population in Japan, 7571 people (Statistics Bureau of Japan n.d.). This indicates a relatively large community concentrated in a specific geographic area. The members of this community are primarily from the three cities, Gaziantep, Kahramanmaraş, and Adıyaman, and include not only ethnic Kurds but also Turks, many of whom maintain strong kinship and hometown ties. Although making a distinction between ethnicity and nationality among Turkish citizens is complex, Kurds are typically described as an ethnic group with historical roots spanning Iraq, Iran, Türkiye, and Syria, unified by a shared language and culture (Husain and Ahamed 2022, p. 4241). In the context of Türkiye, Husain and Ahamed (2022) argue that it is more accurate to define the "other" of Kurdish ethnic

[1] For detailed information regarding provisional release status, see *Kari hōmen [Regarding provisional release]* (Ministry of Justice Immigration Services Agency n.d.-b): https://www.moj.go.jp/isa/08_00050.html

[2] For detailed information regarding the supervisory measures system, see *Kanrisochi ni tsuite [Regarding supervisory measure]* (Ministry of Justice Immigration Services Agency n.d.-a): https://www.moj.go.jp/isa/08_00045.html

identity as "Turks," referring to ethnicity, whereas "Turkish" denotes citizenship according to the Turkish Constitution (p. 4241). Taking both the ethnicity–nationality discussions and the self-identification of informants into account, I use the term *Turkish immigrants* to refer to individuals from Türkiye who hold Turkish citizenship, regardless of ethnic background. The terms *Turkish Kurds* or *Kurdish* refer specifically to Turkish citizens who identify themselves as members of the Kurdish ethnic community.

The Turkish immigrants, particularly the Kurdish community, in Kawaguchi, have attracted considerable public attention, with increasing hate speech/acts and neighborhood conflict narratives, often fueled by false claims in the news and social media platforms (NHK 2024). Despite growing public attention on them over the years, academic studies on Turkish immigrants in Japan remain limited. Existing research primarily focuses on asylum seekers, particularly Turkish Kurds (Ahmadian 2024; Fujibayashi 2020; Nakajima 2003; Tsuchida 2018), as well as transnational marriages and gender (Igarashi 2014), bi-cultural child rearing (Takeshita 2016), and intercultural communication (Katayama 2021). Studies examining the local agency of immigrants and their practices for promoting peacebuilding or integration into the host country are relatively scarce.

Building on the theoretical discussions above and addressing existing gaps in academic knowledge, this research examines the case of Turkish immigrants in Kawaguchi through the lens of everyday peace and agonistic peace. While it focuses on neighborhood conflict narratives and hate speech/acts directed primarily at Turkish Kurds, it aims to offer a broader perspective on the issue by including Turkish immigrants from diverse ethnic backgrounds. By doing so, it aims to explore how groups involved in conflict narratives can counter hate speech, cultivate mutual recognition, and produce counter-narratives to prevailing conflicts.

7.3 Methodology

This research draws on data collected during my fieldwork in Kawaguchi from October 2023 to May 2025 through participant observation, semi-structured interviews with Turkish immigrants and key informants, and passive observation of public social media accounts of key actors.

From October 2023 to May 2025, I conducted field observations through participant observation as a part-time immigrant consultant and interpreter working with immigrants in Kawaguchi, typically once to three times a week. Additionally, I participated in various community events, such as Newroz celebrations, casual lunch/dinner meetings, weddings, and cultural exchange events. As a Turkish woman who is fluent in both Japanese and Turkish, I started volunteering and working as an immigrant consultant and interpreter in February 2020 to assist the Turkish immigrants in Kawaguchi. After a brief break, I resumed this work in October 2022. My years-long professional engagement and personal involvement with the field enabled me to build trust within the community, deepen my knowledge, and observe the changing dynamics in the field firsthand. Although this research benefits from my professional experiences and networks in the field, the individual cases and confidential information obtained through work are not used as research material due to privacy and ethical concerns. I include fieldnotes from events that I attended and informal conversations as part of participant observation data, only with the acknowledgment of the involved parties.

During my fieldwork, I engaged in informal conversations with approximately 500 immigrants from Türkiye. My contact with many community members emerged organically through professional or volunteer activities. I established connections with key informants through academic engagements, workplace interactions, or personal introductions. The community representatives introduced me to other immigrants who are willing to participate in this research and invited me to community events. As a female researcher, I was able to build close connections with many female immigrants who often welcomed me into their houses for interviews, followed by additional hours of informal conversations and shared meals. In contrast, scheduling interviews with men was more challenging due to gender-related dynamics and their long working hours, resulting in a higher proportion of female participants in the final sample.

From December 2024 to May 2025, I conducted twelve semi-structured interviews. Participants were selected through purposive and snowball sampling methods, with careful attention to diversity in ethnicity, community roles, family, and socio-economic background. The interview participants include Turkish immigrants who reside or have resided in Kawaguchi and key informants, such as immigrant community representatives, and members of other pro-immigrant and immigrant-led associations. The key informants are indicated in bold letters. Table 7.1 summarizes the main information regarding the informants and details of their background.

The interviews were conducted in Japanese or Turkish, either face-to-face (n = 11) or online via Zoom (n = 1), based on participants' preferences. Informed consent was obtained from all participants in the form of written consent (n = 10), recorded verbal consent (n = 1), or email confirmation (n = 1), depending on their individual circumstances. I recorded all interviews with participants' permission and transcribed using a combination of Microsoft Word's dictation feature and manual transcription. To protect participants' privacy, I saved all interview files under pseudonyms in a password-protected folder on my computer. I translated the interviews into English, excluded identifiable information from the data presented in this chapter, and used pseudonyms or assigned letters to maintain the anonymity of informants or other individuals identified in the interviews.

In addition, I employed netnographic observation to supplement the ethnographic data, taking fieldnotes through passive observation in line with Kozinets' (2010) framework, which identifies observational, archival, and elicited data as the three core methods of netnographic data collection (p. 98). This involved monitoring public accounts, particularly those of community representatives and individuals who are identified by informants as active actors in shaping conflict narratives and hate speech, on platform X, where these narratives were especially prevalent.

Table 7.1 Interview participants

Pseudonym	Sex	Age	Nationality	Ethnicity	Visa status	Occupation/community position	Years spent in Japan/Kawaguchi
Maria	Female	57	Japanese	Japanese	Japanese citizenship	Former municipality employee/pro-immigrant association worker	57/50
Nergis	Female	40	Turkish	Turkish	Designated activities visa obtained via asylum application	–	2/2
Aras	Male	24	Turkish	Kurdish	Designated activities visa obtained via asylum application	–	2/2
Deniz	Female	18	Turkish	Turkish	Designated activities visa obtained via asylum application	Student	1/1
Arai	Female	29	Japanese	Chinese	Japanese citizenship	Former municipality employee	–/20
Haruka	Female	26	Japanese	Japanese	Japanese citizenship	Student/municipality volunteer	–/6 years in surrounding areas
Emir	Male	43	Turkish	Kurdish	Long-term visa	Lecturer/restaurant owner/community representative	15/2
Nudem	Female	36	Turkish	Kurdish	Provisional release	Former immigrant association volunteer/community representative	19/19
Ela	Female	42	Turkish	Kurdish	Designated activities visa obtained via special grant[a]	Former immigrant association volunteer/community representative	19/19
Duygu	Female	39	Turkish	Turkish	Designated activities visa obtained via work	Factory worker	6/6
Zeynep	Female	27	Turkish	Kurdish	Provisional release	–	5/5
Begüm	Female	23	Turkish	Turkish	Designated activities visa obtained via asylum application	–	3/3

[a] For detailed information regarding this special grant, see *Sōkan kihi-sha no uchi honpō de shusshō shita kodomo no zairyū tokubetsu kyoka ni kansuru taiō hōshin ni tsuite [Policy on special residence permission for children born in Japan among deportation avoiders]* (Ministry of Justice Immigration Services Agency n.d.-c): https://www.moj.go.jp/isa/content/001419300.pdf

For data analysis, I primarily employed **content analysis** to identify recurring themes within participants' narratives, particularly those referencing perceived triggers of neighborhood conflicts. To enhance the depth of the analysis, I cross-referenced these emergent themes with conflict-related narratives found on social media. Additionally, I used **thematic analysis** to explore both immigrants' experiences of hate speech and acts, as well as individual and collective strategies of everyday peace described during the interviews.

7.4 Neighborhood Conflict Narratives and Experience of Hate Speech

Turkish immigrants in Kawaguchi, mainly the Turkish Kurds, have been made the main group around which neighborhood conflict narratives started in social media and some online newspaper groups, such as Sankei News, in 2023 (The Sankei Shimbun 2023a, b). I categorize the key triggers of the conflicts between the immigrant community and the host community articulated in these narratives into two levels: the communal level and the institutional level. The communal level refers to guidelines for the use of public spaces, where violations typically result in little or no formal penalties. The institutional level refers to regulations enforced by official authorities, such as municipal, prefectural, or national governments, where violations can lead to significant legal consequences. These are summarized in Table 7.2:

As an example of conflict narratives, Aras explains how the simple act of immigrants gathering in front of the convenience stores represented a trigger of the conflict:

> In these kinds of days (referring to public holidays and weekends), our people gather there with the purpose of seeing each other. They are meeting each other because everyone is each other's relative. But, they are recording these, sharing them, saying that foreigners gathered again, and they are here again. They are making us look bad, against their own people. Even to the Japanese people, who like us… They degrade us.
>
> …

Table 7.2 Key triggers of disputes in conflict narratives

Communal level	Institutional level
Noise in apartments and public places, in late and early hours	Sexual harassment
Exhaust popping at nights	Traffic rules
Listening to loud music at nights	Damaging public goods
Unhygienic use of space	Littering
Gathering in front of the convenience stores and chatting as a group	Not protecting garbage disposal rules regarding collection days or separation of garbage and recycling materials

> They say the Japanese are scared of you. We are human! Excuse me, but we are not animals, so they would be scared of us. If we were lions... If you bring a lion there, I would be scared, too, but we are not lions. We are humans, too.
>
> …
>
> They are just talking. They buy some drinks and sit in the corner in groups of 3–4 people and chat. They talk about their jobs, what they did on the construction site, or their work is delayed. It is not like they are talking about whether we are going to beat the Japanese or conquer this country. But, they think we are going to take this country away from them. We are just here to earn. We could not make it in our country and decided to come here, to go back after we had some savings. But they degrade us.

Aras explains how the convenience store serves as a socializing place to simply catch up with friends, family members, and acquaintances for many immigrants; however, with neighborhood conflict narratives, they feel that their existence has turned into an "object of fear," perceived as a threat to the host community and country. As a result, they now face hate acts such as people recording them without permission and sharing.

These types of acts are relatively new, as many immigrants indicated during the interviews. With the increasing publicity of Turkish immigrants in Kawaguchi in 2023, online conflict narratives and hate speech have entered the offline world, and immigrants started to feel these in their daily lives (The Asahi Shimbun 2024a, b; Yasuda 2024). The extent of the experience of hate speech depends on various factors such as gender, occupation, language skills, access to social media, public appearances of the person, and ethnicity. Having Japanese skills increases both the chances of dialogue and the exposure to hate speech, especially through various media platforms. Many of the immigrants with limited language skills indicate that they cannot follow Japanese public and social media because of the language barrier, but that they are concerned about rising hate speech, which they know from their relatives or friends who can understand the language. Occupations with more public attention, such as demolition work on construction sites or service work in restaurants, also affect the amount of exposure to hate speech. People who are working in demolition work face stigmatization more stemming from the perception of them being the main subjects of troubles, such as industrial garbage disposal issues. People in service work, mainly in the food sector, face more direct hate speech and acts in their daily lives and their workplace, due to them being out in public with their identity revealed. Some establishments, mostly the ones in the public eye, receive phone calls from anonymous people repeatedly, disturbing their business. Emir mentions these calls he receives in his restaurant, demanding unreasonable takeout orders, such as asking the restaurant, which is in Tokyo, for a delivery to Hiroshima without giving any address or simply blaming the owners and workers for being "criminals." Ethnicity is also an important factor in affecting the extent of hate speech, since many narratives explicitly refer to "Kurds"; however, in the case of hate acts and other discriminatory acts immigrants directly encounter, it is the whole community of Turkish immigrants that is targeted. Emir gives a recent fight in front of a convenience store, recorded and shared on social media, as an example:

> I think this is something against all foreigners. I mean, it is not about just an ethnic group, not about just Kurds. This is not just about Turks either. For example, you just choose a certain group, like Kurdish people. But the target is not the Kurds. Because they also know. What can two thousand Kurds even do? I mean, there is reason, there is logic.
>
> …
>
> Turks are affected, too. You know, recently they had a fight with the Japanese. One of them was Turk. In terms of ethnicity, I mean. That Japanese, the one who takes the video, thinks he is Kurdish. "What Kurd?" he says. He's had it up to here. He was thinking that "Damn you, saying Kurds, Kurds; enough of it already," and they fought with each other.

As explained above, the extent of hate speech and acts immigrants face depends on various factors; however, in what ways do they encounter it? Hate speech and acts take both direct and indirect forms in immigrants' lives. Interviewees described hate speech demonstrations, hostile attitudes, avoiding contact, and people changing their route when they see someone from the immigrant community. More extreme cases include being threatened with calling the police despite the absence of a criminal act, being recorded in public places without consent and shared on social media, and being threatened with violence. The houses of Turkish immigrants are labeled as "Kurdish houses," as if there is a witch-hunt where their doors are marked (140journos 2025, 19:05). In this way, their way of living is placed under constant surveillance by certain individuals and groups. Many Turkish immigrants in Kawaguchi use social media as a platform for community building, solidarity, and self-representation. During my fieldwork, I observed the high use of Instagram, especially by middle-aged women, and TikTok by the younger generation of Turkish immigrants. While acknowledging the use of these platforms, Ahmadian (2024) analyzes the role of social media, particularly Facebook, in the identity representation of Turkish Kurds in Japan as a virtual platform for interacting with other community members (pp. 267-268). At the same time, with the rise of neighborhood conflict narratives and hate speech/acts, immigrants' use of these social media platforms turned into a means of digital surveillance. Some individuals befriend immigrants on social media using fake accounts, posing as members of the same community, then edit their posts, videos, and photographs, resharing them with real name tags. Ela, who has lived in Kawaguchi on provisional release status for 18 years and recently received a visa, shares her own experiences of these hate acts and the feeling of being under constant surveillance:

> The racists were coming to the station and protesting at us. (Saying) "Go away, go away, go away." 2–3 months ago, it was banned by a court order, so they cannot come now.
>
> …
>
> I was really scared. They were coming there every Sunday. Helicopters were flying over us. Lots of police officers were going there.
>
> …
>
> Besides that, I had hesitations whenever I went to a market. "What will happen to me?" "Will someone say something to me?" "What will happen if I go outside?" They reflected that thing to us so much. We were crushed under it like this.
>
> …
>
> He shares the photograph I shared 9 years ago, crossing me out with a red writing. Maybe you know, the red writing is very bad in Japan. The red sign. You know, (it was) that

> racist! A! The post I shared 9 years ago. In a park during Newroz, with my family. He says, "This is this person, these are her children." There are our names, too. He says "terrorist." He says joining a Newroz celebration means (I am a) terrorist.

All of these hate acts, appearing as a result of continuing hate speech targeting an ethnic community and neighborhood conflict narratives, are a reality in the everyday life of Turkish immigrants in Kawaguchi. The next part will explain the individual and collective peacebuilding practices immigrants initiate under these conditions.

7.5 Immigrant Strategies for Building Peace

The findings of this research indicate that immigrant-led peacebuilding practices in Kawaguchi can be considered in the forms of individual practices and collective practices. Individual practices refer to the methods that individual members of the immigrant community choose to implement to keep the peace in their daily lives. Collective practices, on the other hand, refer to peacebuilding through networks, activities of community representatives, and practices of immigrant-led associations and their cooperation with other local actors.

7.5.1 Individual Strategies for Everyday Peace

Individuals' choice of how to react to a dispute, problem, or the possibility of it occurring depends on various factors. Timing and the conditions of encountering a dispute with another person are important in terms of whether individuals will choose methods to calm the situation. At night and when people are within a group, conflicts tend to escalate more quickly. Ethnic/social capital and having relatives in the area or acquaintances from the hometown provide a network of conflict resolution and give its members confidence to prefer methods prioritizing de-escalation of the conflict, since the other members, with more experience and language skills, can play the role of mediators and facilitate dialogue when they are called for help. In addition, positive future aspirations regarding immigrant life in Japan and the possibility of long-term residency with a legal status create positive stimulation for encouraging tolerance and looking for non-violent methods.

While the practices people adopt differ when they face a conflicted situation, I categorize the individual practices aiming to keep the peace in everyday settings into two: practices focusing on preventing conflicts and practices aiming to manage conflicts once they occur. Figure 7.1 summarizes these practices, along with practices that can function as both:

Firstly, as summarized in the figure, individual practices for conflict prevention usually center around protecting oneself from getting involved in disputes and

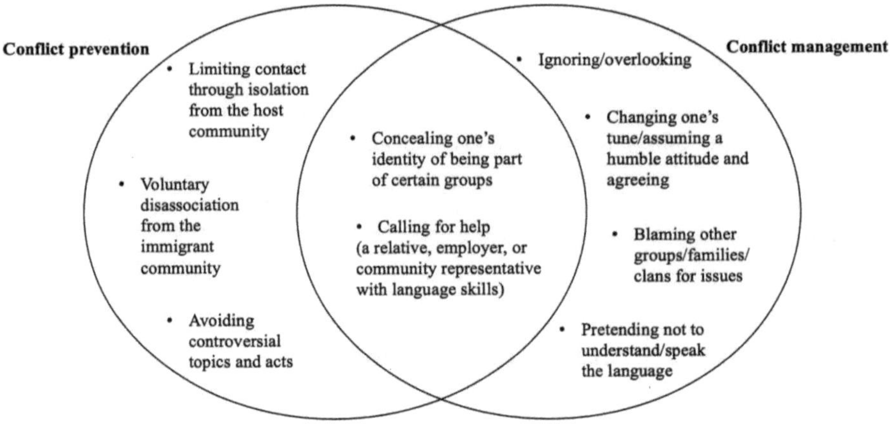

Fig. 7.1 Individual peacebuilding practices in everyday settings

problems by limiting contact with the host community, voluntarily disassociating oneself from the immigrant community that is being targeted, and avoiding controversial topics and acts.

Isolation from the host community has both voluntary and involuntary aspects. Living close to each other and not feeling the need to have contact with the host community or other immigrant communities is a choice for some immigrants. Many Turkish immigrants in Kawaguchi share strong ethnicity, kinship, and hometown connections, and many of them are in Japan with their nuclear or extended family members. These connections form the foundation of their networks that function both as a support system and a socializing mechanism. For example, Nudem compares her relationship with her community and the host community:

> I am now known as a person who can speak Japanese, who has lived in Japan for many years, and who is well integrated here, but from my perspective, I cannot say that I have socially well-connected with the Japanese. We did cultural events together, but that does not count, because it is like an event, a business. So, people from my community are more like my social connections. Closer to me. This is the connection, the habitat, through which I feel more like myself.

In this context, living as a community is a potential choice for Nudem because this is where she feels she belongs. On the other hand, there is also a need to think about what the involuntary aspects of this "choice" are. In the case of Turkish immigrants, these aspects include difficulties with adapting to the culture of a new community, language barriers, and residential segregation due to systematic barriers. Nergis explains how having limited Japanese skills affects her relationship with her neighbors:

> The attitude of the people around me was very good. I did not observe any negative attitude. They even come to my house for a housewarming, saying, "Welcome to our neighborhood." My Turkish friends, I mean. But with Japanese, I could not really do much. I exchange greetings with the people in my building, but besides this, we have not had any conversations. It is because I do not know their language. Actually, I really want to have a conversation

with them, but... I am angry at myself for not learning Japanese. Why shouldn't I? I am angry at myself. If I learn Japanese, at least I will have a chance to chat with them.

While living in the same building with Japanese residents allows immigrants a chance to get in contact, this is usually very limited because of the residential segregation of immigrant families due to systematic barriers in housing markets and identity-based discrimination resulting from conflict narratives. Begüm mentions the problems they had when they were looking for a house to move out of their relatives' house, where they lived when they first arrived in Japan:

> We had so many problems. They do not rent to Turkish people easily. They say, "They do not keep it clean." "They keep it dirty, they yell, they scream," they say. "There is conflict. Too much noise." How can I explain? We had a friend going to Türkiye, my husband's friend. Nyukan (immigration office) was sending him, (saying) "go away." We just took the keys from him. We said, "We will enter this house like this, or they won't give it to us." We moved in. Waited 4–5 months and then went to the real estate agency. We said, "He went to Türkiye, and we want to take over this house."

Nudem also shares her similar experiences of house search:

> Now, I live with my husband and my daughter in a rental house. Approximately 50-year-old building.
>
> ...
>
> My siblings also live there, next door. At that time, when I was looking for a house to rent, I explained our situation to the owner of the house.
>
> ...
>
> The companies did not want to become a guarantor (for us). With the owner, these guarantor companies were the way to rent a house. The company was like that because I did not have a juminhyo (certificate of residency), a visa. Also, they did not give us houses without a job permit, etc. I directly went to the owner of the house and explained our situation. "They do not trust us; they do not rent us houses." I can find a Japanese guarantor if you want. Will you rent us a house? (The owner) trusted my word and rented the house. Since then, he has been the house owner for all of us.

As illustrated in these narratives, many immigrants have difficulties renting a house in the neighborhood, often due to conflict narratives or systematic barriers such as a lack of a residency card or job permit. As a result, despite creative coping strategies, many families and individuals have to live in the same buildings or areas that accept tenants with Turkish nationality. This form of residential segregation highlights the involuntary aspects of their isolation from the host community.

In addition, some immigrants adopt voluntary dissociation from the immigrant community as a strategy for self-protection and to avoid conflicted situations. Deniz illustrates this approach in her description of her social preferences:

> My friends are usually from school. Besides that, like I told you before, I don't have any relations with others, especially from my nationality. ... I have my friends, foreigners... Korean, Chinese... I have good relationships with them.
>
> ...
>
> It might be because of the complaints on social media that I mentioned before, or the warnings they receive. I do not think that (being warned) is nice. Since I don't think it is nice, I don't want to be close to those people. I am generalizing, and probably they are not all like that. But I just prefer not to get close.

From a similar perspective, Duygu compares her experiences in Kawaguchi with those in her new neighborhood:

> If you ask about Kawaguchi, the reaction of the Japanese was very different because there were many Turkish people. For example, because there are people from our community there… Fights, men gathering in front of convenience stores and drinking, our children speaking loudly in parks and fighting… There are the things they (the Japanese) disapprove. I know this because I have lived there. The people there were always on the edge. But, if you ask the place I live now, you already know because you came here. I live in an apartment.
>
> …
>
> This is a place where the Turkish do not live, so they (the Japanese neighbors) notice when I come back from work and greet me. For example, they give responses to me when I am tired, asking me whether I am tired, in Japanese, of course. So, I am very pleased with this environment.

As the quotes above indicate, both Deniz and Duygu disassociate themselves from the groups at the center of neighborhood conflict narratives, and by keeping their distance, they avoid disputes or problems with other residents. Distancing helps them to manage the stigmas and stereotypes associated with the immigrant community, providing a partial way out from being targeted by conflict narratives and hate speech/acts. Thus, this voluntary self-disassociation functions as a mechanism for preventing conflicts in daily life for them.

Secondly, the practices for managing the conflict are usually centered on attitudes of ignoring the dispute or pretending not to understand the language, agreeing with everything the other side argues, and blaming other groups for the issue. These practices are mainly for dealing with issues on that site and protecting oneself from further negative outcomes.

Finally, practices such as concealing one's identity as being part of the targeted groups and calling someone with language skills for help are common practices that are used both before there is a dispute and after it starts. In the following quote, Aras gives an example of how he managed a conflicted situation and protected himself from violence by concealing his identity:

> When I first came to Higashi-Kawaguchi, I saw a group of about seven or eight people, aged between 25 and 28. They stopped me, walking on the street. They asked me where I was from, and I said I was "half" (biracial). I was afraid, you know. I thought, "They are asking me this, and what are they going to do to me?" I told them that my mother is Italian, and my father is Turkish. … I asked them, "What are you going to do with Turkish people?" They said, "We will beat them." I asked, "Why?" And they said, "There are too many of them, we want them out of our country." … And, they said, "We are not interested in you. You are not fully Turkish. We are looking for Kurds and Turks to beat them."

This narrative shows how concealing their identity allows immigrants to manage the first impressions when they start a conversation and mitigate possible negative responses stemming from identity-based discrimination or violence.

In addition, calling someone for assistance helps immigrants de-escalate conflicts caused by miscommunication. Those called upon are often relatives or friends from the same immigrant community who possess stronger language skills. In some cases, Japanese volunteers, in-laws, or employers also provide linguistic and social support. For example, Zeynep describes the disputes between her family and their neighbors:

> My child has a constipation problem. We were even using medicine for it for a while. Because my child was crying (due to this problem), they made a complaint to the police.
>
> …
>
> Police came and asked. "Do you beat her? Do you do anything bad to her?" I said, "No, we were just playing ball."
>
> …
>
> I am wondering, "Why did they call the cops instead of just asking us?". And it affected us. Our upstairs neighbor, how can I say this, is a bad person, Japanese. Banging here and there. One time, he came to our door, complaining about us a lot. I don't know what, but he kept saying things, yelling, screaming. I told my husband to close the door because we did not even understand what he was saying. He might even be swearing at us or something else.

As these examples indicate, the language barrier and inadequate communication often lead to misunderstanding and further escalation of disputes. Zeynep feels like her neighbors' misunderstanding of her child's crying as a case of domestic violence could be resolved if they communicated. She also explains the situation where she and her husband could not respond to their neighbor because of the language barrier. In the same interview, she also adds that at these times, she often relies on her Japanese friends, who are volunteers or researchers, or her other relatives with Japanese skills to help her communicate. In this way, calling for help functions as both a conflict management and prevention method.

7.5.2 Peacebuilding Through Networks, Associations, and Community Representatives

In contrast to individual practices, collective peacebuilding efforts are carried out through networks, immigrant-led associations, and community representatives. These initiatives emerge from complex dynamics shaped by social and ethnic capital, as well as kinship and hometown ties. They aim to establish protective systems and transform dominant conflict narratives. Such collective structures form the foundation of support mechanisms that help immigrant community members navigate everyday challenges, particularly during periods of conflict or tension.

To start with, immigrant networks play a central role in building defense mechanisms against hate speech and discriminatory acts through solidarity. One such mechanism, aiming to counter digital surveillance, involves information distribution when identified individuals enter the areas where immigrants live or hang out. To warn each other, immigrants send messages indicating the whereabouts of individuals who have shared photographs and information about immigrant community members on social media before, with the intent to target them. Other mechanisms involve the collective monitoring of these identified individuals' public social media accounts, sharing information about any recent posts that mention members of the immigrant community, and alerting others about fake accounts if they are identified.

Emir explains how he identified two fake social media accounts associated with one of these individuals:

> To build peace again, women need to actively participate in life, to take a role in social activities, but when women do that, they share their pictures. This is disgusting. They open an account pretending to be a woman and send a friend request to the social media accounts of women. They don't know who it is, but they check and see common friends, so they accept. For example, I found two accounts like that. It was *B*.
>
> ...
>
> The photograph of the man he put on the account was a photo of a man from India. This is so weird. For example, he sends friend requests from these accounts. After that, a Muslim woman, again. Puts a Turkish flag, writes in Turkish, etc. Later, I sent these to the women. They found the man and blocked him. This is what we are doing to restrain these.

Secondly, community representatives and immigrant-led associations serve as mediators and facilitators of communication between immigrants and the local government. Maria, a former municipal worker, shares one of her experiences illustrating how this mediator role functions:

> I was working in the municipality's Promotion of Civic Cooperation Section. In Maekawa. Do you know? There is a place called Maekawa where many Kurds lived. And, the town council consulted us, saying that they are troubled. "We don't know what to do. Please, warn them." We said we would organize a meeting with people from the town council and got in touch with *C*, and his association, I think it was the Kurdish Cultural Association. We asked him to gather the Kurdish people living in that area, and we had a meeting together. At that time, we brought garbage samples we had at the Promotion of Civic Cooperation Section. It was the garbage we collected at our homes, but we brought it and said, "Let's do a game of separating garbage." This is how you teach these. Japanese people enjoyed this, too. And Kurdish people also said they enjoyed it. So, after all, meeting face-to-face like this, doing it together, and teaching is the most effective.

Furthermore, the immigrant-led associations and community representatives carry on practices to raise public awareness against issues immigrants have, such as supporting art projects or joining academic conferences, or giving public speeches. They are involved in taking direct measures to counter anti-immigrant hate speech and demonstrations targeting Kurds in Japan by using the legal system. As a response to a petition from an immigrant association, the Saitama District Court on November 21, 2024, ordered a halt to a protest targeting Kurdish people in Japan, ruling that the demonstration amounted to hate speech (Asada 2024; The Japan Times 2024). In addition, through various channels, such as social media accounts, they warn immigrants against the issues turning into disputes with the host community and ask them to refrain from these types of conduct, such as gathering in front of convenience stores. They also contribute to community building by organizing neighborhood patrols for security and trash-picking events (Japan Kurdish Cultural Association 2024). These activities show consideration for "issues" or "triggers" that appear in the neighborhood conflict narratives. All these practices aim for the de-escalation of the conflict narratives by highlighting the active role of immigrants, who are often portrayed as the main subjects of these narratives, in peacebuilding efforts in their everyday lives.

Complementing these initiatives, immigrant-led associations and community representatives also play a crucial role in developing counter-narratives for anti-immigrant hate speech through various activities. Through social media activity, public appearances on other media platforms, giving interviews to news channels, and participating in discussions on television and YouTube channels, the members of the associations and community leaders are actively engaged in the production of narratives to counter neighborhood conflict narratives and anti-immigrant hate speech rising with it (Abema Prime 2023; Abema Prime 2024). These activities for producing counter-narratives, along with representation of the community in public media, are important for reshaping the existing relations between the immigrant and the host community toward legitimization. Strömbom (2012) argues that the promotion of alternative narratives of "self" and "other," as well as history of the groups of people considered as the "other," is closely related to their recognition process, which can lead to peaceful societies through acceptance of diversity instead of perceiving it as a threat (p. 176, 185). Public appearance of the Kurdish community representatives and active participation in the narrative-making process in the Japanese media allows them to replace or alter the prevailing conflict narratives by promoting their own. Many media organizations, such as NHK and Asahi News, have been providing space for the promotion of counter-narratives for anti-immigrant hate speech and neighborhood conflict narratives centering around Turkish Kurds in Kawaguchi. News articles based on interviews with immigrants and their experience of hate speech raise their voice and increase the publicity of their claims (Akasaka and Sunami 2025; The Asahi Shimbun 2024a, b). A recent NHK documentary, Hate and Reality: Kurds in Kawaguchi, The Truth (2025), also exemplifies the use of public media as a participatory platform, allowing community members to counter dominant conflict narratives by sharing their own perspectives through the documentary-making process.

Finally, cooperation among local actors, such as pro-immigrant groups, civil society organizations, and individual Japanese activists, has an important role in developing countermeasures against anti-immigrant hate speech. These collaborations establish protective mechanisms, including organizing counter-demonstrations through messaging apps. Maria, for instance, cites a LINE group that mobilizes activists against planned hate rallies, even though she does not participate directly. Ela, a witness to these demonstrations, recounts her experience during a cultural event:

> I had an encounter with hate speech once. Like I said, I get scared easily. I was very scared. A Kurdish artist, *D*, came.
>
> …
>
> At Kawaguchi Station, a mini-concert was organized. A small hall for 50 people. We did not know the racists would come that day. I told you before, but no one was telling us these things. We were trying to keep it quiet so that no one would go there and retaliate. We went there. Got out of the car. Parked it. There were voices. "Go away, get out, filthy foreigners, filthy Kurds!" I said, "What is going on? I won't go there." I panicked. They (other family members) laughed at me. There was a group on the upper part of the station, defending us. The ones who did not want us were in the park, with the police in front of them. And we were there for the concert. So, (I thought) how will this work?

Ela's narrative reflects the fear and uncertainty caused by hate speech, alongside the solidarity networks that offer vital support. Together, Ela's and Maria's experiences highlight how cooperation among local actors plays a key role in mobilizing against anti-immigrant demonstrations and countering hate speech.

7.6 Discussion

Immigrant-led peacebuilding practices play a crucial role in fostering neighborhood peace. These practices not only offer immediate responses to conflict but also hold the potential to reshape dominant neighborhood conflict narratives over time. Some individual peace practices in everyday settings increase dialogue and develop mutual understanding among parties involved in the disputes. Collective peacebuilding practices, on the other hand, mainly focus on countering hate speech/acts and developing counter-narratives for neighborhood conflict narratives regarding Kawaguchi and the community of Turkish Kurds. In addition, cooperation with other actors, such as Japanese volunteers and activists, provides support to these immigrant-led initiatives. Although these practices positively contribute to fostering mutual recognition and understanding, this research identifies two key factors that must be addressed to enhance the impact of peacebuilding efforts.

Firstly, while some individual peacebuilding practices aim to prevent conflicts or resolve them through increasing dialogue and understanding, others, in the context of everyday settings, usually center around practices based on avoiding contact with other communities or limiting it. While these practices function as protective mechanisms for individuals who face identity-based discrimination and threats, they also reduce encounter spaces in which mutual understanding and recognition can be fostered. To transform relations and promote peace by disrupting the neighborhood conflict narratives, taking measures to support immigrants to overcome language barriers and limited encounter spaces is crucial. In addition, there is also a need to consider the issues of residential segregation when considering limited encounter spaces. Uslaner (2010) argues that residential segregation lowers the levels of trust, and only if diversity is supported by interrelations among different groups, levels of social cohesion and trust among these groups increase (pp. 1–2). Similarly, the residential segregation of Turkish immigrants in Kawaguchi hinders the trust-building process by dividing the spaces of immigrants and the host community and limiting their encounters.

Secondly, and most importantly, the findings of this research indicate that the primary driver of conflict narratives regarding Kawaguchi and Turkish Kurds lies in their precarious visa and residency status in Japan. Being asylum seekers who are not recognized by the government, and living with short-term visas, or under provisional release, places immigrants in a disadvantaged position, not only compared to Japanese citizens but also to many other immigrant communities. In the example above, Zeynep, when she was explaining the disputes she had with her Japanese neighbor, later added that she had initially wanted to report her neighbor to the police for coming to their door and yelling at them, but she was scared as a person who is on provisional release status. She says that "I was told not to do it by the Turkish interpreters helping me because it would get us into huge trouble. Because he is Japanese." This illustrates how precarious legal standing can limit immigrants' ability to assert their rights and protect themselves, further exacerbating their vulnerability in conflict situations. Moreover, it negatively affects their peacebuilding activities and integration process, leading them to further isolation. Ela and Nudem explain how they were involved in various community activities, such as organizing Japanese language lessons for Turkish immigrants or cultural exchange events to promote friendship and understanding among their community and Japanese residents. Still, they had to stop their activities because both of their husbands were detained by the immigration office despite having lived on provisional release for years, and along with many other community members, they were frightened to continue their activities under this climate of uncertainty. Furthermore, having a Kurdish community tied with strong ethnicity, hometown, and kinship networks also fuels identity insecurities in the host community, strengthening their position as "enemies" in conflict narratives that present their existence with criminalizing arguments, such as calling them "terrorists," "criminals," or "fake refugees" threatening Japanese residents. These criminalizing discourses serve to delegitimize immigrant practices and diminish the legitimacy of their efforts to construct counter-narratives.

7.7 Conclusion

This chapter examined the role of immigrant-led individual and collective peacebuilding practices in terms of countering hate speech, discriminatory acts, and neighborhood conflict narratives surrounding Turkish immigrants, particularly Turkish Kurds, in Kawaguchi. The findings highlight several critical areas requiring attention and improvement to support immigrant communities in fostering sustainable peace and social cohesion.

Firstly, it is necessary to support immigrants' local agency through legal frameworks that recognize immigrants' status, affirming the legitimacy of their actions. Immigrants are engaged in various activities to promote peace in the neighborhood and disrupt conflict narratives; however, their current visa and residency status in Japan undermine their efforts toward mutual recognition and understanding, since dominant conflict narratives often delegitimize them by depicting immigrants as "terrorists" or "criminals" who abuse the asylum system. Secondly, addressing systemic barriers that lead to residential segregation is crucial for fostering encounter spaces and promoting open dialogue among diverse groups. Furthermore, establishing support systems for language learning is of great importance for overcoming language barriers to facilitate direct communication.

Ultimately, by illustrating how the concepts of everyday peace and agonistic peace can be applied in migration contexts, this research points out that empowering immigrant communities through equitable legal recognition and inclusive systems is essential for building sustainable peace in diverse societies.

Acknowledgments The author would like to express her sincere gratitude to all the participants of this study for generously sharing their time and personal stories. She is especially grateful to her academic advisor, Prof. Katsuma Yasushi, for his unwavering support and insightful guidance throughout this research. She would also wish to thank the anonymous reviewers for their valuable feedback on an earlier draft of this chapter, which greatly contributed to improving this work.

References

Abema Prime. (2023, July 8). *Kurudojin: Naze shōtotsu ga? 100-ri rantō wa dema? Netto de wa kōgeki mo? Hiroyuki to kangaeru nanmin shinsei & nyūkanhō kaisei* [Kurds: Why they clashed? Was the 100-people scuffle a demonstration? Attacks on the internet? Discussing refugee applications & immigration law reform with Hiroyuki] [Video]. YouTube. https://www.youtube.com/watch?v=Eo2FKyKgIIA

Abema Prime. (2024, February 23). *Kurudojin: Kuni ni ire no yōbō? Kyōsei sōkan o motomeru? Rantō sawagi mo? Naze Nihon o mezasu? Kyōsei wa dekiru? Tōjisha & Kawaguchi-shi gi to kangae* [Kurds: Unusual request to the government? Demands for forced repatriation? Scuffle disturbance? Why do they aim to come to Japan? Is coexistence possible?] [Video]. YouTube. https://www.youtube.com/watch?v=RmspZPPzQUw

Aggestam, K., Cristiano, F., & Strömbom, L. (2015). Towards agonistic peacebuilding? Exploring the antagonism–agonism nexus in the Middle East peace process. *Third World Quarterly, 36*(9), 1736–1753. https://doi.org/10.1080/01436597.2015.1044961

Ahmadian, S. (2024). Kurdish diaspora in Japan: Navigating Kurdish identity and activism on social media. In L. Momesso & P. Ivanova (Eds.), *Refugees and asylum seekers in East Asia: Perspectives from Japan and Taiwan* (pp. 257–280). Springer Nature Singapore. https://doi.org/10.1007/978-981-97-2867-1_10

Akasaka, T., & Sunami, K. (2025, March 23). Hate speech bringing unwanted focus on Japan's Kurdish community. *Kyodo News*. https://english.kyodonews.net/news/2025/03/1bca3426674d-feature-hate-speech-bringing-unwanted-focus-on-japans-kurdish-community.html#google_vignette

Asada, T. (2024, November 22). Saitama court blocks anti-Kurd protest, calling it hate speech. *The Asahi Shimbun*. https://www.asahi.com/ajw/articles/15519482

The Asahi Shimbun. (2024a, April 20). Hate speech escalates against Kurds living close to Tokyo. *The Asahi Shimbun: Asia & Japan Watch*. https://www.asahi.com/ajw/articles/15238743

The Asahi Shimbun. (2024b, June 5). Social media posters abroad fuel hatred of Kurds in Japan. *The Asahi Shimbun: Asia & Japan Watch*. https://www.asahi.com/ajw/articles/15285385

August, V. (2022). Understanding democratic conflicts: The failures of agonistic theory. *European Journal of Political Theory, 23*(2), 182–203. https://doi.org/10.1177/14748851221120120

Fujibayashi, H. (2020). When public opinion drives national asylum policymaking: The case of Kurdish asylum seekers in Japan. *Asia & the Pacific Policy Studies, 7*(2), 204–216. https://doi.org/10.1002/app5.301

González Martín, D., Lauge Hansen, H., & Parra Grondona, A. (2022). A case for agonistic peacebuilding in Colombia. *Third World Quarterly, 43*(6), 1270–1287. https://doi.org/10.1080/01436597.2021.1993812

Göker, Z. G., & Çelik, A. B. (2022). Women's dialogic encounters: Agonistic listening and emotions in multiple-identity conflicts. *Third World Quarterly, 43*(6), 1251–1269. https://doi.org/10.1080/01436597.2021.1977621

Husain, S., & Ahamed, M. F. (2022). Place & power: A study of Kurdish identity in Turkey. *GeoJournal, 87*(5), 4235–4249. https://doi.org/10.1007/s10708-021-10498-z

Igarashi, Z. M. (2014). *Intimacy crossroads: Turkish-Japanese transnational marriages through lenses of culture, religion and gender* [Doctoral dissertation, Waseda University]. https://waseda.repo.nii.ac.jp/records/24852

Japan Kurdish Cultural Association. (2024, November 2). Community initiatives. https://www.nihonkurdish.or.jp/en/community-initiatives/

The Japan Times. (2024, November 22). Saitama court bans demonstrations over anti-Kurd hate speech. *The Japan Times*. https://www.japantimes.co.jp/news/2024/11/22/japan/crime-legal/kurds-hate-speech/

Katayama, N. (2021). *Zainichi Kurudojin komyuniti ni okeru ibunka-kan komyunikēshon no kenkyū—"Wakari aeru Nihongo komyunikēshon" ni mukete* [A study on intercultural communication in the Kurdish community in Japan: Toward "mutually understandable Japanese communication"] [Doctoral dissertation, University of Tsukuba]. https://doi.org/10.15068/0002005425

Kawaguchi City Official Site. (2025, February 4). *Gurafu de miru Kawaguchi no jinkō* [The population of Kawaguchi in graphs]. https://www.city.kawaguchi.lg.jp/soshiki/01020/010/toukei/12/43956.html

Kozinets, R. V. (2010). *Netnography: Doing ethnographic research online*. SAGE Publications.

Lehti, M., & Romashov, V. (2022). Suspending the antagonism: Situated agonistic peace in a border bazaar. *Third World Quarterly, 43*(6), 1288–1306. https://doi.org/10.1080/01436597.2021.1962274

MacGinty, R. (2014). Everyday peace: Bottom-up and local agency in conflict-affected societies. *Security Dialogue, 45*(6), 548–564. http://www.jstor.org/stable/26292931

MacGinty, R. (2021). *Everyday peace: How so-called ordinary people can disrupt violent conflict*. Oxford University Press.

Ministry of Justice Immigration Services Agency. (n.d.-a). *Kanrisochi ni tsuite* [Regarding supervisory measure]. https://www.moj.go.jp/isa/08_00045.html

Ministry of Justice Immigration Services Agency. (n.d.-b). *Kari hōmen* [Regarding provisional release]. https://www.moj.go.jp/isa/08_00050.html

Ministry of Justice Immigration Services Agency. (n.d.-c). *Sōkan kihi-sha no uchi honpō de shusshō shita kodomo no zairyū tokubetsu kyoka ni kansuru taiō hōshin ni tsuite* [Policy on special residence permission for children born in Japan among deportation avoiders]. https://www.moj.go.jp/isa/content/001419300.pdf

Mouffe, C. (2000). *The democratic paradox*. Verso.

Nakajima, Y. (2003). *Shingetsu no yoru ga akeru toki: Kita Kurudisutan no hitobito* [When the night of the new moon dawns: The people of Northern Kurdistan]. Shinsensha.

NHK. (2024, August 31). "Yōgisha wa fuhō imin" nise jōhō ga Igirisu de bōdō ni Nihon de wa… ["Suspect is an illegal immigrant": Fake news leads to riots in the UK—False claims about Kurds also surge in Japan, especially linked to Kawaguchi City]. *NHK News*. https://www3.nhk.or.jp/news/html/20240831/k10014563531000.html

NHK. (2025, April 5). *Feiku to riaru Kawaguchi Kurudojin shinsō* [Hate and reality, Kurds in Kawaguchi, the truth] [Video]. NHK. https://www.nhk.jp/p/etv21c/ts/M2ZWLQ6RQP/episode/te/8ZVZ4PZ41V/

140journos. (2025, May 15). *Uzakdoğu sorunu* [The Far East problem] [Video]. YouTube. https://youtu.be/aepLFK6Hxmk?si=fAoZ_tUtRW1isE1s

Richmond, O. P. (2010). Resistance and the post-liberal peace. *Millennium, 38*(3), 665–692. https://doi.org/10.1177/0305829810365017

Richmond, O. P. (2020). *Peace in international relations* (2nd ed.). Routledge.

The Sankei Shimbun. (2023a, July 30). Reiwa mo sansei shita Kawaguchi shigikai "Kurudojin" ikensho [Reiwa also supported the Kawaguchi City Council's "Kurdish" statement—Not reported by the media]. *Sankei News*. https://www.sankei.com/article/20230730-H6PIW4LZ25O7LBZKRX6LT7TY34/

The Sankei Shimbun. (2023b, October 14). Zainichi Kurudojin 2-sei "Kekkyoku wa oya shidai" 10-dai zenhan de danshi wa kaitai-gyo, joshi wa kekkon [Second-generation Kurdish residents in Japan: "It all depends on the parents"—Boys enter demolition work, girls marry in early teens]. *Sankei News*. https://www.sankei.com/article/20231014-TP4MHHCHPRP2TLEMTT3PT4TNTE/

Scott, J. C. (1985). *Weapons of the weak: Everyday forms of peasant resistance* (1st ed.). Yale University Press. https://doi.org/10.12987/9780300153620

Scott, J. C. (2008). *Domination and the arts of resistance: Hidden transcripts*. Yale University Press. https://doi.org/10.12987/9780300153569

Scott, J. C. (2013). *Decoding subaltern politics: Ideology, disguise, and resistance in agrarian politics*. Routledge. https://doi.org/10.4324/9780203095041

Shinko, R. E. (2008). Agonistic peace: A postmodern reading. *Millennium, 36*(3), 473–491. https://doi.org/10.1177/03058298080360030501

Shinko, R. E. (2022). A critical (re)reading of the analytical significance of agonistic peace. *Third World Quarterly, 43*(6), 1399–1407. https://doi.org/10.1080/01436597.2022.2042681

Statistics Bureau of Japan. (n.d.). *Zairyū gaikokujin tōkei* [Statistics on foreigners registered in Japan]. e-Stat (Portal Site of Official Statistics of Japan). https://www.e-stat.go.jp/stat-search/files?page=1&layout=datalist&toukei=00250012&tstat=000001018034&cycle=1&year=20240&month=12040606&tclass1=000001060399

Strömbom, L. (2012). Thick recognition: Advancing theory on identity change in intractable conflicts. *European Journal of International Relations, 20*(1), 168–191. https://doi.org/10.1177/1354066112439217

Strömbom, L., & Bramsen, I. (2022). Agonistic peace: Advancing knowledge on institutional dynamics and relational transformation. *Third World Quarterly, 43*(6), 1237–1250. https://doi.org/10.1080/01436597.2022.2054797

Takeshita, S. (2016). Social and human capital among Japanese-Turkish families in Japan. *Asian Ethnicity, 17*(3), 456–466. https://doi.org/10.1080/14631369.2015.1062071

Tsuchida, C. (2018). Causal relationship between choice of applying for refugee status and building an ethnic community: Case study of Kurdish applicants from Turkey in Japan. *Journal of Human Security Studies, 7*(2), 95–112. https://www.jahss-web.org/single-post/2018/10/06/Journal-of-Human-Security-Studies-Vol7-No2-Autumn-2018

Uslaner, E. M. (2010). Segregation, mistrust and minorities. *Ethnicities, 10*(4), 415–434. https://doi.org/10.1177/1468796810378325

Yasuda, K. (2024, October 4). Japanese hate groups targeting Kurdish community. *Nippon*. https://www.nippon.com/en/in-depth/d01048/

Part III
Institutional and Governance Pathways

Chapter 8
Incorporating Policy Process Analysis into Environmental Peacebuilding: The Coupling of Environmental and Foreign Policies in Soviet–Japanese Relations

Balinskaia Aleksandra

8.1 Introduction

Interstate dialogue flows through interconnected economic, political, and social channels that, under favorable conditions, bolster both peace and sustainable development. The United Nations 2030 Agenda for Sustainable Development formalizes this concept by urging states to enhance institutions and protect the environment as mutually reinforcing objectives. However, during periods of tension, these channels constrict: hard-security or commercial priorities overshadow ecological concerns, official talks stall, and at times, even basic diplomatic communication ceases. Yet, in the long run, neglecting shared ecological risks undermines both the capacity of states to sustain constructive relations and the broader socioeconomic resilience of the societies they govern.

The growing scholarship on the peace–sustainability nexus warns that sidelining ecological stress destabilizes societies and erodes possibilities for durable peace (Sharifi et al. 2021; Simangan et al. 2021). Within this scope, Environmental Peacebuilding (EP) scholarship identifies joint resource management as a less politicized entry point for engagement that can prevent, mitigate, or resolve conflict (Floyd and Matthew 2013; Ide et al. 2021).

Cooperation on issues such as pollution control or habitat protection mobilizes scientific expertise, civil society momentum, and multilevel governance to reframe bilateral agendas in future-oriented, non-zero-sum terms. Since the 1960s, an expanding web of transnational norms and instruments has encouraged precisely

B. Aleksandra (✉)
Graduate School of Human and Socio-Environmental Studies, Kanazawa University, Kanazawa, Japan

A. Sharifi et al. (eds.), *Frameworks and Practices Towards a Sustainable and Peaceful Future*, World Sustainability Series,
https://doi.org/10.1007/978-981-95-2310-8_8

this type of collaboration (Corry 2014; Woodhouse 2009), suggesting that ecology can provide both the substance and the symbolism required for sustainable dialogue.

Recent research demonstrates that, even in fragile settings, environmental agendas can intersect with broader governance processes and open diplomatic pathways that seemed foreclosed (Balinskaia 2024). By addressing both tangible problems (such as shared watershed management and climate-risk mitigation) and symbolic ones (readiness to cooperate on low-politics issues), environmental policy can loosen the grip of traditional security paradigms and create space for incremental confidence-building (Conca and Dabelko 2002). Understanding how states weave ecological considerations into foreign-policy decision-making is vital, as it reveals the institutional logics, domestic pressures, and societal demands that determine whether environmental concerns remain peripheral or become strategic engagement tools.

This chapter scrutinizes such a transformation in the late Cold War Soviet–Japanese relationship, an understudied yet instructive "least likely" case (Van Evera 1997). Despite unresolved territorial claims, the absence of a post-war peace treaty, and Japan's close alignment with the United States, the two governments concluded their first comprehensive accord on environmental protection in 1991.

Why did ecological cooperation, of all issues, break through decades of mutual suspicion? Answering this puzzle highlights pathways to engagement that standard foreign policy analysis and mainstream EP have thus far overlooked and speaks directly to contemporary debates on multilateral environmental governance.

To disentangle how this window opened, the study applies the Multiple Streams Framework (MSF; Herweg et al. 2018; Kingdon 1995) to a dual-domain setting. It traces a domestic stream, in which ecological issues rose on the Soviet agenda, alongside an international stream, in which evolving geopolitics increased the appeal of cooperation with Japan. Mapping the intersections between these two streams constitutes a new extension of MSF in the environmental peacebuilding literature.

Integrating insights from EP and policy process research leads to the hypothesis that a policy window opens when a domestically ascending environmental agenda intersects with a favorable shift in the foreign-policy stream. Accordingly, the study asks: Under what conditions does the coupling of two distinct policy domains, domestic environmental policy and international security strategy, enable rival states to institutionalize ecological cooperation? A case-specific sub-question sharpens the puzzle: Why were the USSR and Japan able to conclude an environmental-protection agreement in 1991 despite the absence of a peace treaty?

Addressing this question yields three contributions. Theoretically, the analysis demonstrates the analytical value of policy process tools, especially the MSF, for EP scholarship by showing how stream coupling clarifies the timing and mechanisms of interstate environmental initiatives. Empirically, it offers the first systematic account of Soviet–Japanese environmental diplomacy based on evidence that captures perspectives rarely reflected in existing research. Practically, it distills lessons for today's policymakers who seek low-politics avenues of engagement within the broader peace–sustainability nexus and emerging regimes of multilateral environmental governance. The subsequent chapters detail the theoretical background, methodology, and historical trajectory of Soviet–Japanese relations, and application of MSF to the case, and conclude with findings, limitations, and future research directions.

8.2 Bridging Environmental Peacebuilding and the Multiple Streams Framework

8.2.1 Strengths and Weaknesses of the Environmental Peacebuilding Framework

Environmental peacebuilding views ecological issues as opportunities for cooperation in hostile relationships, challenging the assumption that environmental matters are inherently politicized or securitized (Conca and Dabelko 2002). EP scholarship suggests that environmental concerns can sometimes be strategically reframed, or "unbundled," from entrenched conflicts (Ide et al. 2021).

Environmental issues may appear less politicized when climate adaptation or cross-boundary pollution control is framed as a transboundary, non-zero-sum challenge, thereby encouraging cooperation. This reframing does not render environmental agendas apolitical, but highlights their dual nature. They can intensify political tensions and reinforce power structures (Barnett 2001; Trombetta 2008); nevertheless, they can equally serve as shared, urgent concerns around which adversaries build gradual trust.

It is important to distinguish between environmental cooperation and environmental peacebuilding. While the former refers to joint initiatives on transboundary ecological issues, such as data sharing or watershed monitoring, EP involves integrating environmental management into broader conflict prevention, resolution, or recovery processes (Ide and Detges 2018; Krampe 2017). Not all environmental cooperation leads to peacebuilding.

Ide et al. (2021) emphasize three core dimensions of environmental peacebuilding: security; livelihoods and economic recovery; and politics and social relations. These dimensions complement peace-research arguments that environmental initiatives can evolve from functional cooperation, through confidence-building, to a more profound structural transformation of intergroup relations (Conca and Dabelko 2002).

Despite its growing prominence, EP research has been critiqued for several weaknesses. First, although studies and reviews examine the causal mechanisms that explain under what conditions environmental cooperation leads to peace, these mechanisms still need to be more clearly articulated and studied (Ide 2019; Johnson et al. 2021; Balinskaia 2024). Second, endogeneity persists: environmental projects frequently arise where a baseline of trust already exists, blurring whether peace is an outcome or a precondition of cooperation (Dresse et al. 2019). Third, EP has been criticized for frequently downplaying power asymmetries. Technocratic environmental initiatives can depoliticize conflicts and empower dominant actors while marginalizing others (Aggestam 2015; Verweijen and Hoffmann 2024). Such depoliticization risks masking ongoing structural violence by presenting environmental programs as neutral and purely technical. These critiques suggest that a stronger theoretical framework explaining the policy process leading to EP is needed to clarify the dynamics within the field and enhance its explanatory rigor.

8.2.2 *Environmental Peacebuilding Through the MSF Lens: A Theoretical Synthesis*

The MSF offers a useful heuristic to address some of these challenges in the EP literature. Originally developed to explain policy agenda setting in the United States (Kingdon 1995), the MSF can help to address this gap by illustrating how problems, potential solutions, and political factors interact to elevate or marginalize specific policy proposals in an environment where decision-making is influenced by ambiguity, limited time, incomplete information, shifting preferences, and fluid participation (Herweg et al. 2018). The MSF tries to explain the policy process by suggesting the concept of three independent streams: problems (how an issue is framed), policies (the wide range of potential solutions proposed by experts, policymakers, and other stakeholders), and politics (shifts in public opinion, election outcomes, leadership changes, etc.; Howlett et al. 2015).

The convergence of these streams opens a policy window, usually seized by policy entrepreneurs who invest time, credibility, and resources to couple a problem with a viable solution under favorable political conditions (Herweg et al. 2018). For instance, the 1999 earthquakes in Greece and Turkey catalyzed an unexpected diplomatic rapprochement, partly because political entrepreneurs seized the opportunity to reframe disaster assistance as a basis for rapprochement (Blavoukos and Bourantonis 2012).

Applying MSF to EP highlights how environmental issues can move from peripheral to strategic concerns. A small but growing set of empirical studies illustrates how the framework may sharpen EP analysis. In post-war Bosnia–Herzegovina, Budimir (2022) argues that lethal air pollution remained off the national agenda because the politics stream, dominated by ethnopolitical rivalry and donor-driven reconstruction priorities, failed to couple with recognized problems and available technical fixes, thereby identifying a specific blockage rather than invoking a generic "lack of political will." Analyzing the US–Mexico restoration of the Colorado River Delta, Koebele (2021) traces how nongovernmental entrepreneurs kept habitat-restoration proposals alive in the policy stream until drought-related focusing events and a receptive US administration altered the politics stream, implying that cooperative ventures can precede trust when streams converge. Examining two forest-governance programs in the Congo Basin, Nago (2021) shows that international aid agencies sometimes act as policy entrepreneurs. Where donor activism diverged from national priorities, collaboration appeared to entrench inequalities, whereas closer alignment coincided with more substantive policy change.

These studies indicate three ways in which MSF can enrich EP scholarship. First, by tracing stream congruence and incongruence, MSF helps specify the causal mechanisms that enable or impede ecological cooperation, moving the debate beyond post hoc attribution. Second, sequencing convergence allows researchers to probe whether cooperation is a precursor to, or a consequence of interparty trust, addressing longstanding endogeneity concerns. Third, the framework's focus on entrepreneurs and policy windows foregrounds the politics of agenda control, revealing how ostensibly technical environmental initiatives may reproduce underlying power asymmetries. While the empirical base remains limited, these findings point to a promising research agenda that combines MSF's process orientation with EP's substantive focus on the peace–environment nexus, offering a more nuanced understanding of how, when, and for whom environmental cooperation can foster sustainable peace.

Moreover, MSF allows scholars to explore how domestic and international dynamics interact. Domestic shifts, such as rising public concern over pollution or the emergence of a national environmental movement, can elevate ecological issues within policy agendas. At the same time, external frameworks, such as multilateral environmental agreements or foreign-policy shifts, can open additional windows of opportunity. In environmental peacebuilding, the coupling of streams across two policy domains, domestic environmental policy and foreign affairs, offers a productive perspective on transboundary cooperation in conflict-prone settings (Schlaufer and van den Dool 2024).

Finally, MSF has proven adaptable to various governance settings, including nondemocratic regimes. While originally rooted in pluralist democratic assumptions, recent scholarship shows that the MSF can be applied in authoritarian or hybrid contexts, with appropriate modifications. In such systems, policy streams may not be fully independent, and elite actors often play multiple roles, as both problem definers and policy entrepreneurs (Herweg et al. 2022; Schlaufer and van den Dool 2024). In centralized systems such as the late Soviet Union, informal elite networks and technocratic institutions played critical roles in elevating ecological concerns and aligning them with strategic diplomatic goals. Although state-level or elite actors in authoritarian and hybrid regimes seldom resemble the archetypal policy entrepreneur, their gatekeeping control over agendas and institutional access is often decisive for initiating policy change (Herweg et al. 2022; Schlaufer and van den Dool 2024).

Thus, coupling MSF with EP clarifies the mechanics of environmental policy change in fragile and authoritarian settings and advances theoretical integration between the two frameworks. It provides a way to move beyond normative appeals to the environment as a tool for peace, offering a systematic account of how, when, and why environmental issues gain traction within conflict-prone or diplomatically stalled relationships.

Initially designed for domestic policymaking, MSF's core concepts, uncertainty, the coevolution of distinct streams, and temporal policy windows, also apply to foreign policy. Foreign decisions, influenced by external factors, must align with domestic constraints such as electoral considerations, bureaucratic interests, ideological shifts, and societal demands. Applying MSF to international relations requires adapting the framework to account for multiple governance layers and diverse domestic and international influences (Howlett et al. 2015).

Foreign policy requires balancing domestic agendas with international imperatives. States must address external events such as global environmental challenges and geopolitical tensions while managing internal pressures such as public opinion, elite factionalism, institutional inertia, and economic constraints. This highlights the role of policy entrepreneurs and key decision-makers, such as dominant leaders or institutions, in facilitating foreign policy shifts (Blavoukos and Bourantonis 2012). MSF's focus on problem framing, solution availability, and political receptivity is especially valuable in foreign policy scenarios where longstanding disputes, such as territorial claims or security dilemmas, have stalled traditional negotiation channels.

8.3 Methodology

This study employs a qualitative research design to examine the factors leading to the 1991 Soviet–Japanese environmental agreement. The Soviet-centered focus is intentional: whereas Tokyo's overriding concern was the territorial dispute, Moscow deployed various forms of cooperation as a diplomatic instrument during ideological reorientation and external opening. Analyzing that strategic choice illuminates pathways to environmental peacebuilding that may be obscured in more symmetrical or data-rich cases. Although the Soviet–Japanese episode is not a data-abundant case, its limitations, low mutual trust, and fragmentary evidence make it a proper contrast case against which stronger EP examples can be compared later (Gerring 2016). The asymmetrical, low-trust 1991 Soviet–Japanese environmental agreement can provide a meaningful case to test EP's confidence-building logic and to examine MSF processes across domestic and transnational policy streams.

Primary materials include Soviet legislative acts, Politburo and government minutes, reports, memoirs, diplomatic memoranda, and press coverage from 1985 to 1992. All publicly available periodicals of that time (*Pravda, Izvestiia, Pravitelstvenny Vestnik, Kommersant)* were fully reviewed for the specified period. On the Japanese side, articles were retrieved from *Asahi Shimbun* and *Nikkei Shimbun*, supplemented by files in the Diplomatic Archives of the Ministry of Foreign Affairs of Japan that cover preparations for the April 1991 summit.

Documents were selected based on themes related to environmental policymaking, Soviet–Japanese relations, internal Soviet reforms (Perestroika and Glasnost), and the evolving international environmental agenda. Emphasis was placed on sources that highlight the roles of key political figures such as Gorbachev,

Shevardnadze, Yakovlev, and Chernyaev, who, within the authoritarian regime, controlled decision-making processes (Checkel 1993; Gorbachev 1995). In total, the press corpus comprises approximately 90 Soviet and 20 Japanese articles retained after full-text screening.

Each item identified via keyword search was logged in a source database; a full reading determined inclusion or exclusion. Whenever available, original-language versions were used; the author translated Japanese items without official English summaries. Secondary literature and memoirs provided contextual framing and cross-checks.

To complement archival and published material, a semi-structured interview was held in July 2024 with Professor Kazuhiko Togo, head of the Soviet Section in Japan's Ministry of Foreign Affairs (1988–1992). The interview explored Japanese perceptions of Soviet proposals and the role of environmental issues in bilateral discussions. Written informed consent was obtained; the transcript is stored securely and is available to qualified researchers upon request. Several former Russian officials were approached, but they declined to participate; consequently, the interview is treated as a supplementary, not decisive, source.

During the review, notes, annotations, and preliminary codes were created, focusing on environmental policy, foreign policy decision-making, and the framing of bilateral cooperation. The analysis utilized qualitative content analysis methods as described by Mayring (2015) and adapted to policy research contexts (Creswell and Poth 2018). This involved inductive coding of themes related to environmental agendas, political entrepreneurship, and the intersections of MSF streams. The codes were then organized into categories aligned with MSF logic, identifying how environmental challenges were framed as problems, which policy solutions were proposed, and how the political stream affected the feasibility of environmental cooperation.

Coupling evidence was sought where documents linked at least two streams. A concise chronology anchors the process tracing: 1985 (Gorbachev's election), 1986 (the Chernobyl disaster), 1988 (the CPSU Conference; the creation of the USSR State Committee for the Environment), 1988 (the creation of a bilateral working group for a Gorbachev visit to Tokyo), and April 1991 (the Tokyo Joint Statement).

Triangulation adopted a multidimensional strategy, combining data, methodological, and (where feasible) investigator checks (Tzagkarakis and Kritas 2023). Each key claim was corroborated, where possible, by contemporaneous Soviet press, Japanese press, or MOFA files, and memoirs or interview testimony. Discrepancies were adjudicated by privileging contemporaneous documentary evidence over retrospective accounts. The author conducted all coding; to reduce single-coder bias, provisional findings were discussed with external experts on Soviet foreign policy, whose published work is cited in the chapter.

Additionally, the Soviet focus aligns with the research question, as the USSR initiated environmental cooperation as a diplomatic tool during a period of internal ideological reorientation and international opening. Examining Soviet domestic political processes and leadership priorities is therefore essential. While a fully symmetrical analysis would equally explore the Japanese side, this study emphasizes the Soviet strategic choice to prioritize environmental issues.

Several limitations must be acknowledged. Archival access is uneven: many MOFA and Russian MFA records remain partially closed, and post-Soviet institutional instability has led to partial document loss. Newspaper coverage may reflect official narratives; memoirs are subject to hindsight bias. The author is a Russian-speaking researcher with limited Japanese proficiency; unofficial English abstracts and self-translations were used for Japanese sources lacking authorized translations. These constraints are partially offset by the breadth of Soviet press material and the cross-national press comparison. Findings should, therefore, be read as historically grounded but interpretive, illustrating how MSF can structure inference under evidentiary uncertainty.

Finally, while the MSF is widely applied in policy studies, its application to international relations remains somewhat open-ended. The conclusions are necessarily interpretive and grounded in the interplay of historical documents, scholarly analysis, and theoretical frameworks.

8.4 Changes in the Environmental Policy of the USSR and Soviet–Japanese Environmental Cooperation in the Second Half of the Twentieth Century

8.4.1 The Environmental Policy of the USSR Before 1985

The USSR mobilized its vast natural reserves to meet ambitious economic targets, relegating ecological safeguards to a secondary role (Josephson et al. 2013). From the 1960s through the late 1980s, rapid, loosely regulated expansion, including massive irrigation withdrawals that shrank the Aral Sea, produced acute environmental degradation (Ponomareva 1999; Sokolov et al. 2001; Weiner 1999).

A national monitoring apparatus existed, but bureaucratic fragmentation and a rigid command economy blunted enforcement; pollution data were often suppressed to preserve the state's industrial image (Josephson et al. 2013). Mounting scientific alarm nonetheless fueled grassroots campaigns against megaprojects such as the proposed Siberian rivers diversion to Asia and the Baikal pulp mill projects (Sedaitis and Butterfield 2019).

During the 1970s, the USSR, echoing the global environmental movement, expanded its system of nature reserves and promulgated emission standards, but weak enforcement and overriding growth targets rendered most measures largely symbolic (Corry 2014; Weiner 1999).

Internationally, Moscow sought reputational gains: it endorsed the 1972 Stockholm Declaration and signed several multilateral and bilateral accords, including a wide-ranging agreement with the United States, to project an image of ecological responsibility (Nikitina 1995; Sokolov et al. 2001). However, these efforts were often strategic, aimed at countering Western criticism rather than implementing genuine domestic reforms (Josephson et al. 2013). Implementation lagged as environmental commitments remained subordinate to geostrategic and economic priorities, and many cooperative projects quietly stalled when political tensions resurfaced (Robinson 1988).

8.4.2 Gorbachev's "Liberating" Environmental Policy

Mikhail Gorbachev's accession in March 1985 broke the pattern of geriatric, status-quo leaders and signaled an appetite for systemic change (Taubman 2018). As General Secretary of the CPSU Central Committee and later President of the USSR, he advanced a policy of "new thinking," seeking to de-escalate Cold War tensions and recast Soviet diplomacy around openness and cooperation (Chernyaev and Weber 2010). Crucially, he acknowledged that the USSR could neither remain insulated from global currents nor ignore mounting domestic grievances, including ecological degradation (Gorbachev 1995).

Political and economic reforms coincided with the Chernobyl disaster (26 April 1986), which exposed the systemic failures of Soviet environmental management and triggered unprecedented public outrage (Gorbachev 1995; Larin et al. 2003). Scientists and citizens demanded accountability, making environmental degradation a front-line issue (Sedaitis and Butterfield 2019; Zatravkin and Vishlenkova 2021). In response, Gorbachev created the State Committee for Environmental Protection (Goskompriroda) in 1988, elevating ecology within the domestic policy hierarchy (Josephson et al. 2013; Ponomareva 1999; Weiner 1999).

Several close associates reinforced this agenda. Alexander Yakovlev championed a holistic vision linking "nature-individual-society-humanity-planet" (Yakovlev 1991). Eduard Shevardnadze wove environmental themes into foreign policy, arguing that security and ecology were interdependent (Shevardnadze 1991). Fedor Morgun, and later his successor Nikita Vorontsov, Chairman of the USSR State Committee for Nature Protection and subsequently the first and last USSR Minister for Environmental Management and Protection, worked to unify fragmented environmental agencies and proposed transboundary-reserve models, aiming to foster multinational cooperation (Larin et al. 2003; Weiner 2024).

Gorbachev's personal commitment to environmental issues linked domestic reform to foreign-policy outreach (Gorbachev 2000; Taubman 2018; Yablokov 2007). The openness policy led to the first large-scale environmental protests, forcing the leadership to address environmental-protection institutions in response to public criticism (Gorbachev 1995; Sedaitis and Butterfield 2019).

Externally, the late-1980s thaw encouraged the USSR to condemn the arms race (e.g., endorsement of the 1983 "nuclear winter" findings) and to frame ecology as a shared human value (Rindzevičiūtė 2020). State media began promoting universal human values and environmental ethics (Larin et al. 2003; Weiner 1999), and Moscow hosted the Politburo of the CPSU Central Committee (1986) and The Global Forum on Environment and Development (1990). The latter's declaration urged all nations to unite in safeguarding the planet (The Global Forum on Environment and Development 1990).

8.4.3 *From Political Stalemate to Agenda Broadening: Soviet–Japanese Relations, 1950s–1991*

Soviet–Japanese interaction after 1951 was characterized by a persistent asymmetry: modest but steady economic ties, timber, fisheries, and marine-biological research (Ivanov and Smith 1999; Josephson et al. 2013; Weiner 1999), coexisted with a frozen political dialogue shaped by Cold War bipolarity and the unresolved territorial claims. The absence of a peace treaty constrained high-level contacts and blocked sensitive policy areas, notwithstanding a brief détente in 1971–1973 that followed Prime Minister Tanaka's Moscow visit (Simon 1974). With the return of US–USSR tensions in the late 1970s, the bilateral agenda again narrowed to low-politics trade, and substantive negotiations ceased until 1985.

The arrival of Mikhail Gorbachev, reform-minded and attuned to new thinking, reconfigured this landscape (Taubman 2018). Within weeks of his election, Japanese Prime Minister Nakasone met the new leader at Chernenko's funeral, signaling Japan's willingness to test the opening despite unresolved territorial claims (Torkunov et al. 2015). Foreign Minister Shevardnadze soon added repeated visits that promoted scientific, cultural, and economic initiatives (Politburo of the CPSU Central Committee 1986). In the late 1980s, Moscow therefore deliberately shifted the focus to lower-stakes areas such as trade, cultural exchange, and environmental cooperation in order to keep the dialogue alive while the sovereignty dispute remained deadlocked. Tokyo responded with the concept of *kakudai kinko* (expanded equilibrium), arguing that non-territorial cooperation should proceed wherever possible (Shimotomai 2019).

These maneuvers culminated in Gorbachev's state visit of 16–18 April 1991. The Joint Statement adopted fifteen sectoral accords, including a wide-ranging environmental-protection agreement that established exchanges of experts, joint research, and pilot projects (Panov 1992). Although domestic turbulence curtailed Soviet flexibility and the USSR collapsed eight months later (Shimotomai 2019), the summit institutionalized a mechanism for continued cooperation, illustrating how environmental issues could serve as a politically acceptable bridge even in the absence of a peace treaty (Kuzminkov 2006).

8.4.4 *Soviet–Japanese Cooperation in the Field of Environmental Protection*

Environmental questions long occupied the periphery of Soviet–Japanese relations: apart from a 1973 convention on migratory birds and routine fishery accords, ecological ties were minimal (National Biodiversity Strategy of Japan 1973). The first years of Gorbachev's leadership brought only ad hoc gestures such as the Japanese relief after Chernobyl (1986) and the Armenian earthquake in 1988 (Zhukova 2017), but no structured dialogue.

Momentum shifted in 1989 when Foreign Ministers Shevardnadze and Nakayama agreed to draft a bilateral environmental-protection treaty by year-end (Shimbun 1989; Panov 1992). Tokyo's leaders, Takeshita, Kaifu, and Miyazawa, were promoting a global "environmental leadership" agenda (Ivanov and Smith 1999), and METI's new Green Aid programs targeted Russia and other East Asian neighbors. In June 1990, USSR Minister of Nature Protection Vorontsov headed a delegation to Japan, proposing a Kuril–Hokkaido transboundary reserve; the plan won scientific backing but was shelved by Japan's foreign ministry (Weiner 2024). In the run-up to Gorbachev's visit, both sides mapped joint research on permafrost methane, marine pollution, and hazardous-waste recycling.

Addressing the Japanese Diet on 17 April 1991, Gorbachev emphasized the importance of environmental cooperation, highlighting urgent environmental threats and the need to redirect resources toward protecting the Earth (Pravda 1991). The ensuing Joint Statement contained fifteen accords; the environmental agreement covered air and water pollution control, biodiversity, climate research, and mechanisms for expert exchange (Panov 1992). Diet minutes framed it as the first shift "from technical aid to genuine environmental partnership," citing Soviet ecological crises and a joint pledge to tackle global warming (National Diet of Japan 1991).

According to Professor Togo, the pact stood out as "purely positive... free of the political undertones shadowing most other agreements" and flexible enough to be activated or paused without cost. Despite the USSR's collapse eight months later, Russia inherited the treaty and convened five meetings of the Joint Committee on Environmental Protection between 1994 and 2006, starting with a joint marine ecology survey in the Sea of Japan (Takahashi 2000; Josephson et al. 2013).

Although secondary to the unresolved peace-treaty dispute, the 1991 accord demonstrated that environmental policy could operate as a low-risk bridge, sustaining dialogue and institutional memory when higher-stakes negotiations stalled.

8.5 Discussion

8.5.1 Soviet–Japanese Environmental Agreement: Two Overlapping Policy Processes

Explaining the 1991 environmental accord requires tracking two policy processes that evolved largely in parallel but converged at a single diplomatic moment. The first unfolded inside the USSR, where Chernobyl and a cascade of pollution scandals propelled ecology from a technocratic sideline to a salient governance problem; the second played out in Soviet–Japanese relations, where hard-security talks were stalemated, yet both capitals were searching for a less politicized avenue for dialogue. Table 8.1 summarizes how the MSF helps to locate the critical junctures in each arena.

Table 8.1 USSR environmental policy and Soviet–Japanese 1991 agreement policy processes

MSF streams	Soviet domestic environmental policy	Soviet FP Toward Soviet–Japanese cooperation
Problem Stream	• Environmental degradation became undeniable, culminating in the 1986 Chernobyl disaster, which exposed systemic flaws in policy and governance • New policy allowed scientists, citizens, and journalists to publicly criticize environmental mismanagement, elevating ecological damage as a salient issue demanding state attention	• The bilateral relationship was overshadowed by unresolved territorial disputes and the absence of a peace treaty, which hindered progress in traditional diplomatic spheres • Environmental concerns presented a nontraditional problem space that could bypass heavily securitized territorial debates and engage both sides on less confrontational ground, such as marine biodiversity and air pollution
Policy Stream	• The establishment of the State Committee for Environmental Protection (1988) and related reforms generated solutions such as stricter pollution controls, resource conservation measures, and organizational restructuring • Domestic institutions and expert communities developed a pool of technical and administrative proposals to address ecological degradation, making environmental policy a legitimate governance issue	• Pollution control, climate research, and biodiversity protection emerged as potential bilateral activities that were technically feasible, scientifically grounded, and consistent with global environmental norms • Japanese officials in the Soviet Section of the Ministry of Foreign Affairs advanced the "expanded equilibrium" proposal, reinforcing cooperative options
Politics Stream	• Political reforms under Gorbachev's leadership loosened ideological rigidity, legitimized public demands, and encouraged responsiveness to societal concerns, including environmental stewardship • State leaders recognized that improving environmental management served domestic reform goals and enhanced the USSR's legitimacy	• The shifting geopolitical climate created openings for diplomatic overtures that were less burdened by ideological confrontation • By incorporating various forms of cooperation, including environmental initiatives, influential Soviet actors leveraged improved political conditions to propose dialogue that complemented ongoing but stalled territorial negotiations, signaling goodwill and modernization efforts

By early 1991, these streams aligned almost simultaneously in both domains. Domestically, environmental urgency was undisputed, a comprehensive treaty draft had cleared Soviet vetting, and reformist elites viewed external ecological cooperation as proof of "new thinking" and crucial for maintaining legitimacy, and advancing reforms (Weiner 1999).

Internationally, Gorbachev's approaching visit, Japan's "expanded equilibrium" strategy, and attention to environmental issues created an opening for low-risk engagement. The synchronized alignment constituted a bilateral policy window that entrepreneurial actors such as Shevardnadze moved quickly to exploit. Their coupling of problem salience, off-the-shelf solutions, and favorable politics produced the fifteen-point Joint Statement of 18 April 1991, whose environmental agreement institutionalized joint research, expert exchanges, and pilot mitigation projects (Panov 1992).

Viewing the episode through overlapping MSF lenses clarifies the causal mechanism. Domestic reframing of ecology generated credible policy content; détente-era foreign policy supplied timing and diplomatic demand. Only their intersection converted a previously marginal issue into a viable confidence-building channel. This episode demonstrates how environmental and foreign-policy streams can reinforce one another to create cooperative outcomes even in the absence of trust or a formal peace treaty (Shimotomai 2019; Weiner 1999).

8.5.2 The Role of High-Level Actors as "Policy Entrepreneurs"

A key insight from the MSF is the crucial role of policy entrepreneurs in linking the streams. Traditionally viewed as peripheral actors (Mintrom 2019), later work shows that, in centralized or authoritarian systems, those entrepreneurs can be elite insiders, often members of the very top leadership, because only they command the authority needed to realign a tightly controlled politics stream (Schlaufer and van den Dool 2024; Wu and Wu 2025).

The Soviet–Japanese accord exemplifies this pattern. Domestically, Mikhail Gorbachev prioritized state issues more than ever before: public acknowledgment of Chernobyl, reevaluation of environmental governance, and open recognition of the previous leadership's neglect provided a credible politics stream, while greater freedom of public speech transformed ecological degradation into a nationally significant problem (Gorbachev 2000; Yablokov 2007). He recognized severe pollution from the defense industry, declining life expectancy (Taubman 2018), and the collapse of Soviet agriculture, heightening his awareness of environmental risks (Larin et al. 2003). As General Secretary, Gorbachev pursued a peaceful transition to openness and democracy, raised ecological awareness, and promoted transparency in environmental data and decision-making (Yablokov 2007).

Shevardnadze's role as Foreign Minister was equally instrumental. By integrating environmental policy into diplomatic strategies and advocating for cooperation with Japan, he effectively coupled domestic problem recognition with international opportunities. His diplomatic engagements signaled to both Soviet and Japanese policymakers that environmental cooperation could serve broader foreign-policy goals.

Alexander Yakovlev provided the ideational bridge, articulating a "nature-individual-society-planet" narrative that linked new Soviet domestic policy to global responsibility (Yakovlev 1991). The simultaneous activism of these elite actors, combined with a receptive Japanese stance, aligned the domestic and foreign streams and opened the April 1991 policy window.

MSF research also warns that once a window closes and entrepreneurial sponsors depart, agenda space quickly shifts to new priorities (Herweg et al. 2015; Kingdon 1995). The disintegration of the USSR removed the principal Soviet sponsors. While Russian leaders continued to endorse environmental goals rhetorically, practical efforts remained limited by the prioritization of economic growth and strategic aims (Sosnovskikh 2023). Although Russia honored the treaty and convened five Joint Committee meetings between 1994 and 2006 (Josephson et al. 2013), cooperation dwindled to isolated ventures, such as submarine-reactor decommissioning (TASS 2019) and NOWPAP marine research (2015), which were insufficient to maintain salience (Hønneland and Jørgensen 2003).

The Soviet–Japanese case, therefore, confirms two propositions emerging from authoritarian MSF studies (Schlaufer and van den Dool 2024): first, top leaders frequently perform the entrepreneurial role because only they can synchronize the streams under hierarchical control; second, without their continued sponsorship, stream alignment unravels, and even formal agreements become fragile. Consequently, Russian–Japanese environmental relations have remained structurally precarious: later shocks that heightened ecological salience created moments of problem recognition but did not generate new bilateral initiatives, because no actors of comparable rank stepped forward to re-couple the streams. In the absence of high-level policy entrepreneurs, prospective windows for renewed cooperation opened only fleetingly and closed unused, leaving the 1991 accord an isolated and ultimately unsustained episode.

8.5.3 Process-Oriented Contributions to Environmental Peacebuilding

The Soviet–Japanese episode offers an illustration of how a process-oriented lens can enrich EP theory. First, it supports the idea that environmental cooperation may occur even before positive dynamics appear in the bilateral relationship (Dresse et al. 2019): a confluence of urgent ecological problems, ready practical solutions, and evolving political receptivity, once aligned, can spark genuine goodwill and dialogue, even when core disputes remain unresolved. Over time, such early collaboration could have served as a durable base for deeper trust, but it had to be tested.

Second, the case highlights the decisive role of elite agency in settings where civil society channels are weak. Rather than grassroots actors, it was senior Soviet and Japanese officials whose authority and initiative fused the separate streams of domestic reform and interstate diplomacy. This underscores the importance of integrating a more explicit account of power asymmetries and agenda control into EP frameworks, recognizing that, in many conflict-affected or authoritarian contexts, environmental entrepreneurs will be found at the highest levels of government.

Third, by treating domestic policy shifts and foreign-policy openings as two interlocking domains, the episode points EP scholars toward a dual-domain approach: cooperation emerges most robustly when internal reforms and diplomatic overtures evolve in tandem.

Finally, demonstrating the viability of this process in a highly centralized regime extends EP's analytic toolkit to contexts far beyond pluralist democracies. MSF's adaptability to authoritarian settings offers EP a tested method for diagnosing when, how, and by whom environmental peacebuilding opportunities arise, and why they so often prove fragile without sustained elite sponsorship.

Together, these insights suggest that coupling MSF's stream-convergence diagnostics with EP's substantive focus on ecological cooperation can yield more precise, empirically grounded accounts of environmental peacebuilding.

8.6 Conclusion

In April 1991, environmental dialogue emerged as a channel for Soviet–Japanese discussions, which had previously been pushed to the diplomatic sidelines in favor of more urgent matters. Domestic pressure (Chernobyl-related public outcry and new Soviet institutions) combined with an external norm shift (growing global environmental diplomacy) to raise ecology in both capitals; the streams met when political leaders decided the timing was advantageous. For Moscow and Tokyo, an environmental accord offered a low-stakes, future-oriented issue on which material gains (pollution abatement, technology transfer) and reputational benefits could be secured without conceding territory, thus serving strategic as well as ecological interests. Reframing pollution control and biodiversity protection as shared tasks allowed both sides to sidestep zero-sum logic. Once top-level entrepreneurs synchronized the problem, policy, and politics streams, environmental provisions became the least costly, most workable avenue for signaling goodwill.

Rare alignment of ecological urgency, ready policy proposals, and political willingness created a short-lived window of opportunity. Through an MSF analysis, this study shows how senior decision-makers on both sides seized that window to sign an environmental accord despite making no progress on the Kuril dispute. These findings support the main theses of EP scholarship that, under appropriate conditions, even marginal environmental issues can foster rapprochement.

The application of MSF to the 1991 Soviet–Japanese episode offers several contributions to environmental peacebuilding. First, by mapping the evolution and convergence of problem, policy, and politics streams, the framework identifies precisely where the obstacle lies—in this case, the absence of a receptive political opening, and shows how senior leaders in both countries partially removed that barrier. Second, the timing of events reveals that environmental cooperation itself helped generate new confidence. Third, the case highlights the crucial role of elite agency as "policy entrepreneurs": high-level figures in authoritarian or highly centralized contexts possess the authority to couple these streams, underscoring power asymmetries that can easily be overlooked. Finally, by treating domestic reforms and interstate dimensions as two overlapping policy domains, this study suggests that MSF can provide a rigorous, multi-level lens for analyzing how internal governance shifts must align with diplomatic openings if environmental peacebuilding is to succeed. Together, these insights sharpen EP's causal precision, broaden its applicability beyond pluralist settings, and offer a structured roadmap for both scholars and practitioners seeking to design or assess environmental initiatives in conflict-affected contexts.

A few important limitations should be kept in mind. First, the reconstruction leans heavily on Soviet-centered sources and lacks comprehensive interviews, which means the Japanese side of the story remains underexplored. Second, gaps in both Russian and Japanese archives make it difficult to map every step of the decision-making process with complete confidence. Third, focusing mostly on the late 1980s and early 1990s means it is not yet clear how durable the 1991 environmental agreement has proved under post-Soviet Russian leadership.

To address these gaps, future research could provide a more balanced examination by equally exploring Japanese motivations, policy responses, and strategic considerations. Extending the timeline into the 2000s and beyond would show whether the 1991 accord shaped later cooperation and allow researchers to assess its long-term impact on Russian–Japanese environmental dialogue. Applying the combined MSF with EP framework to other low-trust pairings, such as India–Pakistan or Israel–Jordan, would test its usefulness in different conflict settings. Finally, mixed-methods work, including quantitative network analyses of the key actors involved, could offer deeper insight into what it takes to align problem, policy, and politics streams across varied geopolitical contexts.

Policymakers and practitioners can draw some practical lessons from the Soviet–Japanese case. Environmental cooperation has the potential to contribute to dialogue even when security issues remain frozen, but only when all the conditions have come together for the corresponding policy window to open. In highly centralized or authoritarian settings in particular, the backing of top leadership is indispensable: only senior champions possess the authority to orchestrate these streams and keep momentum alive. Second, setting up permanent inter-ministerial task forces or envoy positions can help sustain momentum beyond a single summit or agreement. Third, weaving ecological initiatives into wider reform efforts (for example, linking pollution control to economic development or public health) makes it easier to extend brief policy windows into longer-term cooperation. By paying close attention to these factors, governments facing deep mistrust can turn environmental initiatives into stepping-stones for gradual trust-building and more durable peace.

Funding The author received no financial support for the research, authorship, and/or publication of this chapter.

References

Aggestam, K. (2015). Desecuritisation of water and the technocratic turn in peacebuilding. *International Environmental Agreements: Politics, Law and Economics*, *15*(3), 327–340. https://doi.org/10.1007/s10784-015-9281-x

Asahi Shimbun. (1989, September 28). Soviet side proposed that Gorbachev visit Japan in 1991 at a meeting of Japanese and Soviet foreign ministers. *Asahi Shimbun*. https://xsearch.asahi.com/kiji/detail/?1722336008343

Balinskaia, A. (2024). Interstate environmental peacebuilding: A systematic review. *Environment and Security*. https://doi.org/10.1177/27538796241287366

Barnett, J. (2001). *The Meaning of Environmental Security: Ecological Politics and Policy in the New Security Era*. https://api.semanticscholar.org/CorpusID:154748533

Blavoukos, S., & Bourantonis, D. (2012). The Greek–Turkish Rapprochement in the 1990s. *Government and Opposition*, *47*(4), 597–617. http://www.jstor.org/stable/26350282

Budimir, A. (2022). Environmental Issues in Post-Conflict States and the Multiple Streams Model. *Anali Hrvatskog Politološkog Društva*, *19*(1), 159–175. https://doi.org/10.20901/an.19.10

Checkel, J. (1993). Ideas, Institutions, and the Gorbachev Foreign Policy Revolution. *World Politics*, *45*(2), 271–300. https://doi.org/10.2307/2950660

Chernyaev, A. S., & Weber, A. B. (Eds.). (2010). *Responding to the challenge of time: Foreign policy of perestroika: Documentary evidence. Based on the records of M. Gorbachev's conversations with foreign leaders and other materials*. Ves Mir.

Conca, K., & Dabelko, G. D. (2002). *Environmental Peacemaking*. Johns Hopkins University Press. https://doi.org/10.56021/9780801871924

Corry, O. (2014). The Green Legacy of 1989: Revolutions, Environmentalism and the Global Age. *Political Studies*, *62*(2), 309–325. https://doi.org/10.1111/1467-9248.12034

Creswell, J. W., & Poth, C. N. (2018). *Qualitative Inquiry & Research Design* (4th ed.). SAGE.

Dresse, A., Fischhendler, I., Nielsen, J. Ø., & Zikos, D. (2019). Environmental peacebuilding: Towards a theoretical framework. *Cooperation and Conflict*, *54*(1), 99–119. https://doi.org/10.1177/0010836718808331

Floyd, R., & Matthew, R. (Eds.). (2013). *Environmental Security*. Routledge. https://doi.org/10.4324/9780203108635

Gerring, J. (2016). Case Study Research: Principles and Practices. In *Strategies for Social Inquiry* (2nd ed.). Cambridge University Press. https://doi.org/10.1017/9781316848593

Gorbachev, M. (1995). *Life and reforms* (First Edition). Novosti. https://www.gorby.ru/gorbachev/memoirs/

Gorbachev, M. (2000, September 5). *Gorbachev about Environment*. Living On Earth. https://www.loe.org/shows/segments.html?programID=00-P13-00037&segmentID=8

Herweg, N., Huß, C., & Zohlnhöfer, R. (2015). Straightening the three streams: Theorising extensions of the multiple streams framework. *European Journal of Political Research*, *54*(3), 435–449. https://doi.org/10.1111/1475-6765.12089

Herweg, N., Zahariadis, N., & Zohlnhöfer, R. (2018). The Multiple Streams Framework: Foundations, Refinements, and Empirical Applications. In *Theories of the Policy Process* (pp. 17–53). Routledge. https://doi.org/10.4324/9780429494284-2

Herweg, N., Zahariadis, N., & Zohlnhöfer, R. (2022). Travelling Far and Wide? Applying the Multiple Streams Framework to Policy-Making in Autocracies. *Politische Vierteljahresschrift*, *63*(2), 203–223. https://doi.org/10.1007/s11615-022-00393-8

Hønneland, G., & Jørgensen, A.-K. (2003). *Implementing international environmental agreements in Russia*. Manchester University Press. http://www.jstor.org/stable/j.ctt155j8v6

Howlett, M., McConnell, A., & Perl, A. (2015). Streams and stages: Reconciling Kingdon and policy process theory. *European Journal of Political Research*, *54*(3), 419–434. https://doi.org/10.1111/1475-6765.12064

Ide, T. (2019). The Impact of Environmental Cooperation on Peacemaking: Definitions, Mechanisms, and Empirical Evidence. *International Studies Review*, *21*(3), 327–346. https://doi.org/10.1093/isr/viy014

Ide, T., Bruch, C., Carius, A., Conca, K., Dabelko, G. D., Matthew, R., & Weinthal, E. (2021). The past and future(s) of environmental peacebuilding. *International Affairs*, *97*(1), 1–16. https://doi.org/10.1093/ia/iiaa177

Ide, T., & Detges, A. (2018). International water cooperation and environmental peacemaking. *Global Environmental Politics*, *18*(4), 63–84. https://doi.org/10.1162/glep_a_00478

Ivanov, V. I., & Smith, K. S. (Eds.). (1999). *Japan and Russia in northeast Asia: partners in the 21st century*. Westport, Conn.

Japan and the Soviet Union to Conclude Environmental Agreement, First Investigating Frozen Methane Gas. (1991, March 25). Nikkei Shimbun.

Johnson, M. F., Rodríguez, L. A., & Quijano Hoyos, M. (2021). Intrastate environmental peacebuilding: A review of the literature. *World Development*, *137*, 105150. https://doi.org/10.1016/J.WORLDDEV.2020.105150

Josephson, P., Dronin, N., Mnatsakanian, R., Cherp, A., Efremenko, D., & Larin, V. (2013). *An Environmental History of Russia*. Cambridge University Press. https://doi.org/10.1017/CBO9781139021043

Kingdon, J. W. (1995). *Agendas Alternatives and Public Policies*. HarperCollins College Publishers.

Koebele, E. A. (2021). When multiple streams make a river: analyzing collaborative policymaking institutions using the multiple streams framework. *Policy Sciences*, *54*(3), 609–628. https://doi.org/10.1007/s11077-021-09425-3

Krampe, F. (2017). Toward sustainable peace: A new research agenda for post-conflict natural resource management. *Global Environmental Politics*, *17*(4), 1–8. https://doi.org/10.1162/GLEP_a_00431

Kuzminkov, V. (2006). Gorbachev's Foreign Policy toward Japan during 1985-1991. *Kobe Law Journal*.

Larin, V., Mnatsakanyan, R., Chestin, I., & Shvarts, E. (2003). *Nature Conservation in Russia: From Gorbachev to Putin*. Interprogress.

Mayring, P. (2015). Qualitative Content Analysis: Theoretical Background and Procedures. In A. Bikner-Ahsbahs, C. Knipping, & N. Presmeg (Eds.), *Approaches to Qualitative Research in Mathematics Education: Examples of Methodology and Methods* (pp. 365–380). Springer Netherlands. https://doi.org/10.1007/978-94-017-9181-6_13

Mintrom, M. (2019). *Policy Entrepreneurs and Dynamic Change*. Cambridge University Press. https://doi.org/10.1017/9781108605946

Nago, M. (2021). The role of aid agencies within environmental cooperation in Congo Basin: facilitators or policy entrepreneurs? *Development Studies Research*, *8*(1), 135–146. https://doi.org/10.1080/21665095.2021.1937255

National Biodiversity Strategy of Japan. (1973). *Laws related to water and soil conservation*. https://www.biodic.go.jp/biolaw/wata_so/index.html

National Diet of Japan. (1991, April 23). *4th plenary session of the 120th National Diet*. https://kokkai.ndl.go.jp/txt/112004006X00419910423/0

Nikitina, E. (1995). *National Implementation of International Environmental Commitments: A Review of Soviet Literature*. WP-95-026. https://pure.iiasa.ac.at/id/eprint/4566/

NOWPAP. (2015). *Meeting Report of the 13th NOWPAP DINRAC Focal Points Meeting*.

Panov, A. N. (1992). 不信から信頼へ—北方領土交渉の内幕 *[From distrust to trust: Inside the Northern Territories negotiations]*. Simul Press.

Politburo of the CPSU Central Committee. (1986). *USSR Politburo Session on East Asia*.

Ponomareva, V. I. (Ed.). (1999). *Ecology and Power, 1917–1990*. International Democracy Foundation.

Pravda. (1991, April 17). *Chronicle of foreign policy events in the USSR*. Pravda. https://www.gorby.ru/archival/expocenter/vneshpolitika/show_29328/

Rindzevičiūtė, E. (2020). Soviet Policy Sciences and Earth System Governmentality. *Modern Intellectual History*, *17*(1), 179–208. https://doi.org/10.1017/S1479244318000161

Robinson, N. A. (1988). *The U.S.-U.S.S.R. Agreement to Protect the Environment: 15 Years of Cooperation*.

Schlaufer, C., & van den Dool, A. (2024). Policy processes in authoritarian settings. *Review of Policy Research*, *41*(6), 860–864. https://doi.org/10.1111/ropr.12634

Sedaitis, J., & Butterfield, J. (2019). *Perestroika from Below* (J. B. Sedaitis & J. Butterfield, Eds.). Routledge. https://doi.org/10.4324/9780429301353

Sharifi, A., Simangan, D., Kaneko, S., & Virji, H. (2021). The sustainability–peace nexus: why is it important? *Sustainability Science*, *16*(4), 1073–1077. https://doi.org/10.1007/s11625-021-00986-z

Shevardnadze, E. (1991). Forum: Governments alone won't turn the world green – Eduard Shevardnadze looks at ways of ensuring the triumph of political ecology. *New Scientist*.

Shimotomai, N. (2019). The Rise to Power of Mikhail Gorbachev and the Policy of "Expanding Equilibrium." In *A History of Russo-Japanese Relations* (pp. 443–458). BRILL. https://doi.org/10.1163/9789004400856_022

Simangan, D., Virji, H., Hendrix, C., Islam, M., Kaneko, S., Ma, Y., Mechler, R., Pangotra, P., Peters, K., Sharifi, A., & Shams, S. H. (2021). A co-designed heuristic guide for investigating the peace-sustainability nexus in the context of global change. *Sustainability Science, 16*(4), 1097–1109. https://doi.org/10.1007/s11625-021-00970-7

Simon, J. (1974). Japan's "Ostpolitik" and the Soviet Union. *The World Today, 30*(4), 161–169. http://www.jstor.org/stable/40394773

Sokolov, V., Jäger, J., Pisarev, V., Nikitina, E., Ginzburg, A., Goncharova, E., Cavender-Bares, J., & Parson, E. A. (2001). Turning Points: The Management of Global Environmental Risks in the Former Soviet Union. In *Learning to Manage Global Environmental Risks*. The MIT Press. https://doi.org/10.7551/mitpress/4173.003.0013

Sosnovskikh, S. (2023). Environmental Policies in Russia. In *Environmental Finance and Green Banking* (pp. 169–208). Routledge. https://doi.org/10.4324/9781003206194-10

Takahashi, W. (2000). *Environmental Cooperation in Northeast Asia.* https://www.iges.or.jp/en/pub/environmental-cooperation-northeast-asia/en

TASS. (2019). *Japan to invest 1.16 billion rubles in creation of radioactive waste storage center in Primorye.* https://tass.ru/obschestvo/6024472

Taubman, W. (2018). *Gorbachev. His Life and Times.* Simon & Schuster.

The Global Forum on Environment and Development. (1990) *Global forum on environment and development [Conference declaration].* https://www.osti.gov/biblio/5626813

Torkunov, A. V., Iokibe, M., Streltsov, D. V., & Grishachev, S. V. (2015). Russian–Japanese relations in the format of parallel history. MGIMO University Press & Association of Japanologists.

Trombetta, M. J. (2008). Environmental security and climate change: Analysing the discourse. *Cambridge Review of International Affairs, 21*(4), 585–602. https://doi.org/10.1080/09557570802452920

Tzagkarakis, S. I., & Kritas, D. (2023). Mixed research methods in political science and governance: approaches and applications. *Quality & Quantity, 57*(1), 39–53. https://doi.org/10.1007/s11135-022-01384-y

Van Evera, S. (1997). *Guide to Methods for Students of Political Science.* Cornell University Press. http://www.jstor.org/stable/10.7591/j.ctvrf8bm7

Verweijen, J., & Hoffmann, K. (2024). Dangerous environments: environmental peacebuilding's technomoral imaginary and its power-knowledge effects. *Ecology and Society, 29*(3). https://doi.org/10.5751/ES-15391-290323

Weiner, D. R. (1999). *A Little Corner of Freedom Russian Nature Protection from Stalin to Gorbachev.* University of California Press.

Weiner, D. R. (2024). Nikolai Nikolaevich Vorontsov's credo. *Studies in the History of Biology, 16*(3), 86–95.

Woodhouse, K. M. (2009). The Politics of Ecology: Environmentalism and Liberalism in the 1960s. *Journal for the Study of Radicalism, 2*(2), 53–84. https://doi.org/10.1353/jsr.0.0008

Wu, Y., & Wu, X. (2025). Navigating the streams of power: applying the multiple streams framework in China's authoritarian regime. *Policy Sciences, 58*(1), 189–219. https://doi.org/10.1007/s11077-024-09561-6

Yablokov, A. (2007, October 17). *Heroes of the Environment: Michail Gorbachev.* Time. https://content.time.com/time/specials/2007/article/0,28804,1663317_1663319_1669882,00.html

Yakovlev, A. (1991). *Note from A.N. Yakovlev to M.S. Gorbachev "On the draft of the CPSU Program."*

Zatravkin, S., & Vishlenkova, E. (2021). Deterioration of Soviet people's health in an era of stagnation. *Problems of Social Hygiene, Public Health and History of Medicine.* https://doi.org/10.32687/0869-866X-2020-29-2-359-368

Zhukova, E. (2017). Foreign aid and identity after the Chernobyl nuclear disaster: How Belarus shapes relations with Germany, Europe, Russia, and Japan. *Cooperation and Conflict, 52*(4), 485–501. https://doi.org/10.1177/0010836717710529

Chapter 9
Access to Justice as a Guarantee of Rights and Sustainable Development: The Case of Argentina

Federico Gustavo Bartalini

9.1 Introduction

Access to Justice is more than merely a legal guarantee; it's a crucial mechanism of modern liberal democracies based on the rule of law that ensures that rights are enforced in reality. This chapter aims to provide an in-depth analysis of how Access to Justice contributes to achieving Sustainable Development Goal (SDG) 16 and its ripple effect on other goals, such as reducing poverty, ensuring gender equality, and environmental protection. This chapter will adopt a socio-legal approach based on mixed methods. A qualitative analysis will be made based on laws, treaties, and case studies while delving into statistical analysis of Access to Justice issues in Argentina. This chapter will focus on the relationship between SDGs and Access to Justice and how that is reflected by analyzing some key jurisprudence and recent public policies. By the end, we will explore the findings and formulate concrete, actionable public policies that can guarantee that both human rights and sustainable development goals can be achieved.

9.2 Sustainable Development Goals and Access to Justice

The SDGs were adopted by the UN General Assembly in 2015 with the primary objectives of eradicating poverty, protecting the planet, and achieving peace and security for all people by 2030. These goals are divided into 17 specific targets. While each of them is crucial for their intended purpose, their understanding and enforcement should be integrated with each other. Therefore, the commitment made

F. G. Bartalini (✉)
Hiroshima University, Higashihiroshima, Japan

A. Sharifi et al. (eds.), *Frameworks and Practices Towards a Sustainable and Peaceful Future*, World Sustainability Series,
https://doi.org/10.1007/978-981-95-2310-8_9

by all participating States was to work on these targets collectively. Naturally, the priorities vary across regions due to differing needs and political interests, leading to uneven progress among the goals. However, the UN emphasizes that these objectives are interconnected. Therefore, progress (or setbacks) in one area will impact others. For instance, progress in the first goal, "No Poverty," will likely lead to improvements in health and well-being, as well as in decent work and economic growth.

This work focuses on the importance of Access to Justice, which is a target of SDG 16. While often overshadowed by urgent goals like reducing inequalities or ending hunger, Access to Justice is critical because it deals with making Human Rights, and the rest of the goals, a concrete reality and not merely text on paper. Advancing this goal will have a ripple effect, positively influencing other SDGs such as poverty reduction (SDG 1), health and well-being (SDG 3), gender equality (SDG 5), and reduced inequalities (SDG 10).

The UN has also explicitly stated that progress on the SDGs must be balanced to avoid achieving specific goals at the expense of others. For example, public policies that drive economic growth and reduce social inequality but simultaneously harm the environment cannot be considered sustainable development. (United Nations 2015)

In this regard, this chapter understands the goal of "Peace, Justice, and Strong Institutions," not only as an end but also as a fundamental tool to achieve the other goals. Within this goal, several essential targets guide state policies toward its fulfillment. These include reducing all forms of violence and mortality, ending abuse and exploitation of children, combating illicit financial flows and organized crime, eliminating corruption, building accountable and transparent institutions, ensuring inclusive decision-making, increasing the representation of developing countries in global governance, providing legal identity for all (primarily through birth registration), ensuring public access to information and fundamental freedoms, and strengthening institutions to prevent terrorism and crime. This work is particularly relevant to promoting the rule of law at national and international levels and guaranteeing equal access to justice.

The following sections will analyze Access to Justice as a fundamental right, exploring its legal foundation and necessity in all legal systems. We will also examine key cases in Argentina's jurisprudence highlighting the importance of a socially conscious approach to justice, demonstrating its critical role in achieving broader societal goals.

9.3 Access to Justice

9.3.1 Basis of Access to Justice

The existence of the Right of Access to Justice can be inferred in any legal system just by the existence of the rule of law, which is essential for developing human rights and for the very existence of those rights. There are various conceptions of what the rule of law entails, ranging from minimalist views that focus solely on the protection of rights, guarantees, and obligations at a formal level to more comprehensive views that include specific Human Rights, particularly Social, Economic, and Cultural Rights, as an integral part of the system. For the sake of brevity, we will adopt a basic definition based on Tamanaha's minimalist Rule of Law.

According to Tamanaha (2012), the Rule of Law means that citizens and public officials are subject to and act according to the law. While this definition may appear overly simplistic, as it neither requires a democratic political system nor guarantees Human Rights, it provides a helpful starting point. A state may operate under the rule of law yet still face significant issues such as health crises, income inequality, or environmental degradation. Thus, while the rule of law is foundational, it is insufficient by itself to create a desirable or justice-minded society. For that to be realized, additional elements, such as the protection of Human Rights, are needed.

Tamanaha characterizes the Rule of Law with three key components: the supremacy of law, the limitation of government by law, and the principle of formal legality, often expressed as "the rule of law, not of man." The first element ensures that the law binds public officials and that they must follow legal procedures to modify their actions. The second element, formal legality, specifies the nature of laws: they must be public, general, non-retroactive (with some exceptions), and uphold the principles of equal treatment and non-discrimination. Finally, the third element emphasizes that citizens and government officials, particularly judges, must abide by the law. Judges must make decisions based on legal rules rather than personal preferences or wishes. This highlights the importance of judicial independence, the non-interference of judges in political matters, and the role of other legal professionals, such as lawyers, in advising and guiding individuals and institutions according to the Rule of Law. Importantly, Tamanaha also stresses that for the rule of law to function effectively, it must be legitimized by the population and culturally accepted.

The Right to Access Justice is a fundamental right that serves as the mechanism for guaranteeing the enjoyment of all other rights. To ensure this right, the state must provide the necessary means for all individuals to freely exercise their rights and find swift and effective remedies for any violations, regardless of their circumstances. This becomes problematic when there is a lack of trust in the rule of law or when it is viewed negatively by the public. In such cases, a weak Right to Access to Justice can prevent legal norms from being effectively enforced, to the detriment of citizens, further eroding confidence in the rule of law and making it even harder to establish a foundation for development. In fact, according to one study (Latinobarometro Report 2024), trust issues with the judicial system are a pervasive problem in all of Latin America, with only 22% of Argentinian respondents reporting they trust the legal system.

While statistical data provide a valuable snapshot of social realities, it often fails to reflect the situation of vulnerable communities, who face the most significant barrier to enforcing their rights. Quantitative indicators may not be enough to depict the complex realities of groups that are not reached by the usual normative approach to legal issues. Statistics are a powerful tool, but they are insufficient unless they are contrasted with a social-legal analysis of these situations. The following section will analyze Argentina's positivist situation before moving to a more socially grounded approach to this issue, with the purpose of giving the best reflection of complex lived experiences.

9.3.2 Access to Justice in Argentina

9.3.2.1 Normative Dimension

There are multiple definitions of what constitutes Access to Justice. While the focus of this paper is to see Access to Justice mainly through the lens of sustainable development and how it can affect other SDGs, the most basic definition and dimension of this right is the purely positivist approach. To this end, we will first look at the legal standing of this right in Argentina.

Several norms explicitly protect the Right of Access to Justice, while others do so tacitly. In the national Constitution of Argentina, the Right of Access to Justice is not explicitly guaranteed; the closest is the right to due process in Article 18 of said norm. However, it can be inferred by Article 33 of the constitution under the reading of the whole document, including the preamble. Nevertheless, it is found explicitly in the Constitution of the Autonomous City of Buenos Aires (from now on referred to as CABA), in its article 6, explicitly establishing that there will be no economic barriers to Access Justice and that it will provide tools to facilitate access to Justice. In turn, article 108 of the said norm establishes the obligation of the legislative and executive branches to guarantee access to justice, especially ensuring that it is economically accessible and resolved reasonably.

Argentina is a party to many Human Rights treaties, of which the Right to Access Justice can be found in several instruments. First, however, it is fundamental to address the peculiarities of the Argentine system, which is only monist from an international rights perspective, understanding the national system of law and the public international law system as a single, unified system, in contrast to dualism, which views them as separate systems (Ferreira and Ferreira-Snyman 2014). This means that all the Human Rights treaties to which Argentina is a party are directly applicable without needing internal regulation. Additionally, with the constitutional reform of 1994, several human rights treaties were given Constitutional hierarchy, part of Article 75 subsection 22 of the National Constitution.

Further, while not directly enforceable, the Brasilia Rules on Access to Justice for Vulnerable Persons contain a great deal of information on this topic, developing the issues that must be treated as a condition of vulnerability and possible remedies for them. These non-binding rules were accepted during the XIV Ibero-American Judicial Summit held in 2008. Although these are not binding for the participating countries, they represent a concrete example of interest in improving Access to Justice, especially for vulnerable people. The rules themselves are principles of action, basic ideas, or recommendations that aim to generate greater guarantees and better exercise of rights. The spirit of these rules is to prevent rights in the states from being recognized formally instead of being easily and efficiently exercised by the population (Langford 2009).

A modern international instrument is the Escazú Agreement on Environmental Protection, which was ratified by Argentina in 2020 and effective in 2021. This Treaty, currently signed by 17 Latin American countries, focuses on protecting access to justice for environmental rights and recognizing the rights of the current and future generations to a safe and healthy environment. It also guarantees access to public information, promotes community participation in environmental matters, and establishes sustainable development as a Right that can be judicially petitioned (Peña Chacón 2020).

Argentina also has many institutions that deal directly or indirectly with Access to Justice and the promotion of Human Rights. One of those is the Ombudsman Office (*Defensoria del Pueblo*), whose role is to protect and promote human rights and provide free legal services and mediation in some instances. It also has legal standing to initiate legal processes and support interested parties, which happened in several important cases and will be developed later in the Matanza-Riachuelo case.

As we will develop in the next section, Argentina also has many modern approaches to Access to Justice, not only limited to access to the judiciary. The compulsory civil mediation and labor conciliation processes are examples of such policies. These norms establish these processes as mandatory before starting any civil or labor judicial process, enabling the parties to reach faster and better solutions to the controversy.

9.3.2.2 Socio-legal Approach

Oftentimes, Access to Justice is equated to Access to the Judiciary. The latter means accessing courts without any barriers and requesting that the courts take actions to guarantee the fulfillment of the rights that assist individuals. Said judiciary procedure must be non-discriminatory and allow legal guarantees such as due process and proper defense (Cappelletti and Garth 1978). However, the right of Access to Justice is even broader. It implies that the person can guarantee his rights and guarantees, regardless of whether he must do so through a judicial or administrative body or achieve a self-composition of rights. This distinction is essential since it is in this difference that the possibility of generating a much broader degree of satisfaction with these rights lies in making access to justice more efficient (de Santos 2002).

In summary, the right of Access to Justice can be defined by this approach as a fundamental right whose function is the prerogative of guaranteeing the enjoyment of the rights that assist a person. In turn, to ensure this right, the state must facilitate all the means so that any person, regardless of their particularities, can freely exercise them and find a quick and efficient solution to their violation. This state obligation can be divided into two distinct subcategories: a commitment not to do or to allow doing and a positive commitment to do. Implementing the Right of Access to Justice as an obligation of the state can be argued in two ways: a negative and a positive obligation.

The negative obligation is configured not only as allowing people to exercise their right to access justice freely but in a broad sense; it is also understood as allowing individuals to administer this autonomously. This concept implies that the state must necessarily allow people to distribute their justice without interference from third parties, whether human beings, legal entities, or even the state in its various forms. This broad conception of what is understood as the negative obligation of the state concerning Access to Justice may seem to strongly clash with republican principles, especially with the principle of separation of powers and the state as the only party entitled to resolve disputes between individuals. However, the purpose of this principle is protective, so that by using public order as a means, the state may still reserve the monopoly of solving certain disputes, notably criminal, tax, and those in which the state is a party (Rawls 1971). In turn, this public order must be understood as transversal to the rest of the relationships that may generate a conflict (such as civil or commercial conflicts), laying the foundations of non-transactional rights, that is, those that the parties cannot negotiate in their self-composition of justice.

Those mentioned above must be understood from the point of view of promoting the self-composition of conflicts since these methods have multiple advantages concerning the traditional administration of justice exclusively by judicial power. This does not imply, however, a renunciation or a banning of the judicial power, but merely that of creating more attractive options to resolve controversies that entail a benefit for both the inhabitants of the nation and the state by implying an improvement in efficiency and a significant reduction in costs that can be used to guarantee other rights more comprehensively (Fiss 1984).

The positive obligation of the state to act must be understood essentially as a facilitator. Currently, the Right to Access Justice is understood as a protective right, mainly by breaking down the various barriers so that people can access the courts regardless of their conditions, and justice is administered to them in an almost paternalistic way. However, the state is not currently focused on giving people the tools to solve their conflicts in the way they see fit. It can be verified that the judiciary continues to be the hegemonic tool to solve conflicts, and the public policies that are primarily carried out aim to facilitate access to the judiciary. Although, in recent years, mediation has been seen as a good alternative to court-based adjudication, its main objective is to create a residual option for traditional justice (Menkel-Meadow 2004). Some even understand that mediation has the purpose of "decongesting courts," an accessory function (Sander and Goldberg 1994).

This does not mean that policies that facilitate access to the judiciary are not necessary or desirable. Still, they are not efficient enough to achieve significant progress in the protection of rights, particularly from a sustainable development approach for vulnerable populations, as will be assessed later in this paper. I believe that policies for access to the judiciary should be continued and improved as much as possible, but at the same time, tools should be provided to people so that they can resolve their conflicts in an autonomous and flexible way. This, however, cannot be achieved in the short term since it implies teaching a generation as a whole conflict management skills that, unfortunately, many do not possess today. Although this can be partially alleviated through training and information campaigns, the definitive solution is to incorporate into the compulsory education of every child the necessary tools to manage their conflicts and administer justice. At the same time, more conflict-buffering spaces must be created whose function will be to help people find solutions rather than to dispense justice. In this way, the judiciary will position itself as a last resort in conflicts between individuals, imposing a solution only when the parties cannot do so together.

9.3.2.3 Barriers to Access to Justice

By employing a purely doctrinal reading of the legal system, we can understand how it deals with legal cases and solutions, but that is often insufficient to understand how it can be effectively enforced, especially for vulnerable groups. What is called by Gargarella (2002) as the "Latin American paradox" is at the crux of these issues. This referenced paradox paints the picture of very generous law systems (especially constitutions) but with significant difficulties for people, particularly vulnerable populations, to enforce them.

Particularly in the field of Access to Justice, as De Stefano (2012) outlines, various informal institutions act as obstacles to achieving effective justice. Following the cited author, we will use the term "barriers" to describe these challenges, as it aligns with the intended terminology.

The first significant barrier is the length of judicial proceedings. Due to the high volume of litigation within the Argentine system, inadequate resource allocation, and other systemic issues, judicial processes tend to take much longer than reasonable. The phrase "justice delayed is justice denied" is commonly cited and aptly describes the current situation. Many people avoid going to court simply because of the time involved in judicial procedures, which represents a significant shortcoming in the administration of justice. Even if all other barriers to Access to Justice are overcome, if the process remains excessively slow, it cannot be considered true justice (De Stefano 2012).

The second barrier is economic. Many individuals lack the financial resources to initiate a legal process or to secure an adequate defense if they are on the receiving end of litigation. This issue can be alleviated by free legal representation or the constitutional provision in Buenos Aires that allows individuals to litigate without incurring expenses. However, there remains a pressing need for a practical solution for more minor, neighborhood-level disputes. Despite numerous legislative proposals, no concrete law has yet been enacted to address this issue (De Stefano 2012).

Another significant impediment is the geographical centralization of courts. This is not trivial, as nearly all individual legal claims must be filed in the central judicial district. This results in additional time and financial costs, which are especially burdensome for those in vulnerable situations who may lack the resources to exercise their rights. In cases involving small claims, the time and costs required to make the claim can often exceed the value of the claim itself. Additionally, the limited working hours of the judiciary present another challenge. It is the only essential public service that operates just six hours per day, does not function on weekends, and has a 45-day annual recess, making it the service with the fewest working days. Simply eliminating the judicial recess would not fully resolve this issue, as it would place undue demands on self-employed legal professionals, making it difficult to fulfill (De Stefano 2012).

Another barrier to full Access to Justice is the complexity of judicial language. Legal terminology can often be incomprehensible, even to professionals well-versed in the field. While it is essential to use technical language in legal matters, there is room for simplification and more neutral vocabulary that does not compromise the clarity or precision of communications, particularly in notifications and court summons. Legal education is crucial in this regard, as well as simplification of terminology (De Stefano 2012).

In addition to the previous section, a barrier in all Latin America is the distrust of the population of the legal system. This directly affects people in vulnerable positions and often makes them not even attempt to go to the judiciary because of a distrust of its adequate functioning or a perception of corruption. When low-income citizens or vulnerable populations are asked for their opinion on the courts, most respond that either the judges were often bought by the rich or there were two very different justices available, one for the rich and one for the poor (Gargarella 2002).

9.4 SDG 16, Access to Justice and Public Policies

9.4.1 Measuring Access to Justice

Accurately measuring the sub-dimensions of Access to Justice requires multiple data sources. Administrative data is vital for generating indicators on institutions, legal processes, and services, including alternative dispute resolution and citizen interactions with entities like police, prosecutors, courts, and ombudsman offices. To be useful for policymakers, analysts, and media, this data must have a clear purpose and be consistent, precise, and based on standardized concepts and procedures. However, a major limitation is its failure to reflect individuals excluded from these institutions. Additionally, duplicated or inconsistent records are common, for instance, a single victim may suffer multiple offenses, or multiple people may be arrested for one crime. Someone might also consult several law firms for the same issue. Improving the quality and coverage of this data requires stronger inter-institutional coordination across the judicial sector.

During its 46th session in March 2015, the Statistical Commission under the UN Economic and Social Council created the Praia Group on Governance Statistics to contribute to the establishment of international standards and methods for the collection of statistics on the main dimensions of governance and to collaborate with UN bodies and other organizations dealing with specific aspects of governance statistics. The Manual indicates that most statistical governance collection sources are based on non-official sources and are more complementary than substitutes for information from official State sources. The Manual brings up the example of The Netherlands, given that in 2006, its statistics agency carried out consultations with more than 30 organizations producing civil society information that is an unofficial source and, following the standards recommended by the Manual, generated a large number of supplementary information going from covering 30% of the SDG indicators to 51% in 2018.

In the case of Argentina, universities and nonprofit organization also play an important role in the gathering of data and improving the existing statistics. As will be explained in the next section, *Universidad Catolica Argentina* (Catholic Argentinian University) provides significant reports on social inequality and vulnerable populations. *Asociacion Civil Por la Igualdad y la Justicia (civil association* for equality and justice) is a nonprofit organization that also works intensely for the protection of Human Rights. While they do not focus on producing statistics of Access to Justice, their focus is empowering vulnerable populations to enforce their Rights, providing legal education to low-income and vulnerable populations, and providing legal assistance (Asociación Civil por la Igualdad y la Justicia (ACIJ) 2025).

While there is certainly a diversity of sources to get information on Access to Justice in Argentina in order to measure it, sometimes the information can be disorganized since the methodologies of the reports and surveys tend to vary a lot. To this end, in April 2025 a network of coordination of Access to Justice was launched (*Red Federal de Acceso a la Justicia)*, with the purpose of enhancing the articulation of public institutions among themselves and with NPOs when it comes to producing and organizing data and exchange technical cooperation (Consejo de Procuradores y Fiscales Generales de la República Argentina 2025).

9.4.2 Argentina's Reports on SDGs and Access to Justice

In 2017, Argentina submitted a report to the United Nations Office of the High Commissioner for Human Rights (OHCHR) on its efforts and progress in achieving the Sustainable Development Goals (SDGs). Argentina underscored its vision of the SDGs as both transversal and indivisible, emphasizing a strategy to address all objectives simultaneously. The country linked its national "eight government objectives" to the SDGs, aligning public policy goals with the global agenda. Argentina aimed to progress on SDG 16 through several key objectives: National Productive Agreement, Sustainable Human Development, Fight against Drug Trafficking and Improvement in Security, Institutional Strengthening, Modernization of the State, and Intelligent Insertion into the World.

The Argentine report highlighted the Judiciary's crucial role in promoting citizen participation, improving access to information and justice, encouraging the use of alternative dispute resolution mechanisms, and fostering a more human-centered justice system. The report emphasized that achieving SDG 16 would require stronger interconnections between the rule of law and development, positioning the justice system as a central player in realizing the SDGs. Argentina's focus on developing Access to Justice was primarily through the Judiciary, even though civil society and specialized administrative agencies also offered promising avenues for progress. According to the report, only 37% of civil organizations were aware of the SDGs, but 85% expressed interest in receiving more information, pointing to an opportunity to enhance Access to Justice through these organizations. Figure 9.1 below shows that SDG 16 was the fourth most frequently addressed goal among civil society initiatives (United Nations 2017).

In 2017, Argentina also underwent an examination by the UN Human Rights Council. The subsequent report detailed recommendations and the state of human rights protection in the country, with SDG 16 receiving the most attention, 27% of the recommendations. Significant progress was noted in several areas related to Access to Justice, especially in domestic violence cases. It is also of note that in the year 2019, Argentina was one of the five highest-ranked countries on the UN Human Rights Committee's ranking for the implementation of selected recommendations,

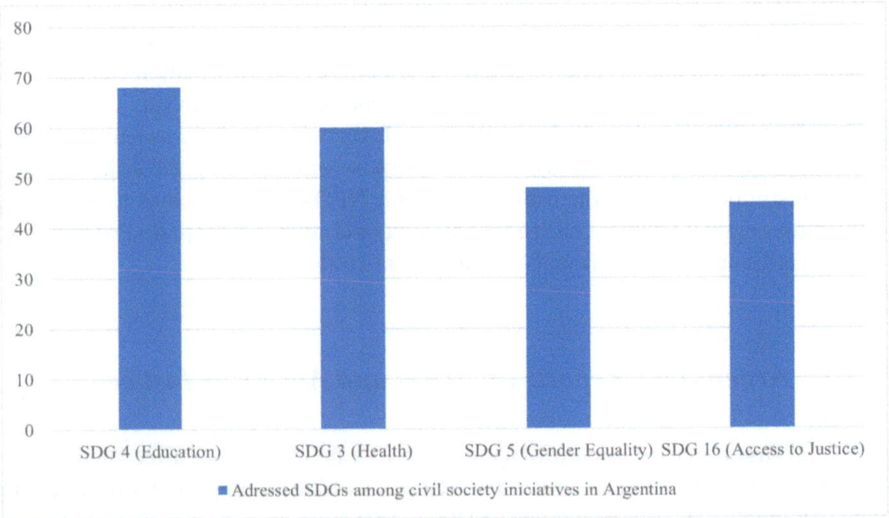

Fig. 9.1 SDG 16 as the fourth most frequently addressed goal among civil society initiatives in Argentina. Author's own work, based on United Nations (2017)

noting the case of a ruling of the supreme court of the Province of Tucuman, which freed a young woman after her abortion was criminalized (United Nations 2017).

In 2019, within the framework of the United Nations Development Program and taking SDG 16 as a guide, the National Ministry of Justice and Human Rights, together with the Social Debt Observatory of the Universidad Católica of Argentina, developed a report on Access to Justice in conditions of social inequality for the years 2018–2019. Said document highlights the link between poverty and Access to Justice insofar as: "the impossibility of accessing justice has a disproportionately negative effect on people of social vulnerability, reducing economic opportunities and increasing the chances of keeping them in conditions of structural poverty." The report was based on a broad reading of objective 16.3 to focus on civil justice needs on the basis that the former are those experienced by the majority of the population and, in particular, especially by people experiencing poverty and groups at greater risk of being left behind. In this sense, it analyzed access to civil justice in line with the methodological developments of the unsatisfied legal needs in a transversal way to the situation of multidimensional poverty to deepen the studies of justice with a development approach. It was based on surveys such as those proposed by the Praia Group Manual with specific questions on legal problems, legal needs satisfaction with the results obtained, and unsatisfied legal needs complaints and gender violence. Likewise, the factors studied are related to gender, educational level, and socio-economic level. In relation to legal problems, understanding different types of issues that have been difficult to solve in the last two years that can lead to legal issues, the report concludes that the lower the socio-economic level (very low) or the socio-occupational stratum (marginal worker), there are more family-type problems or conflicts or in the sphere of social rights. The opposite occurs at a higher economic-social level (middle-high) or socio-occupational stratum (professional medium). It has a greater incidence in terms of economic problems or as consumers and in the sphere of insecurity. Figure 9.2 shows an exceptionally high rate of dissatisfaction with the legal service obtained, regardless of whether the process was successful (Observatorio de la Deuda Social Argentina, Universidad Católica Argentina 2019).

9.4.3 Recent Public Policies

In recent years, many public policies have been put in place to improve the situation of human rights. As explained previously, one of the major problems is not the inexistence of rights but the difficulty in exercising them, as demonstrated by the high rate of dissatisfaction with the judicial and administrative system. To that end, many of the initiatives that deal with human rights are listed as SDGs that facilitate Access to Justice. We will proceed to explain some of the initiatives, giving priority to the ones that have been praised and mentioned by the UN Report regarding the rights of indigenous communities, which will be explained in the next section under the Lhaka Honhat case.

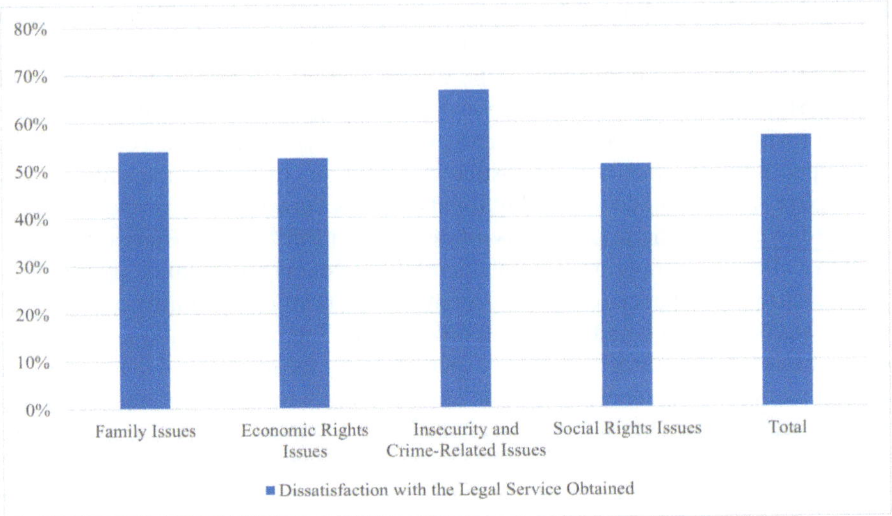

Fig. 9.2 Percentage of dissatisfaction with the legal service obtained, regardless of case outcome. Author's own work, based on Observatorio de la Deuda Social Argentina, Universidad Católica Argentina (2019)

Regarding gender violence, the dial-144 hotline was highlighted by the UN Human Rights Council for offering around-the-clock assistance managed by an interdisciplinary team. The "Victims Against Violence" program, which provides support for sexual assault and violence victims, was also praised. This program includes line 137 (available in Buenos Aires and Misiones) for family and sexual violence cases and line 0800-222-1717 for child sexual abuse cases, both of which are available 24/7 and managed by psychologists and social workers. In urgent situations in Buenos Aires and Misiones, a Mobile Team accompanies victims to hospitals or to file complaints.

For people with disabilities, two key initiatives were highlighted by the report: the National Assistance Program for Persons with Disabilities in their Relations with the Administration of Justice (ADAJUS) and the Electoral Accessibility Program. ADAJUS provides training and support to judicial operators, prison officers, security forces, public officials, and civil society organizations to improve accessibility for persons with disabilities within the justice system. In practice, this creates the obligation for public servants to attend yearly training sessions in order to better accommodate people with different kinds of disabilities.

Lastly, the same report highlighted two major initiatives that represent Argentina's commitment to advancing SDG 16: the creation of the Access to Justice Centers (CAJs) and the policies under the Justice 2020 plan. The CAJs, spread across the country, aim to facilitate Access to Justice for the most vulnerable groups, historically the most marginalized, in exercising their rights. These centers offer legal and psychosocial support, raise awareness of legal rights, and engage in community mediation on issues such as family, housing, social security, disability, labor, and neighbor disputes. The Justice 2020 program, promoted by the Ministry of Justice and Human Rights, provides a platform for institutional dialogue and is designed to create, implement, and evaluate policies that produce socially relevant outcomes. It also promotes swift and reliable conflict resolution and plays a vital role in raising awareness of citizens' rights through publications and outreach efforts. The next section will analyze the effect that the pandemic had on the issue and which concrete public policies were enacted.

9.4.4 Access to Justice During COVID-19

As highlighted by Rossi (2020), during the COVID-19 pandemic, the obstacles of Access to Justice were exacerbated by the near-total closure of public offices during the strict period of lockdown, followed by a slow and gradual reopening of service and complaint channels. By November 2020, many administrative offices for claims and complaints were still operating at a minimal capacity, further undermining the exercise of essential rights.

One particularly serious consequence of the pandemic was the total interruption of trials for crimes committed against humanity by the military junta. This pause caused significant social harm, as it blocked Access to Justice for victims and their families and hindered the exercise of the *Right to Truth and Memory*. Only in May 2021, some courts resumed these trials, starting to hold virtual hearings or, where necessary, face-to-face hearings with social distancing measures in place. This marked a significant achievement in terms of justice, not only because it demonstrated a commitment to continue addressing issues of profound importance to the Argentine people but also because it proved that judicial processes could adapt to the constraints imposed by the pandemic. This success set a precedent that could potentially be replicated in other courts and administrative bodies across the country.

Rossi (2020) also pointed out that, unlike other crimes, incidents of violence against women, particularly femicides, escalated during the confinement period. Reports indicate 1280 cases of intrafamily violence in the first 4 months of the lockdown, increasing the rate 28% nationwide and 48% in the City of Buenos Aires (Rochi Monagas et al. 2021). In response, various measures were introduced, such as the automatic renewal of protective orders and the strengthening of non-face-to-face support channels, primarily through telephone services. Nevertheless, these efforts proved insufficient as the number of cases continued to rise. This led multiple organizations and entities to call for the implementation of new, more effective measures to combat this growing issue. These numbers have stabilized by now, but they still show the importance of implementing specific and compelling measures in times of need.

Despite the difficulties of normal judicial functioning due to the intense lockdowns, this period also brought the digitalization of many areas of the conflict resolution system out of necessity. The digital system to file claims online and manage the documents and evidence in judicial processes was available already, but some jurisdictions started to experiment with online hearings and provincial courts with available technology switched to online processes completely (Federal Judicial Center 2020).

Many households in Argentina still do not have their basic needs covered, and that includes the digital tools necessary to be part of online proceedings. Digitalization of these processes is fundamental to enhance Access to Justice, but statistics show that 36% of households do not possess Internet access (Instituto Nacional de Estadística y Censos (INDEC) 2022). Those populations also run the risk of being in an especially vulnerable position, so public policies regarding their Access to Justice should be a priority.

As a lesson from the Covid crisis, today's judicial and mediation processes are held in principle online, with different jurisdictions providing different online systems or platforms. Legal information, including law, regulations, rulings, and doctrine, is also unified and available online. (The Impact of Covid-19 Measures on the Justice System Report 2025.) Nevertheless, while these innovations represent progress, they also raise questions about the judiciary's role in times of crisis and in advancing the protection of Human Rights. The following section is going to examine how landmark judicial decisions in Argentina can prove that Human Rights and SDGs are not just ink on paper but legal realities that need to be protected by all institutions. Court-based adjudication, with all its faults, is in a great position to play the role of expanding the scope and depth of the protection of rights.

9.5 Important Jurisprudence of Access to Justice in Argentina

To analyze the impact of Access to Justice on people's life and their ability to use it to guarantee all their rights, three thematic cases from Argentina were chosen. They all have very different focuses, but they all contribute to a broader recognition of rights, especially the internationally recognized. As was explained in the previous section, Argentina's system has a strong influence on international law, and that enables it to come up with leading rulings on the protection of Human Rights and sustainable development. The selected cases were chosen based on their different topics but having the overarching theme of enforcing rights in a concrete manner, according to the standards of the SDGs. Each of these rulings not only exemplifies the protection of Human Rights but also illustrates clearly how jurisprudence can serve as a tool to advance the agenda of the SDGs.

The first case analyzed, Ekmekdjian, deals with the application of international norms and the importance of their conception as being directly applicable without the need for extra legislative steps. The second case deals intensively with the recognition of the Right to a Safe Environment and how International Standards can and should be applied by the government, taking active steps to protect the environment and repairing damages caused by the state's negligence in protecting the environment. Finally, the last case involves indigenous populations and their Right to Land, which is of particular interest not only by looking at the Inter-American system of protection of Human Rights and how their standards should be applied by governments but also how public policy plays a fundamental role in vulnerable groups being heard and giving them a democratic voice when it comes to their exercise or rights.

9.5.1 Ekmekdjian V. Sofovich

This case is one of Argentina's most important civil jurisprudence rulings, particularly relevant to Access to Justice and SDGs 10 and 16. Its legacy lies in redefining how international human rights are understood and exercised. Miguel Angel Ekmekdjian, a lawyer and constitutional law professor, initiated the case in 1988 to demand application of international law, specifically the Right to Reply. Before this, he filed a related case against Bernardo Neustadt, a prominent TV host who had made defamatory remarks during an interview with ex-president Arturo Frondizi. Ekmekdjian requested the same airtime to reply, invoking the American Convention on Human Rights. While the court acknowledged the violation of his rights, it denied enforcement, citing Argentina's lack of domestic legislation and prioritizing press freedom. This reflected a dualist approach, where international treaties were not directly applicable unless legislated.

A second case involved TV host Gerardo Sofovich, where similar defamatory comments were made. Though initially dismissed, the Supreme Court ultimately upheld Ekmekdjian's claim and allowed him to exercise his Right to Reply, even without domestic regulation. The ruling marked a turning point: it established that international law could be self-executing in Argentina, challenging the traditional dualist view. This evolution toward monism, treating domestic and international law as a unified system, is central to ensuring effective implementation of human rights and the SDGs.

This marked a new understanding of Argentina's constitutional law, shifting to a monist system in which international law is directly applicable, without the need for laws to be passed regulating the matter. This also incentivized the Constitutional reform of 1994, in which this approach was made explicitly clear, and certain International Human Rights treaties were given Constitutional hierarchy. This can be consulted in Argentina's Constitution Article 75 subsection 22. This also goes hand in hand with upholding an International Standard of Human Rights protection after the return to democracy after the military government in which fundamental rights were extensively violated.

This critical case also focused on the necessity of a Right to Reply to enforce accountability in the media, without that being a limitation on freedom of speech or press, but rather, establishing clear liabilities in the case of infringements. It's clear that the power of the individual is relatively small when compared to prominent media outlets, but being able to exercise a Right to Reply not only helps with self-expression but also works toward reducing inequalities, as seen in SDG 10. Being able to exercise international law, in particular, Human Rights treaties directly, is a massive milestone toward achieving not only objective 16 of the SDGs but also guaranteeing all of them not merely in a formal way but in a concrete and readily exercisable way, particularly reducing inequalities. The next case is going to tackle inequalities head on, and other SDGs associated with it, but the focus is going to be on environmental protection.

9.5.2 Matanza-Riachuelo Case

This case, also known as Mendoza Beatriz Silvia and Others v. the State of Argentina, is a landmark environmental justice case in Argentina that highlights how the legal mechanisms available to citizens can be used to promote adequate environmental protection, as described in SDG 13. The Matanza-Riachuelo basin is one of the most polluted rivers in the world. Many factories drop pollutants into it, some with a high rate of toxicity, which has proven to cause death and illness in the population that lives close to it. Most of this population is also highly vulnerable economically. It does not have proper access to basic services, such as clean water or the gas network, having to rely on unsafe or dangerous options. While Argentina guarantees the Right to a Safe Environment in its constitution and has also signed several international agreements toward this goal, the truth is that in many cases, the vulnerable populations had no way to enforce their rights in practice, mainly due to the usual issues of access to justice but in particular a lack of interest of the government and a lack of coordination between the national and provincial government of the province of Buenos Aires. We will also see the importance of Human Rights Institutions like the Ombudsman office, which do not rely on the current political party in power for their decisions.

This lawsuit was started by the inhabitants of the basin with the help of several environmental organizations. They filed it against several companies and the provincial and national governments. Many reports were made about the situation, including health reports detailing respiratory illnesses, skin diseases, and organ failure due to the untreated sewage and toxic chemicals that were dropped into the river, as well as the economic devastation of the zone not enabling the communities to take part in either fishing or agriculture. The Supreme Court ruled that the Right to a Safe Environment was being violated, and the right of the population to enjoy the river for recreational purposes was also recognized by the court.

In 2008, the court ruled in favor of the plaintiffs on three main points. The first one is ordering monetary reparations for the damage caused by pollution, be it by action (in the case of companies) or inaction (in the case of negligence by the state). This could provide economic reparation to the victims but could not solve the situation that sparked the conflict.

The second point was ordering the government to take immediate action to clean the river and solve the environmental crisis that the population was facing. This was a crucial precedent since it not only recognized the right to a healthy environment but also the responsibility of the state to take concrete action to enforce it. Crucially, the state of Argentina was deemed liable for failing to protect its population from environmental harm. Failing to enforce this protection would imply not only a violation of Environmental Rights but also of their fundamental rights of access to water, food, equality, and economic opportunities, among others.

On a third point and as a true novelty, the Supreme Court ordered the creation of a state control agency called the Authority of the Matanza-Riachuelo Basin (ACUMAR), whose goal was to visualize the adequate fulfillment of the ruling and its objectives. This was an attempt by the ruling to have an enforcing arm that would continue to make reports and analyze the progress of the ruling. This self-governing agency has three main goals: improving the life of the basin's inhabitants, reconstructing the environmental state of land, water, and air before the pollution starts, and preventing further damage to the population and area.

While the river's condition remains far from ideal, progress has been made toward its rehabilitation. Pollution persists due to governmental negligence and ACUMAR's limited resources to enforce its findings. Nonetheless, cleanup efforts are ongoing, and conditions are gradually improving. The basin's residents remain vulnerable, but steps are being taken to address their needs, primarily through the construction of plumbing and sewage systems and the provision of clean water. Additional measures include integrating residents into safer economic zones, expanding health services, and implementing housing programs tailored to the area.

The role of nonprofit organizations and the Ombudsman's Office was critical in securing Access to Justice. These actors not only supported the legal process and covered associated costs but also produced key environmental, urban, and health reports that bolstered the case. Their involvement ensured that vulnerable communities could participate meaningfully in the litigation. The Ombudsman even became a co-plaintiff, reinforcing human rights framing of the case and strengthening its outcome.

While the right to the environment is only partially guaranteed in this case, it's still a benchmark case for the development of this right and the exercise of rights in general. This case directly influenced not only the right to the environment as protected in SDG 13, but also SDGs 14 and 15, regarding submarine and land life, sustainable communities as in SDG 11, reducing inequalities as in SDG 10, protecting economic development as in SDG 8 as well as access to clean water (SDG 6), not the mention SDGs 1 to 3, ending poverty, zero hunger, and health and well-being, which were a critical focus of this case. In the following case we will analyze how these principles are viewed through the lens of multiculturalism, and which should be the role of the state in such cases.

9.5.3 *Lhaka Honhat (Our Land) Case*

The *Lhaka Honhat v. Argentina* case was brought before the Inter-American Court of Human Rights in 2020 and became a landmark in Access to Justice for Indigenous communities, particularly regarding ancestral land rights and equality (SDG 10). Across the Americas, Indigenous populations coexist with communities of European descent, whose customs and legal systems dominate state institutions. While the conquest of the continent is now widely viewed as genocidal, efforts have emerged to protect the rights of native populations. These efforts vary greatly. Countries like Bolivia, Guatemala, Peru, and Mexico have Indigenous populations exceeding 80%, while Argentina and the USA each report only around 1% (World Bank 2015). Despite these differences, Indigenous peoples often live in precarious conditions, with their land claims and customary legal practices frequently unrecognized. As this case shows, legal rulings alone are often insufficient; what truly advances sustainable development are flexible, inclusive public policies.

The *Lhaka Honhat* case is Argentina's most significant intercultural ruling to date. It involved five Indigenous groups representing 150 communities living across 643,000 hectares in Salta Province. After stalled negotiations with the national government, the case was brought to the Inter-American Court, which found Argentina internationally responsible for violating rights to communal property, cultural identity, a healthy environment, and access to clean water and food. However, the court did not uphold claims related to freedom of thought, expression, movement, or legal recognition of the communities.

As was expressed by the court, Argentina lacked proper legislation on communal property, and as such, it was violating the human rights present in international treaties. This was not only an issue with the Right of Property but also of Access to Justice, as the communities lacked the proper tools to guarantee their internationally protected Rights. The state was also considered liable for allowing the burning and deforestation of these areas, which a priori constituted valid under private ownership but was considered an infringement of communal property.

The court requested reparations in two ways. Firstly, the necessary measures must be taken to allow the native communities to enjoy their communal property, working together to find viable solutions. Secondly, to remove fences and agricultural production centers belonging to non-native communities, such as companies or non-native inhabitants. The court prohibited forced evictions for 3 years after the ruling, having to negotiate the relocation of that population peacefully. They also ruled to guarantee the property rights of that population, either by offering lands elsewhere or a proper monetary reparation.

This ruling was received in a very constructive way by Argentina, which had a history of seeing the Inter-American Court of Human Rights rulings as an opportunity to improve the legal system and public policies to guarantee rights better. Following the verdict, the national government created a specific agency to control and execute the ruling under the wing of the Human Rights Ministry. This agency worked together with the national government, the provincial government, the municipal government, and the native communities to better establish the priorities and roadmap to execute the ruling.

Based on that, four main points were established: territorial restitution, improved living conditions for the native population, environmental policies, and support for local initiatives that respect indigenous customs and identities. According to government reports, all actions were carried out through public hearings in local communities, using both Spanish and regional languages depending on the group. Considerable time and effort were invested in securing community approval for public projects, particularly to bring clean water without infringing on communal property rights. Thanks to this democratic and horizontal approach, community consent was obtained in all cases. While parts of the roadmap are still under government discussion, once finalized, it will be presented to the communities for their input and recommendations. Although the process is slow, involving communities is essential for the successful implementation of the court's mandates, as each group's unique customs must be respected. Similar strategies are being considered for other Indigenous communities across the country.

As explained, this case is not only crucial to SDG 16 when it comes to Access to Justice but also plays a fundamental role in reducing inequalities (SDG 10), creating sustainable communities (SDG 11), protecting the environment (SDG 13), ending poverty (SDG 1), zero hunger (SDG 2), health (SDG 3), and access to clean water (SDG 6). It is also crucial for understanding how multicultural governance should take place, working to gap differences, and giving a voice to traditional communities. As was also seen in the Matanza-Riachuelo case, institutional reform is necessary for this, by the creation in that case of the ACUMAR institution and creating effective communication channels in this case.

It's important to reiterate the importance of focusing on access to justice when it comes to approaching SDGs concretely, as it involves the exercise of other rights and strengthens all the objectives altogether. These selected cases are just a glimpse of the interesting development of Argentina in Human Rights, and that's precisely why studying its case is helpful for learning not only the importance of access to justice but also its different angles. The importance of inclusive public policies to this end cannot be understated, as the three cases had very different approaches but the same goal: better guarantee the exercise of rights by the individuals. Institutional reform is also crucial, as we can see through the constitutional reform that modernized the Argentinian constitution, the use of a Human rights protection institution such as the Ombudsman's office, the creation of audit institutions such as ACUMAR, and having direct channels of communication with citizens, giving them an active role in the implementation of public policies. This implies a more democratic and inclusive government that follows the rule of law but also seeks a social approach to law and sustainable development. This shift in understanding about the role of law, not only as formal tool to keep social order but as something that can change social realities is what is truly needed to develop sustainable societies.

9.6 Conclusions: Looking into the Future of Law and Sustainability

By exploring Argentina's approach to Access to Justice, this chapter has emphasized the crucial role that justice plays in advancing societal well-being and achieving Sustainable Development Goals (SDGs), particularly SDG 16. Access to Justice is not only fundamental in upholding the Rule of Law but also serves as the backbone for the realization of other rights. When justice is accessible, citizens are empowered to defend their rights, which leads to progress in related areas such as health, education, and environmental protection. This interconnectedness is at the heart of the SDGs, as advancements in one area often trigger positive outcomes in others. Argentina's legal framework offers a particularly insightful example of how Access to Justice underpins sustainable development.

Despite Argentina's robust constitutional and legal framework, significant barriers hinder the full realization of Access to Justice. Economic inequalities, geographic disparities, and judicial inefficiencies, such as lengthy legal processes and complex legal language, continue to restrict marginalized groups from accessing the justice system. These obstacles disproportionately affect vulnerable populations, including women, Indigenous communities, and low-income individuals. Addressing these systemic challenges is essential not only for ensuring that justice is available to all but also for fostering long-term, equitable development.

The examination of key legal cases within this chapter further illustrates the multifaceted impact of Access to Justice on societal development. Landmark rulings, such as those concerning the right to reply, environmental degradation, and indigenous land rights, demonstrate how the legal system can act as a powerful tool for upholding human rights, protecting the environment, and reducing social inequalities. These cases highlight that effective Access to Justice is not just a theoretical concept but one that produces tangible outcomes. This focus, also called moving law "from books to action," is the key factor in determining the adequate implementation of the development goals and the promotion of Human Rights. Based on the case studies, this chapter would suggest three areas that appear to be key for the development of SDGs in general.

Innovation can bring efficiency into conflict resolution systems, as we have seen looking at the response to the pandemic. Direct access to information and direct communication via smartphone to institutions seems to be a clear road to developing a more effective and inclusive legal system. Although hotlines and online platforms are used extensively for this, the development of specialized AI helpers for legal matters, such as providing precise legal information, would be aligned with SDG 16. In fact, an AI called *Prometea* is currently being used in the prosecutor's office of Buenos Aires, showing impressive results, such as a 937% increase in processing times with a 96% success rate in predicting case outcomes (Corvalan 2020).

Institutions like the Ombudsman office play an important role in Human Rights protection, but currently, it's mainly active in both urban centers, not so much in rural areas. Enhancing this and perhaps doing joint work with Access to Justice

Centers would provide cost-effective solutions to bringing Human Rights protection to remote areas.

Participation in civil society cannot be overstated. Nonprofit organizations and the government help assist vulnerable communities, but these groups are usually seen as objects to be protected in a paternalistic way. This methodology strips empowerment from these communities and does not allow them to reasonably choose how to engage with social issues. Reinforcing local autonomy while enhancing legal pluralism is decisively aligned with the SDGs, as was seen in the Lhaka Honhat case.

In conclusion, Argentina's experience with access to justice offers useful insight into the development of concrete, rights-based public policies for sustainable development. Although Argentina struggles with social inequality, judicial inefficiencies, and systemic barriers affecting vulnerable populations, these challenges are present in many countries worldwide. By examining both the successes and shortcomings of its reforms, other nations can design more resilient and inclusive justice systems that are better aligned with the Sustainable Development Goals.

References

Asociación Civil por la Igualdad y la Justicia (ACIJ). (2025, May). *Personas en situación de calle en la Ciudad de Buenos Aires* [Informe]. ACIJ.

Cappelletti, M., & Garth, B. (1978). *Access to justice: The newest wave in the worldwide movement to make rights effective. Buffalo Law Review*

Consejo de Procuradores y Fiscales Generales de la República Argentina., 2025, *Se crea la Red Federal de Acceso a la Justicia.* Available online: https://consejompra.org/index.php/novedades/421-se-crea-la-red-federal-de-acceso-a-la-justicia

Corvalan, J. (2020). Prometea: Artificial intelligence to transform justice and public organizations. *Revue Internationale de droit des données et du numérique, 6*, 89–102.

De Stefano, J. S. (2012). *Access to Justice: Analysis and Perspectives of the New Challenges*. Eudeba.

Federal Judicial Center. (2020). *Courts and COVID-19: Judiciaries around the world rally and innovate to meet the challenge of COVID-19.* Available online: https://www.fjc.gov/content/348402/courts-and-covid-19-judiciaries-around-world-rally-and-innovate-meet-challenge-covid

Ferreira, G., & Ferreira-Snyman, A. (2014). The Incorporation of Public International Law Into Municipal Law and Regional Law Against the Background of the Dichotomy Between Monism and Dualism. *Potchefstroom Electronic Law Journal, 17*(4), 1471-1496.

Fiss, O. M. (1984). *Against Settlement. Yale Law Journal, 93*(6)

Gargarella, R. (2002, November). Too Far Removed From the People: Access to Justice for the Poor—The Case of Latin America. In *Chr. Michelsen Institute Workshop, United Nations Development Programme Oslo Governance Centre* (Vol. 18).

Instituto Nacional de Estadística y Censos (INDEC). (2022). *Indicadores de las condiciones habitacionales de las viviendas particulares ocupadas: Censo Nacional de Población, Hogares y Viviendas 2022.*

Langford, M. (2009). *Rights-based approaches to development: Concepts, evidence, and policy.* In M. Langford (Ed.), *Social rights jurisprudence: Emerging trends in international and comparative law* Cambridge University Press.

Latinobarometro Report. 2024, Latinobarometro Corporation. Available online: https://www.inep.org/images/2024/TXT/Latinobarometro-Informe_2024.pdf

Menkel-Meadow, C. (2004). *From legal disputes to conflict resolution and human problem solving: Legal dispute resolution in a multidisciplinary context.* 54 Journal of Legal Education, 7–29
Observatorio de la Deuda Social Argentina, Universidad Católica Argentina. (2019). *EDSA Agenda para la Equidad.* Observatorio de la Deuda Social Argentina, UCA.
Peña Chacón, M. (2020). *El Acuerdo de Escazú y La Consagración de los Principios de Progresividad y no Regresión.* Revista Iberoamericana de Derecho Ambiental y Recursos Naturales, (38), Argentina
Rawls, J. (1971). *A theory of justice.* Harvard University Press
Rochi Monagas, M., Maisley, N., Asar, M., Gargarella, R., & Álvarez Ugarte, R. (2021). *Argentina: Legal Response to Covid-19.* In Oxford Compendium of National Legal Responses to Covid-19. Oxford University Press.
Rossi, J. (2020). Access to Justice During the Pandemic: To What Justice For Whom and For What? In J. P. Boholavsky (Ed.), *COVID-19 and Human Rights: The Pandemic of Inequality.* Editorial Biblos.
Sander, F. E. A., & Goldberg, S. B. (1994). *Fitting the forum to the fuss: A user-friendly guide to selecting an ADR procedure. Negotiation Journal, 10*(1)
de Santos, B. S. (2002). *Toward a new legal common sense: Law, globalization, and emancipation* (2nd ed.). LexisNexis Butterworths
Tamanaha, B. Z. (2012). The History and Elements of the Rule of Law. *Singapore Journal of Legal Studies.*
The Impact of COVID-19 Measures on the Justice System Report, 2025, The Alberta Civil Liberties
United Nations. (2017). *Report of Argentina before the UN Human Rights Council.* Retrieved from https://www.ohchr.org/EN/HRBodies/UPR/Pages/ARIndex.aspx
United Nations. (2015). *UN General Assembly Resolution 70/1 (2015) – Transforming our World: the 2030 Agenda for Sustainable Development.* Retrieved from https://www.un.org/en/development/desa/population/migration/generalassembly/docs/globalcompact/A_RES_70_1_E.pdf
World Bank. (2015, April 28). *Justice and the rule of law*

Chapter 10
Institutional Dynamics and Policy Directions for Official Development Assistance and Public–Private Partnership Urban Transport Financing in Metro Manila, Philippines

Martin Angelo T. Millete

10.1 Introduction

Developing economies face significant challenges in building urban infrastructure, particularly in the transportation sector. The Philippines, especially its capital region, Metro Manila, exemplifies this issue. Chronic delays in transportation projects and significant deficiencies in urban planning have brought Metro Manila to the brink of unsustainable urban mobility. The quality of public transportation in the region is considered one of the worst globally (OliverWyman Forum 2022). A report from the Japan International Cooperation Agency (JICA), published in 2014 in collaboration with the National Economic and Development Authority (NEDA) of the Philippines, estimated that residents of Metro Manila incur daily losses of approximately PHP 2.4 billion due to traffic congestion. If effective interventions are not implemented, this figure could rise to as much as PHP 6 billion by 2030 (JICA and NEDA 2014). Urban transport issues have become a defining characteristic of daily life in Metro Manila, resulting in conflicts among policymakers and stakeholders as they strive to have a full understanding of the problem.

Transportation megaprojects, such as rail infrastructure, often encounter persistent issues, including cost overruns, underestimated risks, and shortcomings in benefits. This phenomenon is known as the "iron law of megaprojects" (Flyvbjerg 2009, 2014, 2017; Cantarelli et al. 2010). The challenges are further complicated by the complexities of managing these projects, which involve various factors such as scope, organization, quality, cost, and time (Turner 2006a, b). Metro Manila's transportation infrastructure projects face these challenges. Successfully managing these

M. A. T. Millete (✉)
International Peace and Co-existence Program, Hiroshima University, Higashihiroshima, Japan

A. Sharifi et al. (eds.), *Frameworks and Practices Towards a Sustainable and Peaceful Future*, World Sustainability Series,
https://doi.org/10.1007/978-981-95-2310-8_10

projects requires navigating intricate economic, political, and managerial dynamics, alongside disputes and external factors such as the economic climate and political influences, which frequently lead to delays (Rivera et al. 2020, p. 7). These dynamics underscore the importance of adaptability in leveraging diverse funding sources, such as Official Development Assistance (ODA) and Public–Private Partnerships (PPPs), to ensure the successful completion of projects and effective collaboration (Estache 2010; Pessoa 2008).

This chapter investigates the institutional mechanisms that underlie two primary financing modalities for transportation infrastructure projects: ODA and PPP. It focuses explicitly on ongoing railway projects in Metro Manila, examining two projects funded through ODA (the Metro Manila Subway Project and the North–South Commuter Railway Phase 1) and two funded through PPP (MRT-7 and the Makati Intra-city Subway). By using rational choice institutionalism as a theoretical framework, this research critically analyzes the role of institutional structures and actor behaviors in shaping the design and implementation of these infrastructure projects. It unpacks the governance dynamics, transparency mechanisms, and institutional capacities that distinguish these financing approaches. Given the trade-offs between quality and cost in megaprojects (Estache and Iimi 2012), strict quality requirements may limit the pool of contractors, while lenient standards can lead to cost overruns and delays. This highlights the importance of striking a balance between institutional constraints and project goals.

Although previous research has offered valuable perspectives on infrastructure development, both within the Philippines and globally, many studies focus either on single-country or modality-specific assessments, such as the PPP-centered policy analysis of Llanto (2016) and Capuno (2020) or on broader global finance trends (Estache 2010; Flyvbjerg 2014; Pessoa 2008). While these works significantly contribute to our understanding of financing modalities and governance reform, they often lack comparative, empirically grounded evaluations of how distinct institutional frameworks shape project performance in real time. This chapter addresses that gap by offering a structured, cross-case institutional analysis of four railway projects in Metro Manila, comparing effectiveness, efficiency, and sustainability outcomes across ODA and PPP-funded initiatives.

The primary objectives of this chapter are to analyze the institutional dynamics that govern ODA and PPP-funded railway projects in Metro Manila, evaluate the effectiveness, efficiency, and sustainability of projects funded under each financing modality, and identify governance challenges and propose potential reform strategies for infrastructure development. This chapter also contributes to the scholarly understanding of infrastructure financing in developing countries by positioning urban transport not only as a technical and fiscal challenge but also as a peacebuilding and sustainability imperative. Efficient and inclusive transport systems can foster social cohesion by enhancing access to employment, education, and essential services, particularly for marginalized populations (Estache 2010; Haggard et al. 2008). Chronic congestion and inequitable access to mobility exacerbate urban frustration, stress, and social division (Flyvbjerg 2009; Timberman 2019). Therefore, understanding how financing models influence the delivery of equitable and resilient transport systems is critical to advancing long-term peace and sustainability goals.

In addition to its normative framing, this chapter offers concrete contributions to the academic literature on infrastructure financing in developing countries by providing empirical insights into the complex interactions between institutional frameworks and infrastructure project outcomes, offering a nuanced perspective that views ODA and PPP as complementary financing strategies rather than competing ones and generating policy recommendations aimed at enhancing urban transport infrastructure development. The findings of this chapter have significant implications for policymakers, urban planners, and researchers interested in the intricate institutional mechanisms that shape infrastructure development in environments characterized by complex governance. The scale of megaprojects, political capital, and agency-related conflicts of interest can significantly impact project outcomes (Flyvbjerg 2017). Therefore, understanding these mechanisms is crucial for mitigating risks and enhancing project delivery. This chapter aims to provide valuable insights into these vital aspects by employing rational choice institutionalism.

10.2 Institutional Dynamics in Infrastructure Financing

10.2.1 Rational Choice Institutionalism

Rational choice institutionalism offers a comprehensive analytical framework for understanding how institutional structures influence strategic decision-making in infrastructure development. This theoretical approach provides a nuanced understanding of the interplay between formal rules, informal norms, and the behaviors of various actors within complex institutional environments.

Drawing on North's (1991) foundational work, institutions are viewed as essential structures that organize human interactions, reduce uncertainty in economic activities, and provide a basis for informed strategic decision-making. This perspective recognizes that institutions are not just passive constraints; instead, they are active mechanisms that shape economic and political outcomes. As North (1991) asserted, institutions represent the "rules of the game" that govern human interaction and economic activity.

Extensive research has indicated the critical role of institutional quality in infrastructure development. Strong institutional governance prioritizes public benefit, encourages investment through a consistent rule of law, and fosters a stable regulatory environment conducive to infrastructure development (Haggard et al. 2008; Kyriacou et al. 2019; Percoco 2014; Wang et al. 2019). In contrast, weak institutional frameworks can undermine infrastructure development by deterring investment, fostering inefficiency, and increasing the risks of project delays and corruption (Benitez et al. 2012; Hammerschmid and Wegrich 2017).

Rational choice institutionalism also emphasizes the importance of cost–benefit analyses in policy formulation decisions. This encourages strategic approaches to minimize cost overruns and optimize project implementation (Shepsle 2006). Policymakers should consider the design of policy interventions in conjunction with institutional quality, as both factors significantly impact economic performance (Rodrik et al. 2004). Additionally, both human and social capital must be integral to project design and implementation, particularly in terms of project management, optimizing labor resources, and engaging with communities and stakeholders (Glaeser et al. 2004).

By synthesizing these theoretical perspectives, rational choice institutionalism emerges as a robust analytical framework that goes beyond traditional understandings of institutional dynamics. This approach conceptualizes institutions as dynamic, interactive mechanisms rather than static constraints, offering vital insights into how strategic decision-making occurs in the context of complex infrastructure development. Its strength lies in its holistic examination of formal and informal institutional structures, integrating economic, political, and social dimensions to provide a comprehensive understanding of the outcomes of infrastructure projects. This theoretical framework not only clarifies the complex processes underlying infrastructure development but also offers a strategic lens for policymakers and researchers to diagnose, analyze, and potentially transform the institutional environments that influence large-scale public investments.

10.2.2 ODA and PPP as Complementary Strategies

Both ODA and PPP offer distinct pathways for financing large-scale infrastructure projects in the Philippines. With the country's infrastructure gaps becoming increasingly urgent amid limited resources and technical capacities, these two modalities serve as crucial alternatives to traditional procurement through the national budget. However, selecting the appropriate financing approach demands a nuanced evaluation of several factors: the nature of the project, the technical capabilities of the implementing agency, and the terms and conditions associated with ODA loans or PPP arrangements.

Not all economically viable infrastructure projects—those where societal benefits outweigh costs—are financially feasible. For projects where revenue generation is uncertain or insufficient to recoup costs, a taxpayers-pay scheme, such as ODA, might be more appropriate. Conversely, for projects with clear potential for revenue generation, a users-pay model through PPP may prove more suitable (Bernardo 2017, p. 66). This initial choice, typically made during the project identification phase, is pivotal in shaping the project's overall trajectory. As projects advance through various stages of the development cycle, continual reassessment of these factors is necessary to ensure alignment with strategic goals and evolving conditions (NEDA 2005, pp. 8–9). The chosen financing modality has a significant impact on project funding, management, and long-term sustainability. By selecting ODA or

PPP early in the process, implementing agencies can optimize resource allocation, leverage external expertise, and mitigate potential risks. A well-informed choice helps avoid pitfalls such as project delays, mismanagement, or outright failure.

ODA offers concessional funding with low-interest rates and long repayment terms, making it suitable for projects with high societal benefits but low financial returns. The modality fosters strong diplomatic relationships, facilitates knowledge and technology transfer, and serves as a catalyst for broader economic development. ODA can be particularly useful for sectors such as education, health, and public utilities, where profitability is secondary to societal impact. Despite its benefits, ODA is susceptible to corruption, political interference, and policy misalignment, which can compromise effectiveness (Rivera 2003; Trinidad 2017; Villacorta 2003). Concerns about debt sustainability can arise, particularly if ODA funds are not managed prudently. Bureaucratic red tape and slow approval processes can also delay project implementation. Furthermore, dependence on ODA may undermine local ownership and limit innovation in project execution (Batalla 2018, p. 212; Dohner and Intal 1989, pp. 175–177; Overholt 1986, p. 1143).

ODA refers to financial support provided to countries and territories listed as recipients by the OECD-DAC. It primarily consists of grants and soft loans designed to foster economic growth and improve welfare (OECD n.d.). However, the distribution of ODA often reflects the strategic interests of donor countries, which can lead to political considerations overshadowing the economic policies of recipient nations (Alesina and Dollar 2000; Boone 1996; Burnside and Dollar 2000; Easterly 2003; Hynes and Scott 2013; Maizels and Nissanke 1984). ODA helps bridge infrastructure gaps, supports policy reforms and modernization, enhances regional and global integration, and builds resilience against socio-economic challenges (Sunaga 2004). It also aligns with international development goals, such as the United Nations' Sustainable Development Goals (SDGs), fostering long-term growth and stability (Kitaoka 2016; Ohno 2017; Iacobuta et al. 2022). The diplomatic relationships built through ODA can also open avenues for technical cooperation and policy exchange (McEwan and Mawdsley 2012; Endo and Murashkin 2022). The efficacy of ODA, however, can be undermined by political and economic volatility, overreliance on donor support, public mistrust, and shifting geopolitical priorities (Alesina and Dollar 2000; Bernardo 2017, p. 66; Burnside and Dollar 2000; NEDA 2005, pp. 8–9). Changes in donor policies or international relations can disrupt the flow of assistance and affect project timelines. Additionally, misalignment between donor priorities and local needs may result in projects that are less impactful or unsustainable. Case studies from South Korea and China illustrate the complexity of using ODA as a diplomatic tool. Both countries strategically deployed ODA in Myanmar and the Philippines but faced institutional mismatches and political controversies that hindered their effectiveness (Camba 2021a, 2021b; Kim 2018; Trinidad 2016). These challenges underscore the need for adaptable ODA strategies to navigate the political and institutional realities of recipient countries effectively.

PPP leverages private-sector expertise, transfers risk to private partners, increases infrastructure capacity, and promotes knowledge and technology transfer. This modality can deliver projects more efficiently by tapping into private-sector innovation and efficiency (Canlas 2017; Llanto 2007). PPPs also reduce the fiscal burden on the government by sharing costs with private entities, thereby accelerating infrastructure development. Moreover, the contractual nature of PPPs often imposes discipline and accountability, leading to timely project completion. Nevertheless, PPPs often involve complex negotiations and contract management processes, which can lead to delays. Corruption, political risks, and public skepticism about private-sector involvement in public infrastructure pose additional challenges (Juego 2015; Llanto 2016; Ronda 2006). In some cases, the profit-driven nature of private partners may conflict with public interest, leading to higher user fees or inequitable access to services. Furthermore, the long-term nature of PPP contracts may lock governments into unfavorable terms if not carefully designed (Juego 2015).

The Philippines recognizes two main types of PPP models: availability PPPs, where the private sector guarantees the availability of a public service for a fixed fee, and concession PPPs, which allow the private sector to build, operate, and charge for a service with government approval (PPP Center n.d.). These arrangements fall under the Build-Operate-Transfer (BOT) Law, which defines various risk-sharing and role allocation structures (RA 6957 as amended by RA 7718). Despite their potential advantages, the effectiveness of PPPs depends on proper cost management, governance, risk management, and political stability (Llanto 2007; Joha and Janssen 2010; Sharma 2012; Zhang et al. 2015). Moreover, the financing component of PPPs can also include ODA, demonstrating how these two models can intersect (Pessoa 2008). PPPs attract diverse investments, respond quickly to urgent infrastructure needs, and can serve as a catalyst for economic development and revitalization (Cheng et al. 2021; Fabre and Straub 2021). They can foster competition and innovation in service delivery, resulting in higher-quality outcomes (Carbonara and Pellegrino 2020). Successful PPP projects can also build investor confidence and attract further investments in the country (Osei-Kyei and Chan 2017). The involvement of reputable private partners can enhance project credibility and reduce perceptions of government inefficiency (Brinkerhoff and Brinkerhoff 2011). However, challenges such as political and economic instability, legal and regulatory hurdles, and public distrust can hinder the success of PPP projects (Bernardo 2017, p. 66). Disputes over contract terms, risk allocation, and profit sharing can derail projects, while inadequate regulatory frameworks may fail to protect public interests adequately (Navarro and Llanto 2014; Llanto 2016; Leigland 2018). Additionally, relying on user fees for cost recovery may disproportionately affect low-income populations, thereby raising concerns about equity (Trebilcock and Rosenstock 2015).

Given the challenges associated with ODA and PPP, viewing them as complementary rather than competing strategies is vital. ODA can provide initial funding and political support for large-scale infrastructure projects, while PPPs can leverage private-sector efficiency and innovation in project execution and service delivery (Pessoa 2008; Estache and Saussier 2014). The Philippines' experience highlights the importance of adaptable financing strategies that take into account local institutional capacity, governance, and economic conditions (Navarro and Latigar 2022;

Ito 2018). Ultimately, a balanced approach that leverages the strengths of both ODA and PPP while attending to their respective challenges is crucial for achieving successful infrastructure development in developing countries. Continuous evaluation throughout the project cycle ensures that the chosen modality remains aligned with the country's infrastructure priorities and development vision. By strategically integrating these financing methods, governments can optimize project outcomes, address institutional weaknesses, and promote sustainable growth.

10.2.3 Assessment Framework

This chapter employed an analytical approach rooted in rational choice institutionalism to evaluate the financing modalities of Metro Manila's railway projects, financed either through ODA or PPP. An assessment framework (Table 10.1) is designed to determine whether ODA or PPP financing better serves the railway infrastructure needs of Metro Manila (Millete 2024). The framework provided a structured analysis by focusing on three primary criteria: effectiveness, efficiency, and sustainability. The roles inform these criteria of key actors—the government as the demand-side actor, and funding agencies (for ODA) or private-sector participants (for PPP) as the supply-side actors. This structured evaluation helps assess how institutional constraints and incentives impact the design, execution, and outcomes of railway projects financed through ODA and PPP.

10.3 Key Institutional Developments in ODA and PPP Infrastructure Financing in the Philippines

The successful implementation of transport megaprojects is closely tied to the institutional capacity and governance structures shaped by political and economic history. In the Philippines, ODA and PPP frameworks did not evolve in isolation; they developed through successive policy cycles, external pressures, and administrative reforms under various presidential administrations (Dohner and Intal 1989; Tsunekawa 2014). Early challenges in ODA governance, including the misuse of foreign loans under the Marcos regime, led to the strengthening of accountability mechanisms in later administrations (Rivera 2003). Meanwhile, the growth of the PPP model reflected both global liberalization trends and domestic legal and institutional adaptations (Llanto 2007, 2016). This section provides historical context clarifying how earlier decisions, particularly those concerning foreign borrowing, privatization, and institutional accountability, have shaped the divergent governance characteristics of ODA and PPP financing modalities today. By understanding these trajectories, the analysis in Sect. 10.4 can be more firmly grounded in the structural and historical legacies that continue to affect project performance.

Table 10.1 Assessment framework (Millete 2024)

Criteria	Indicators
Effectiveness	Public benefit • To what extent were the objectives of each project realized? • What methods were employed in assessing the public benefit of each project? • How were stakeholders consulted in assessing the public benefit of each project? • How did changes to project design and implementation affect the public benefit of each project? • What are the flaws of each project that may potentially lead to further benefit shortfalls? • To what extent can each project lessen automobile reliance and reduce traffic congestion? • To what extent can each project deconcentrate population density in Metro Manila?
Efficiency	Costs • What are the causes of cost overruns/underruns? • What decisions under project design and implementation have affected the overall cost? (e.g., bidding, procurement) • What technical features of the project have affected the overall cost? (e.g., rolling stock energy source, track gauge type, station building type, line length) Timeliness • How did timeliness impact the overall cost of each project? • What are the causes of project delays? (e.g., technical, economic, political)
Sustainability	Project life • How does the estimated lifespan of each project compare with their respective payment terms under their financing modality? • What issues on durability did each project encounter? • What are the operations and management (O&M) schemes planned for each project? Environmental impact • What issues did each project encounter regarding their environmental impact? • To what extent can each project improve their respective environments?

During the authoritarian rule of President Ferdinand E. Marcos Sr. (1965–1986), the Philippines saw a significant increase in foreign borrowing to finance large-scale infrastructure projects. This debt-driven approach contributed to widespread corruption and economic mismanagement, ultimately resulting in a balance of payments crisis in 1983 (Dohner and Intal 1989). Notable infrastructure projects, such as the Bataan Nuclear Power Plant (BNPP), became emblematic of misappropriated funds and political patronage (Batalla 2020). Japan emerged as the largest ODA donor to the Philippines, motivated by its strategic interests in regional diplomacy and economic stability. However, corruption scandals related to Japanese ODA-funded projects prompted Tokyo to implement stricter aid policies through the 1992 ODA Charter (Tsunekawa 2014).

The fall of the Marcos regime in 1986, facilitated by the People Power Revolution, marked the beginning of a period of institutional rebuilding. Under President Corazon Aquino, the administration prioritized debt repayment and governance reforms, despite facing limited fiscal resources. Key reforms included the passage of the Local Government Code and the Foreign Investments Act, both of which decentralized governance and promoted foreign investment (Canlas 2017).

President Fidel Ramos's administration (1992–1998) is often credited with stabilizing the economy through market-oriented reforms. Ramos expanded infrastructure investments via privatization and introduced the amended Build-Operate-Transfer (BOT) Law, addressing a severe power crisis through the Electric Power Crisis Act (Llanto 2007). These reforms institutionalized PPP mechanisms, laying the groundwork for sustainable infrastructure development.

Political instability reemerged during the administration of President Joseph Estrada (1998–2001), which was marred by controversies over PPP agreements, including the cancellation of the NAIA-3 contract due to corruption allegations (Juego 2015). The subsequent presidency of Gloria Macapagal-Arroyo (2001–2010) faced similar governance challenges. Her tenure was marked by high-profile scandals involving the NorthRail and NBN-ZTE projects, which reflected systemic issues of political rent-seeking (Batalla 2020; Canlas 2017).

President Benigno Aquino III (2010–2016) emphasized governance reforms and fiscal consolidation. His administration achieved investment-grade ratings but encountered difficulties in rolling out infrastructure, mainly due to bureaucratic inertia and project delays. Controversies, such as the CALAX bidding process and the unconstitutional Disbursement Acceleration Program (DAP), underscored ongoing governance challenges. At the same time, widespread frustrations over poor public transportation service delivery became a significant point of referendum in the 2016 elections (Villamejor-Mendoza et al. 2018).

The election of President Rodrigo Duterte in 2016 brought a bold infrastructure agenda through the "Build, Build, Build" program. His administration aimed to address infrastructure deficits and stimulate economic growth by attracting foreign direct investment. While concerns regarding authoritarian governance persisted, Duterte's infrastructure initiative resulted in significant project completions and new investments (Capuno 2020; Teehankee and Thompson 2016; Timberman 2019).

Taken together, these political transitions reflect a fragmented yet cumulative process of institutional development. Under continued pressure from international donors, particularly Japan, ODA governance has become increasingly codified and externally accountable (Rivera 2003; Tsunekawa 2014). In contrast, PPP governance emerged from liberalization reforms and structural adjustment policies (Llanto 2007; Dohner and Intal 1989) but continues to be susceptible to elite capture and regulatory inconsistencies, as evidenced in recent infrastructure experiences (Batalla 2020; Capuno 2020). These divergent institutional legacies now underpin the transparency, risk allocation, and stakeholder coordination practices examined in the succeeding section.

10.4 Assessment of ODA and PPP-Funded Railway Projects in Metro Manila

This section presents the findings of the assessment of four railway projects in Metro Manila: the Metro Manila Subway Project (MMSP), North–South Commuter Railway Phase 1 (NSCR), MRT-7, and Makati Intra-city Subway. The projects, funded via ODA or PPP as indicated, were analyzed using the framework detailed in Sect. 10.2.3, focusing on three criteria: effectiveness (public benefit), efficiency (cost efficiency and timeliness), and sustainability (project life and environmental impact). Table 10.2 summarizes the evaluation of each project across these dimensions.

10.4.1 Analytical Approach

This chapter applies an original, author-developed framework grounded in rational choice institutionalism to evaluate the institutional performance of railway infrastructure projects financed through different modalities. The framework operationalizes three interrelated criteria—effectiveness, efficiency, and sustainability—through a matrix of performance indicators (see Table 10.1, Sect. 10.2.3). Effectiveness refers to alignment with planning goals and institutional responsiveness; efficiency captures procurement, cost, and implementation performance; and sustainability addresses long-term viability, environmental integration, and stakeholder continuity.

The analysis draws on publicly available project documents, including environmental impact assessments (EIAs/EIS), government briefings, and technical reports from the Philippine Department of Transportation (DOTr), JICA, the Philippine Public–Private Partnership (PPP) Center, and project proponents, as well as scholarly and policy literature (Flyvbjerg 2009; Pessoa 2008; Estache 2010). Where direct data is limited, institutional analysis and triangulation of secondary sources

Table 10.2 Summary of ODA and PPP project findings per assessment framework indicator

		ODA		PPP	
Criteria	Indicator	MMSP	NSCR Phase 1	MRT-7	Makati Subway
Effectiveness	Public benefit	25-km subway reducing travel to 31 min	38-km line reducing travel to 35 min	Key link in northeast Metro Manila	10.1-km subway within Makati City
Efficiency	Cost	PHP 356.9 billion	PHP 149.1 billion	PHP 77 billion	$2.5 billion
	Timeliness	Strategic land use minimized delays	Design integration caused delays	Delayed due to land disputes	Expedited but faces political risks
Sustainability	Project life	40-year life, TOD, emission reduction	35-year life, TOD, emission reduction	Reduced traffic emissions but depot issues	Reduced emissions, fully underground design
	Environmental impact	Seismic risk mitigated	Air pollution reduction	Proximity to protected area	Construction noise minimized

were employed to infer governance dynamics, particularly about transparency, stakeholder engagement, and procedural accountability (Llanto 2016; Capuno 2020; Rivera 2003; Batalla 2020).

This framework enables a structured, comparative application of theory-informed criteria across multiple cases rather than isolated descriptions (Haggard et al. 2008). Although this chapter does not involve new primary fieldwork, its contribution lies in synthesizing fragmented data to reveal how financing modalities shape governance practices and project performance in urban transport. The following subsections demonstrate how this framework was applied to each case.

10.4.2 Effectiveness (Public Benefit)

The effectiveness of each project was evaluated by examining its stated objectives, projected benefits (particularly reductions in travel time and congestion), the application of transit-oriented development (TOD) principles, and the extent of stakeholder engagement during the design and planning phases. Indicators such as passenger capacity, mode shift potential, and urban connectivity improvements were analyzed using the framework introduced in Sect. 10.2.3.

The MMSP, financed by JICA's Special Terms for Economic Partnership (STEP) ODA loan, is designed to enhance connectivity and significantly reduce travel times across Metro Manila. The fully underground 25-km subway aims to cut travel from

2.5 h to approximately 31 min. It addresses the assessment criteria on public benefit by employing methods such as stakeholder consultations and data-driven projections to assess benefits (DOTr and JICA 2019). The project promotes TOD by connecting major business districts like Ortigas and Bonifacio Global City, which supports population deconcentration in Metro Manila. Projections indicate that it will reduce the number of 1300 cars, 220 jeepneys, and 60 buses daily, contributing to reduced traffic congestion (DOTr and JICA 2019). The Philippine Railway Institute (PRI) plays a role in sustaining operational benefits through workforce training. Changes in project design, such as realigning to avoid the West Valley Fault, demonstrate an adaptive approach to addressing safety concerns. Government prioritization of land use minimized right-of-way acquisition delays, ensuring timeliness and strengthening public-benefit delivery (DOTr and JICA 2019; Millete 2024).

The NSCR Phase 1 aligns with the public-benefit criteria by reducing travel times from 85 to 35 min along its 38-km elevated commuter line, directly improving mobility between Bulacan and Manila. The use of TOD principles enhances accessibility and local economic activity along its alignment. Stakeholder collaboration with local government units (LGUs) and DPWH ensures stations support urban development while addressing local needs (JICA 2015). The design's integration with future NSCR phases reflects a long-term strategic approach to connectivity, directly contributing to regional deconcentration and economic growth. Capacity estimates of 2288 passengers per trip highlight its role in reducing reliance on automobiles (JICA 2015; Millete 2024).

The MRT-7 meets public-benefit criteria through its objective to alleviate vehicular traffic along Commonwealth Avenue, accommodating up to 850,000 passengers daily. Its 22.8-km alignment from Quezon City to Bulacan serves as a critical link for commuters, shortening end-to-end travel time to 34 min (PPP Center 2019). Delays caused by land disputes, however, hindered the timely realization of project benefits. Despite this, its potential for reducing traffic congestion remains significant.

The Makati Subway is set to reduce travel time within Makati City from 1.5 h to 15 min, serving up to 700,000 passengers daily. By incorporating TOD strategies, the project aims to decongest intra-city traffic and support sustainable mobility (Lichtel Technologies Inc 2019b). However, disputes over territorial alignments with the EMBO barangays introduce risks that could hinder long-term benefits and necessitate redesigns. Such flaws, if left unaddressed, may result in shortfalls in achieving projected benefits (Ramos 2017).

10.4.3 Efficiency (Cost Efficiency and Timeliness)

The efficiency dimension was assessed using two principal indicators: cost efficiency and timeliness. Cost efficiency was evaluated based on available data regarding original and revised project costs, technical specifications (e.g., track gauge, rolling stock, tunnel construction methods), and procurement decisions. Timeliness was examined through the identification of causes and implications of delays, including political, economic, and technical factors. Where direct project-level cost–benefit analyses were unavailable, this chapter inferred efficiency outcomes from documented project design decisions, implementation timelines, and secondary evaluations by government agencies and development partners.

The MMSP experienced a cost increase from PHP 218.3 billion to PHP 356.9 billion, primarily driven by the addition of new stations and safety-motivated realignments. Despite higher costs, the project's efficiency remains comparatively high due to early right-of-way planning, strategic land use, and the use of tunnel boring machines (TBMs), which enhanced both safety and project speed (JICA 2018; Embassy of Japan in the Philippines 2021). These decisions reflect an institutional capacity to absorb cost escalations while optimizing construction timeframes, as evidenced by the minimization of expropriation-related delays.

The NSCR Phase 1 registered a 42% cost increase attributed to technical upgrades, including elevated viaducts and signaling integration. Although this may suggest inefficiency from a fiscal standpoint, the project adopted forward-looking design enhancements that improve long-term cost-effectiveness. By choosing a standard rail gauge and electrified systems compatible with future extensions, the project demonstrates strategic planning to accommodate future line expansions, albeit at the expense of immediate budget containment, which may result in delays (NEDA 2018, 2020).

The MRT-7 experienced construction delays due to a spate of bureaucratic, legal, and corporate disputes (Camus 2016a, 2016b; DOTr 2017; Valdez 2018). As a result, traffic congestion and poor air quality worsened along Commonwealth Avenue, the major thoroughfare spanning the majority of the MRT-7's right of way. As it turned out, delays have ironically turned the project into part of the very problem it was designed to solve (Gascon 2019; Piedad 2023). Furthermore, the project's use of a third-rail electrification system was a novel technical choice, but it raised project-specific risks, particularly regarding long-term operations and maintenance (O&M) requirements (Gines Jr. 2021).

The Makati Subway achieved cost reductions from $3.5 billion to $2.5 billion through modifications to its alignment. However, territorial disputes between the cities of Makati and Taguig continue to risk additional costs and delays (Lopez 2019; Camus 2023). This matter has been well documented in the public domain (Lalu 2018; Ramos 2017). However, none of the official project documents (Lichtel Technologies Inc 2019a, b), and neither the Makati City Ordinance (No. 2019-A-020) detailing the city government's joint venture agreement with the project's private-sector proponent (City Government of Makati 2019) nor any of the annual reports of the private proponent (Philippine Infradev Holdings Inc 2020, 2021) acknowledged it as a potentially disruptive factor—at worst, a potential cause of project failure.

10.4.4 Sustainability (Project Life and Environmental Impact)

Sustainability was evaluated using two indicators: project life (alignment between infrastructure lifespan and financing terms, durability, and long-term operations and maintenance arrangements), and environmental impact (ecological footprint, emissions reduction, noise and vibration mitigation, and alignment with environmental standards). Given the limited availability of primary environmental audits, this chapter triangulated data from project documents (EIA/EIS), secondary literature, and inferred institutional practices to assess sustainability potential.

The MMSP is designed for a projected lifespan of 40 years and incorporates several of Japan's sustainable practices in underground railway construction. Its underground design, emission reduction goals, and collaboration with the PRI for long-term workforce development reflect a careful approach to ensuring the project's longevity (JICA 2018). The project also addresses environmental risk mitigation, particularly regarding seismic activity, by implementing realignment strategies that demonstrate both institutional responsiveness and design foresight (Department of Transportation [DOTr] and Japan International Cooperation Agency [JICA] 2019). The use of TBMs minimizes disruptions during construction while enhancing efficiency in this complex engineering endeavor (JICA 2018; Embassy of Japan in the Philippines 2021).

The NSCR Phase 1 has a lifespan of 35 years and includes scalability within its operational framework (NEDA 2018). As the system evolves, it has the potential to significantly reduce automobile usage and related emissions, similar to the results seen with the MMSP. Once this project is completed, commuters living and traveling to and from the vertical peripheries of Metro Manila will benefit greatly from efficient rail travel, especially with the planned extensions of the lines. The resulting decrease in dependence on road-based travel could enhance the quality of life for commuters and benefit the environment overall through lower emissions (JICA 2018).

The MRT-7 project is expected to provide environmental benefits, including reduced vehicular emissions. However, there are concerns regarding the depot's location near the La Mesa Watershed, which is a protected area. A major issue is the lack of verifiable information due to the absence of publicly disclosed project documents. This situation highlights a common tendency among private-sector participants in public–private partnerships (PPPs) to prioritize confidentiality to ensure fair competition (Piedad 2023). Unlike projects funded by Official Development Assistance (ODA), private-sector partners who possess project documents are often reluctant to share this information. Furthermore, Section 2 of the Philippine Republic Act No. 10173 (Data Privacy Act of 2012) safeguards the private sector's right to keep project details confidential, regardless of their relevance to public interest. This lack of transparency, stemming from these legal provisions, raises concerns about the project's sustainability.

The Makati Subway minimizes disruptions with its fully underground design and the use of TBMs, reducing surface noise and pollution (Lichtel Technologies

Inc 2019a). While this suggests a high degree of environmental consideration for construction within a highly urbanized area, the project's long-term sustainability is uncertain due to unresolved jurisdictional disputes that may affect maintenance accountability and future network integration (Lopez 2019; Camus 2023). Similar to the MRT-7, its private-sector-driven model restricts access to key project evaluations, inhibiting comprehensive environmental analysis.

10.5 Discussion

The financing and implementation of infrastructure projects in the Philippines through ODA and PPP have become vital strategies for tackling the country's significant infrastructure deficits. The analysis of ODA and PPP-funded railway projects in this chapter has shown that these financing methods display contrasting institutional behaviors due to their different operational frameworks. This divergence has a direct impact on project transparency, efficiency, and long-term sustainability.

As demonstrated in Sect. 10.4, the ODA-funded projects, the MMSP and NSCR Phase 1, benefit from structured governance mechanisms rooted in long-standing bilateral cooperation, stringent documentation requirements, and developmental conditionalities imposed by donors such as Japan. All these have produced positive outcomes in terms of stakeholder engagement, cost management, environmental safeguards, and operational planning. In contrast, the PPP-funded MRT-7 and Makati Subway projects reflect a more opaque institutional structure that prioritizes investor competitiveness, often at the expense of public disclosure, social accountability, and long-term risk mitigation.

Institutional dynamics shape the strategic actions of stakeholders within both ODA and PPP frameworks. The Japanese-funded MMSP and NSCR Phase 1 exemplify how historical experiences with ODA have fostered transparency and cooperation. Decades of managing ODA loans from Japan, based on a shared institutional memory of past mismanagement during the Marcos Sr. era (Rivera 2003; Villacorta 2003; Trinidad 2016), have led to rigorous documentation and open sharing of project details. This transparency facilitates technical, financial, and environmental assessments, reinforcing accountability and enabling effective progress monitoring.

In contrast, the PPP-financed MRT-7 and Makati Subway projects reflect a preference for confidentiality rooted in private-sector competitive practices. The World Bank noted that PPP arrangements typically restrict public disclosure to protect business interests (World Bank Group 2015). In these projects, limited access to essential documents and the unavailability of key informants hamper comprehensive evaluation. Although this approach aims to preserve competitive fairness, it compromises the public's ability to assess the overall benefits of such projects. More importantly, it weakens institutional responsiveness to concerns around equitable access, environmental protection, and public risk exposure.

Beyond transparency, project delays and cost overruns further reveal the influence of institutional frameworks. The ODA-supported MMSP and NSCR Phase 1 have progressed steadily due to strong bilateral cooperation between Japan and the Philippines, facilitated by the Duterte administration's "Build, Build, Build" initiative and supported by transparent implementation practices. Conversely, the PPP-funded MRT-7 and Makati Subway face significant implementation uncertainties, including right-of-way issues, intricate design adjustments, and unresolved jurisdictional disputes. For example, the territorial conflict between Makati and Taguig City, overlooked in the planning of the Makati Subway, demonstrates the risks posed by limited preemptive disclosure and inadequate political contingency planning.

Sustainability considerations also underscore key differences between financing modalities. The project documents for the ODA-funded MMSP and NSCR Phase 1 demonstrate a commitment to comprehensive environmental assessments. These assessments not only ensure compliance with safety protocols but also support the long-term viability and continuity of projects (Tsunekawa 2014). In contrast, both the PPP-funded MRT-7 and Makati Subway projects lack comparable levels of transparency, as their proponents have not made environmental documentation widely accessible. This raises potential concerns regarding ecological sustainability and institutional resilience (Batalla 2020). The divergence reflects systemic issues related to openness, regulatory enforcement, and institutional accountability (Estache and Iimi 2012).

Decisions regarding the financing of transportation infrastructure have significant social and developmental consequences that extend far beyond technical and economic considerations (Pessoa 2008). These projects have enduring spatial and temporal impacts, and therefore require meaningful public scrutiny. While commercial confidentiality in PPPs may be justified to protect private-sector investments, it should not override the public interest. Large-scale transport infrastructure primarily serves the needs of commuters and the broader community (Capuno 2020). Ensuring public accountability and fostering participatory governance requires strong transparency mechanisms, particularly in the form of open access to project documents.

To achieve long-term objectives such as reducing urban inequality, promoting spatial deconcentration, and building social cohesion, infrastructure must be viewed as a public resource that enables dialogue among diverse stakeholders. Consequently, railway systems backed by transparent financing arrangements play a crucial, yet often overlooked, role in advancing peaceful and sustainable urbanization. This perspective encourages a reevaluation of engagement with infrastructure and highlights the crucial role of transparency in shaping an equitable urban future.

The sociopolitical dimensions of these projects further illustrate how institutional contexts shape stakeholder engagement and inter-agency coordination (Haggard et al. 2008). ODA-funded projects exhibit enhanced cooperation among government entities driven by bilateral agreements and development assistance protocols. This structured collaboration promotes smoother project execution and capacity building. By contrast, PPP projects, motivated by profit considerations, often experience fragmented coordination due to competing priorities between government agencies and private investors.

These findings highlight the significant impact of institutional arrangements on infrastructure outcomes. ODA-driven projects, with their transparency-focused frameworks, facilitate more effective evaluation and enhanced public accountability. Meanwhile, the confidentiality embedded in PPP models complicates public-interest assessments despite enabling competitive project delivery. Taken altogether, these underscore the need for adaptive financing approaches that integrate the technical and financial strengths of both models, while explicitly prioritizing institutional reform, transparency, and inclusivity. Policymakers must, therefore, navigate institutional complexity by strengthening regulatory frameworks, enforcing transparency norms, and fostering inclusive project planning and execution.

10.6 Conclusion

This chapter illustrated the complex institutional dynamics that influence the effectiveness, efficiency, and sustainability of ODA and PPP-financed urban transport projects in Metro Manila. By analyzing four railway projects—two funded by ODA and two through PPP—this chapter revealed key differences and synergies between these modalities, highlighting their unique contributions and challenges in addressing Metro Manila's chronic infrastructure deficits.

The findings confirmed that ODA-financed projects, such as the MMSP and the NSCR Phase 1 benefit from well-established frameworks that emphasize transparency, stakeholder collaboration, and adherence to international standards. These characteristics enhance public accountability, foster institutional capacity building, and ensure long-term sustainability. Conversely, PPP projects such as the MRT-7 and the Makati Intra-city Subway demonstrate the private sector's capacity to deliver innovative solutions and mobilize resources efficiently. However, they face governance challenges, including limited public disclosure and political risks, which can undermine public trust and equitable access.

Informed by this comparative analysis, the following policy recommendations directly address the governance challenges and performance gaps identified in Sect. 10.4, aiming to support inclusive, resilient, and sustainable transport development in Metro Manila and similar urban contexts. Firstly, enhancing PPP regulatory frameworks is essential, which entails strengthening the legal and institutional mechanisms governing these partnerships to balance private-sector competitiveness with public accountability. Key actions should include revising procurement policies to mandate greater transparency in project costs, contracts, and environmental assessments. By implementing public-interest safeguards, this governance framework will effectively mitigate risks encountered in the MRT-7 land acquisition issues and the jurisdictional conflict regarding the Makati Subway.

ODA can play a significant role not just as a financing tool but also when leveraged to institutionalize best practices across project planning, implementation, and long-term operational strategies. The PRI, supported by the MMSP, exemplifies how ODA can fund institutional development, and future projects should incorporate similar training, research, and monitoring mechanisms to ensure technical sustainability and foster human capital growth. Additionally, exploring hybrid financing models that combine the strengths of ODA and PPP can achieve optimal results; these innovative financing arrangements can capitalize on concessional loans to meet initial capital requirements while utilizing the efficiency of the private sector for operations and maintenance. However, it is essential that these models include clearly defined governance rules to ensure that risk allocation and public-benefit goals are protected throughout the project cycle.

Strengthening coordination among stakeholders serves as a pivotal recommendation, particularly in the context of institutional fragmentation within PPPs, which often hampers effective implementation and sustainability. To mitigate this issue, it is imperative to enhance mechanisms for inter-agency coordination, dispute resolution, and LGU involvement, ensuring cooperation and alignment among all parties. The Makati–Taguig conflict, which jeopardizes the progress of the Makati Subway, illustrates the costs associated with overlooking these vital linkages. Creating cross-jurisdictional project boards and shared spatial planning protocols can help to reduce political friction and implementation delays. Additionally, the institutionalization of comprehensive monitoring and evaluation (M&E) systems is crucial, as robust M&E frameworks should be standardized across both official development assistance (ODA) and PPP modalities. Continuous tracking of cost efficiency, environmental impact, and delivery timeliness will facilitate early course corrections and enhance project resilience. It is essential to adapt lessons from ODA-funded projects, where development partners require such frameworks, for use in domestic institutions and local PPP oversight bodies.

Prioritizing public engagement and communication is essential, as community consultations and public access to information must be required at all stages of infrastructure planning. Marginalized urban populations often find themselves excluded from PPP negotiations, yet they bear the brunt of transportation inequities. Therefore, inclusive planning processes reduce social grievances, improve legitimacy, and foster a shared vision of mobility justice. Additionally, adopting technology and innovation is vital for enhancing project efficiency and safety; this includes encouraging the use of advanced technologies such as TBMs, seismic mitigation systems, and innovative mobility platforms. However, technical choices should be guided by comprehensive public evaluation frameworks to mitigate the risks associated with untested innovation or cost inflation. Finally, it is critical to recognize infrastructure as a social contract; transport infrastructure must be viewed as a public good with peacebuilding potential rather than merely a growth stimulus. Reliable, equitable, and low-carbon mobility networks can help mitigate urban stressors, promote social inclusion, and foster long-term resilience, making it essential to evaluate financing mechanisms based on their capacity to uphold these democratic and developmental principles.

Through these recommendations, the Philippine government can unlock the full potential of ODA and PPP modalities while addressing their limitations. Integrating transparency, accountability, and inclusivity into project frameworks will ensure more equitable and sustainable urban mobility solutions. Ultimately, this chapter calls for a paradigm shift in the institutional approach to urban transport financing in Metro Manila. Policymakers must embrace adaptive strategies that leverage the complementary strengths of ODA and PPP to overcome existing institutional constraints. This holistic perspective will enable the Philippines to build resilient transport systems that not only enhance urban connectivity but also promote inclusive economic growth, reduce environmental risks, and improve the quality of life for its citizens.

References

Alesina, A., & Dollar, D. (2000). Who gives foreign aid to whom and why? *Journal of Economic Growth, 5*, 33-63. https://doi.org/10.1023/A:1009874203400

Batalla, E. V. C. (2020). Grand corruption scandals in the Philippines. *Public Administration and Policy, 23*(1), 73-86. https://doi.org/10.1108/PAP-11-2019-0036

Batalla, E. V. C. (2018). Bypassing industrial development. In M. R. Thompson & E. V. C. Batalla (Eds.), *Routledge handbook of the contemporary Philippines* (pp. 211-224). Routledge. https://doi.org/10.4324/9781315709215

Benitez, D. A., Estache, A., & Søreide, T. (2012). Infrastructure policy and governance failures. *Chr. Michelsen Institute (CMI) Working Paper, 5*. https://doi.org/10.1007/S11127-009-9448-5

Bernardo, R. L. (2017). The great infrastructure debate. In C. Chikiamco, R. Bernardo, E. de Dios, R. Fabella, & C. Paderanga Jr. (Eds.), *Momentum: Economic reforms for sustaining growth* (pp. 65-68). Foundation for Economic Freedom.

Boone, P. (1996). Politics and the effectiveness of foreign aid. *European Economic Review, 40*, 289-329. https://doi.org/10.1016/0014-2921(95)00127-1

Brinkerhoff, D. W., & Brinkerhoff, J. M. (2011). Public-private partnerships: Perspectives on purposes, publicness, and good governance. *Public Administration and Development, 31*, 2-14. https://doi.org/10.1002/pad.584

Burnside, C., & Dollar, D. (2000). Aid, policies, and growth. *The American Economic Review, 90*(4), 847-868. https://doi.org/10.1257/aer.90.4.847

Camba, A. (2021a). *How Duterte strong-armed Chinese dam-builders but weakened Philippine institutions*. Carnegie Endowment for International Peace. https://carnegieendowment.org/files/202106-Camba_Philippines_final_updated.pdf

Camba, A. (2021b). Sinews of politics: State Grid Corporation, investment coalitions, and embeddedness in the Philippines. *Energy Strategy Reviews, 35*, 1-9. https://doi.org/10.1016/j.esr.2021.100640

Camus, M. R. (2016a, February 5). *SMC given 2 weeks to close funding deal for MRT-7*. Philippine Daily Inquirer. https://web.archive.org/web/20160206045453/https://motioncars.inquirer.net/42962/42962

Camus, M. R. (2016b, February 17). *With funds secured, MRT-7 to run by '19*. Philippine Daily Inquirer. https://business.inquirer.net/207206/funds-secured-mrt-7-run-19

Camus, M. R. (2023, September 7). *Makati subway stalls on Taguig win in territorial row*. Philippine Daily Inquirer. https://business.inquirer.net/419524/makati-subway-stalls-on-taguig-win

Canlas, D. B. (2017). Philippine policy reforms and infrastructure development: a historical account. *Philippine Review of Economics, 54*(2), 61-87. https://econ.upd.edu.ph/pre/index.php/pre/article/view/955/850

Cantarelli, C. C., Flyvbjerg, B., Molin, E. J., & Wee, B.V. (2010). Cost overruns in large-scale transportation infrastructure projects: Explanations and their theoretical embeddedness.

European Journal of Transport and Infrastructure Research (EJTIR), 10(1), 5-18. https://doi.org/10.18757/ejtir.2010.10.1.2864

Capuno, J. J. (2020). Dutertenomics: Populism, progress, and prospects. *Asian Economic Policy Review, 15*(2), 262-279. https://doi.org/10.1111/aepr.12296

Carbonara, N., & Pellegrino, R. (2020). The role of public private partnerships in fostering innovation. *Construction Management and Economics, 38*(2), 140-156. https://doi.org/10.1080/01446193.2019.1610184

Cheng, Z., Wang, H., Xiong, W., Zhu, D., & Cheng, L. (2021). Public–private partnership as a driver of sustainable development: toward a conceptual framework of sustainability-oriented PPP. *Environment, Development, and Sustainability, 23*, 1043-1063. https://doi.org/10.1007/s10668-019-00576-1

City Government of Makati. (2019). *Joint venture agreement with Philippine Infradev Holdings Inc. and the consortium which it represents regarding the project for the construction, establishment, management and operation of a subway system with the City of Makati* [Ordinance No. 2019-A-020]. Retrieved July 7, 2025, from https://www.makati.gov.ph/content/resolutions-and-ordinances/category/infrastructure/67262013-92eb-9d49-90a6-b8ec599d0c51?type=all

Dohner, R., & Intal, P. (1989). Debt crisis and adjustment in the Philippines. In J.D. Sachs (Ed.), *Developing country debt and the world economy* (pp. 169–192). University of Chicago Press. http://www.nber.org/chapters/c7525

Department of Transportation [DOTr]. (2017). *Unified Station/Grand Central Station memorandum of agreement.* https://www.scribd.com/document/342418742/Compromise-Common-Station-MOA-Signed-Copy-Jan-2017

Department of Transportation [DOTr] & Japan International Cooperation Agency [JICA]. (2019). *Metro Manila Subway Project (MMSP) phase 1 environmental performance report and management plan* (Vol. 1). https://eia.emb.gov.ph/wp-content/uploads/2019/08/MMSP-EPRMP_-Executive-Summary_English.pdf

Easterly, W. (2003). Can foreign aid buy growth? *The Journal of Economic Perspectives, 17*(3), 23-48. https://doi.org/10.1257/089533003769204344

Embassy of Japan in the Philippines. (2021, February 5). *Tunnel Boring Machine Arrives for Metro Manila Subway Project*. https://www.ph.emb-japan.go.jp/itpr_en/11_000001_00326.html

Endo, K., & Murashkin, N. (2022). Japan's infrastructure export and development cooperation: the role of ODA loan projects in the 2010s. *Australian Journal of International Affairs, 77*(2), 129-149. https://doi.org/10.1080/10357718.2022.2064972

Estache, A. (2010). Infrastructure finance in developing countries: An overview. *EIB Papers, 15*(2), 60-88. https://hdl.handle.net/10419/45371

Estache, A., & Iimi, A. (2012). Quality or price? Evidence from ODA-financed public procurement. *Public Finance Review, 40*(4), 435-469. https://doi.org/10.1177/1091142111422436

Estache, A., & Saussier, S. (2014). Public-private partnerships and efficiency: A short assessment. *CESifo DICE Report, 12*(3), 8-13. https://www.ifo.de/en/publications/2014/article-journal/public-private-partnerships-and-efficiency-short-assessment

Fabre, A., & Straub, S. (2021). The impact of public-private partnerships (PPPs) in infrastructure, health, and education. *Journal of Economic Literature, 61*(2), 655-715. https://doi.org/10.1257/jel.20211607

Flyvbjerg, B. (2009). Survival of the unfittest: why the worst infrastructure gets built—and what we can do about it. *Oxford Review of Economic Policy, 25*(3), 344-367. https://doi.org/10.1093/oxrep/grp024

Flyvbjerg, B. (2014). What you should know about megaprojects and why: An overview. *Project Management Journal, 45*(2), 6–19. https://doi.org/10.1002/pmj.21409

Flyvbjerg, B. (2017). Introduction: The iron law of megaproject management. In B. Flyvbjerg (Ed.), *The Oxford handbook of megaproject management* (pp. 1-18). Oxford University Press. https://doi.org/10.1093/oxfordhb/9780198732242.001.0001

Gascon, M. (2019, August 1). *Firm urged to explain delay in MRT 7 construction.* Philippine Daily Inquirer. https://newsinfo.inquirer.net/1148580/firm-urged-to-explain-delay-in-mrt-7-construction

Gines Jr., B. (2021, December 17). *MRT-7 gets new train sets*. The Manila Times. https://web.archive.org/web/20220311091136/https:/www.manilatimes.net/2021/12/17/business/top-business/mrt-7-gets-new-train-sets/1826198

Glaeser, E. L., La Porta, R., Lopez-de-Silanes, F., & Shleifer, A. (2004). Do institutions cause growth? *Journal of Economic Growth*, 271-303. https://doi.org/10.1023/B:JOEG.0000038933.16398.ed

Haggard, S., MacIntyre, A., & Tiede, L. (2008). The rule of law and economic development. *Annual Review of Political Science, 11*, 205-234. https://doi.org/10.1146/annurev.polisci.10.081205.100244

Hammerschmid, G., & Wegrich, K. (2017). Infrastructure governance as political choice. In K. Wegrich, G. Kostka, & G. Hammerschmid (Eds.), *The governance of infrastructure* (pp. 21-42). Oxford University Press. https://doi.org/10.1093/acprof:oso/9780198787310.001.0001

Hynes, W., & Scott, S. (2013). *The evolution of official development assistance: Achievements, criticisms and a way forward* (OECD Development Co-operation Working Papers No. 12). OECD. https://www.oecd.org/content/dam/oecd/en/publications/reports/2013/12/the-evolution-of-official-development-assistance_g17a242d/5k3v1dv3f024-en.pdf

Iacobuta, G. I., Brandi, C., Dzebo, A., Donaji, S., & Duron, E. (2022). Aligning climate and sustainable development finance through an SDG lens. The role of development assistance in implementing the Paris Agreement. *Global Environmental Change, 74*, 1–12 https://doi.org/10.1016/j.gloenvcha.2022.102509

Ito, S. (2018). PPP vs ODA revisited: Key issues for PPP infrastructure development in the Philippines. *The Philippine Review of Economics, LV*(1-2), 56-86. https://pre.econ.upd.edu.ph/index.php/pre/article/view/969/867

Japan International Cooperation Agency [JICA] & National Economic and Development Authority [NEDA]. (2014). *Roadmap for transport infrastructure development for Metro Manila and its surrounding areas (Region III & Region IV-A).* https://openjicareport.jica.go.jp/pdf/12149605_01.pdf

Japan International Cooperation Agency [JICA]. (2015). *Ex-ante evaluation (NSCR Malolos-Tutuban).* https://www2.jica.go.jp/en/evaluation/pdf/2015_PH-P262_1_f.pdf

Japan International Cooperation Agency [JICA]. (2018). *Operational rules of Special Terms for Economic Partnership (STEP) of Japanese ODA loans.* https://www.jica.go.jp/Resource/english/our_work/types_of_assistance/oda_loans/step/c8h0vm000053zae9-att/operational_rules.pdf

Joha, A., & Janssen, M. (2010). Public-private partnerships, outsourcing or shared service centres? Motives and intents for selecting sourcing configurations. *Transforming Government: People, Process and Policy, 4*(3), 232-248. https://doi.org/10.1108/17506161011065217

Juego, B. (2015). Elite capture and elite conflicts in Southeast Asian neoliberalization processes. *Inequality, democracy and development under neoliberalism and beyond* (pp. 68–93). IDEAs; CLACSO; CODESRIA. https://www.networkideas.org/tricont/aug2015/IDD.pdf?orden=&id_libro=929&pageNum_rs_libros=0&totalRows_rs_libros=897

Kitaoka, S. (2016). Sustainable Development Goals and Japan's Official Development Assistance policy: Human security, national interest, and a more proactive contribution to peace. *Asia-Pacific Review, 23*(1), 32-41. https://doi.org/10.1080/13439006.2016.1195951

Kyriacou, A. P., Muinelo-Gallo, L., & Roca-Sagalés, O. (2019). The efficiency of transport infrastructure investment and the role of government quality: An empirical analysis. *Transport Policy, 74*, 93-102. https://doi.org/10.1016/j.tranpol.2018.11.017

Kim, H. S. (2018). The political drivers of South Korea's Official Development Assistance to Myanmar. *Contemporary Southeast Asia, 40*(3), 475-502. https://muse.jhu.edu/article/714348

Lalu, G. P. (2018, August 22). *Binay insists: BGC, other barangays are Makati*. Philippine Daily Inquirer. https://newsinfo.inquirer.net/1023897/binay-insists-bgc-other-barangays-are-makati

Leigland, J. (2018). Public-private partnerships in developing countries: The emerging evidence-based critique. *World Bank Research Observer, 33*(1), 103-134. https://doi.org/10.1093/wbro/lkx008

Lichtel Technologies Inc. (2019a). *Makati Public Rail Transport System Project: Project Description*. https://eia.emb.gov.ph/wp-content/uploads/2019/05/Makati-Subway-PD-22May2019.pdf

Lichtel Technologies Inc. (2019b). *Makati Public Rail Transport System Project: Environmental Impact Statement*. https://eia.emb.gov.ph/wp-content/uploads/2019/10/Makati-Public-Rail-Transport-System-Project__EIS10012019.pdf

Llanto, G. M. (2007). Build-Operate-Transfer for infrastructure development: Lessons from the Philippine experience. In *ERIA research project report 2007-2* (pp. 319-359). IDE-JETRO. https://www.eria.org/uploads/media/Research-Project-Report/RPR_FY2007_2_Chapter_11.pdf

Llanto, G. M. (2016). Philippine infrastructure and connectivity: Challenges and reforms. *Asian Economic Policy Review, 11*, 243-261. https://doi.org/10.1111/aepr.12141

Lopez, M. L. (2019, October 29). *Makati subway gets $332M funding from Chinese firms*. CNN Philippines. https://web.archive.org/web/20230604164832/https://www.cnnphilippines.com/news/2019/10/29/Makati-subway-funding.html

Maizels, A., & Nissanke, M. K. (1984). Motivations for aid to developing countries. *World Development, 12*(9), 879-900. https://doi.org/10.1016/0305-750X(84)90046-9

McEwan, C., & Mawdsley, E. (2012). Trilateral development cooperation: Power and politics in emerging aid relationships. *Development and Change, 43*(6), 1185-1209. https://doi.org/10.1111/j.1467-7660.2012.01805.x

Millete, M. A. T. (2024). *The political economy of rail transport financing: A closer look at ODA and PPP-financed railway projects in Metro Manila*. (Publication No. ETDM-9) [Master's Thesis, De La Salle University]. De La Salle University - Animo Repository.

Navarro, A. M., & Latigar, J. S. (2022). *Road and rail transport infrastructure in the Philippines: Current state, issues, and challenges* (PIDS Discussion Paper Series No. 2022-34). Philippine Institute for Development Studies (PIDS). https://www.pids.gov.ph/publication/discussion-papers/road-and-rail-transport-infrastructure-in-the-philippines-current-state-issues-and-challenges

Navarro, A. M., & Llanto, G. M. (2014). *Financing infrastructure in the Philippines: Fiscal landscape and resources mobilization* (PIDS Discussion Paper Series No. 2014-01). Philippine Institute for Development Studies (PIDS). https://www.pids.gov.ph/publication/discussion-papers/financing-infrastructure-in-the-philippines-fiscal-landscape-and-resources-mobilization

National Economic and Development Authority [NEDA]. (2005). *Reference manual on project development and evaluation* (Vol. 1). https://drive.google.com/file/d/0BzhHHzvgVKq8UkNBdEJDZ0xoSGs/view?resourcekey=0-6jbuJqIdKGFPWoxoNXpgxg

National Economic and Development Authority [NEDA]. (2018, November 9). *ICC approves change in scope and cost of North-South Commuter Railway system*. https://depdev.gov.ph/icc-approves-change-in-scope-and-cost-of-north-south-commuter-railway-system/

National Economic and Development Authority [NEDA]. (2020). *Official development assistance portfolio review 2019*. https://depdev.gov.ph/oda-portfolio-review-2019/

North, D. C. (1991). Institutions. *The Journal of Economic Perspectives, 5*(1), 97-112. https://doi.org/10.1257/jep.5.1.97

OECD. (n.d.). *Official development assistance (ODA)*. https://www.oecd.org/dac/financing-sustainable-development/development-finance-standards/official-development-assistance.htm

Ohno, I. (2017). Network-based development cooperation as a way forward for Japan. In A. Asplund & M. Soderberg (Eds.), *Japanese development cooperation: The making of an aid architecture pivoting to Asia* (pp. 159-176). Routledge. https://doi.org/10.4324/9781315407746

OliverWyman Forum. (2022). *Urban mobility readiness index: 2022 report.* University of California, Berkeley. https://www.oliverwymanforum.com/content/dam/oliver-wyman/ow-forum/template-scripts/urban-mobility-index/PDF/Mobility-Index-Report.pdf

Osei-Kyei, R., & Chan, A. P. C. (2017). Factors attracting private sector investments in public–private partnerships in developing countries: A survey of international experts. *Journal of Financial Management of Property and Construction, 22*(1), 92-111. https://doi.org/10.1108/JFMPC-06-2016-0026

Overholt, W. H. (1986). The rise and fall of Ferdinand Marcos. *Asian Survey, 26*(11), 1137-1163. https://doi.org/10.2307/2644313

Percoco, M. (2014). Quality of institutions and private participation in transport infrastructure investment: Evidence from developing countries. *Transportation Research Part A: Policy and Practice, 70*, 50-58. https://doi.org/10.1016/j.tra.2014.10.004

Pessoa, A. (2008). Public–private partnerships in developing countries: Are infrastructures responding to the new ODA strategy? *Journal of International Development, 20*, 311-325. https://doi.org/10.1002/jid.1416

Philippine Infradev Holdings Inc. (2020). *PIHI 17A Annual Report – 2020.* https://www.infra.com.ph/wp-content/uploads/2023/04/pihi-17a-annual-report-2020.pdf

Philippine Infradev Holdings Inc. (2021). *PIHI 17A Annual Report – 2021.* https://www.infra.com.ph/wp-content/uploads/2023/04/pihi-17a-annual-report-2021.pdf

Piedad, M. (2023, May 7). *19 years and counting: The story of MRT-7.* Ibon Foundation. https://www.ibon.org/19-years-and-counting-the-story-of-mrt-7/

PPP Center. (n.d.). *What is PPP?* https://ppp.gov.ph/ppp-program/what-is-ppp/

PPP Center. (2019). *Project Briefer: The MRT-7 Project.* https://web.archive.org/web/20220811091611/https:/ppp.gov.ph/wp-content/uploads/2019/08/UNESCAP-SiteVisit-MRT7-Project-Briefer.pdf

Ramos, M. (2017, November 6). *CA: Taguig owns BGC, not Makati.* Philippine Daily Inquirer. https://newsinfo.inquirer.net/943180/ca-taguig-owns-bgc-not-makati

Rivera, T. C. (2003). The politics of Japanese ODA to the Philippines, 1971-1999. In S. Ikehata & L. Y. Jose (Eds.), *Philippines-Japan relations* (pp. 509-545). Ateneo de Manila University Press.

Rivera, L., Hilario Baguec, J., & Yeom, C. (2020). A study on causes of delay in road construction projects across 25 developing countries. *Infrastructures, 5*(10), 1-16. https://doi.org/10.3390/infrastructures5100084

Rodrik, D., Subramanian, A., & Trebbi, F. (2004). Institutions rule: The primacy of institutions over geography and integration in economic development. *Journal of Economic Growth*, 131-165. https://doi.org/10.1023/B:JOEG.0000031425.72248.85

Ronda, R. A. (2006, April 3). *Government should back out of 'ill-advised expropriation' of NAIA-3, says lawyer.* The Philippine Star. https://www.philstar.com/headlines/2006/04/03/329598/government-should-back-out-145ill-advised-expropriation146-naia-3-says-lawyer

Sharma, C. (2012). Determinants of PPP in infrastructure in developing economies. *Transforming Government: People, Process and Policy, 6*(2), 149-166. https://doi.org/10.1108/17506161211246908

Shepsle, K. A. (2006). Rational choice institutionalism. In R. A. Rhodes, S. A. Binder, & B. A. Rockman (Eds.), *The Oxford handbook of political institutions* (pp. 23-38). Oxford University Press. https://doi.org/10.1093/oxfordhb/9780199548460.001.0001

Sunaga, K. (2004). *The reshaping of Japan's Official Development Assistance (ODA) Charter* (FASID Discussion Paper on Development Assistance No. 3). Foundation for Advanced Studies on International Development (FASID). https://www.mofa.go.jp/policy/oda/reform/paper0411.pdf

Teehankee, J. C., & Thompson, M. R. (2016). Electing a strongman. *Journal of Democracy, 27*(4), 125-134. https://doi.org/10.1353/jod.2016.0068

Timberman, D. G. (2019). *Philippine politics under Duterte: A midterm assessment.* Carnegie Endowment for International Peace. https://carnegieendowment.org/2019/01/10/philippine-politics-under-duterte-midterm-assessment-pub-78091

Trebilcock, M., & Rosenstock, M. (2015). Infrastructure public–private partnerships in the developing world: Lessons from recent experience. *The Journal of Development Studies, 51*(4), 335-354. https://doi.org/10.1080/00220388.2014.959935

Trinidad, D. D. (2016). Institutional mismatch and Chinese aid in the Philippines: Challenges and implications. *Asian Perspective, 40*, 299-328. https://doi.org/10.1353/apr.2016.0013

Trinidad, D. D. (2017). Domestic factors and strategic partnership: Redefining Philippines-Japan relations in the 21st century. *Asian Politics & Policy, 9*(4), 613-635. https://doi.org/10.1111/aspp.12352

Tsunekawa, K. (2014). *Objectives and institutions for Japan's official development assistance (ODA): Evolution and challenges* (Working Paper No. 66). JICA Research Institute. https://www.jica.go.jp/english/jica_ri/publication/workingpaper/wp_66.html

Turner, J. R. (2006a). Towards a theory of project management: The nature of the project. *International Journal of Project Management, 24*, 1-3. https://doi.org/10.1016/j.ijproman.2006.03.002

Turner, J. R. (2006b). Towards a theory of project management: The functions of project management. *International Journal of Project Management, 24*, 187-189. https://doi.org/10.1016/j.ijproman.2006.03.001

Valdez, D. A. (2018, August 7). *Right-of-way issue hounds MRT 7*. BusinessWorld. https://www.bworldonline.com/editors-picks/2018/08/07/178682/right-of-way-issue-hounds-mrt-7/

Villacorta, W. V. (2003). Political relations between Japan and the Philippines during the Aquino and Ramos administrations. In S. Ikehata & L. Y. Jose (Eds.), *Philippines-Japan relations* (pp. 576-608). Ateneo de Manila University Press.

Villamejor-Mendoza, M. F., Baylon, M. S., Cuaresma, J. C., Diola, M. F. L., Florano, E. R., & Sobrepena, A. G. A. (2018). *The performance of the Aquino administration (2010-2016): An assessment*. University of the Philippines National College of Public Administration and Governance (UP-NCPAG). https://ncpag.upd.edu.ph/wp-content/uploads/2018/07/PerfromanceOfAqunioAdmin.pdf

Wang, H., Liu, Y., Xiong, W., & Song, J. (2019). The moderating role of governance environment on the relationship between risk allocation and private investment in PPP markets: Evidence from developing countries. *International Journal of Project Management, 37*, 117-130. https://doi.org/10.1016/j.ijproman.2018.10.008

World Bank Group. (2015). *A framework for disclosure in public-private*. World Bank Group. https://pubdocs.worldbank.org/en/773541448296707678/Disclosure-in-PPPs-Framework.pdf

Zhang, S., Gao, Y., Feng, Z., & Sun, W. (2015). PPP application in infrastructure development in China: Institutional analysis and implications. *International Journal of Project Management, 33*, 497-509. https://doi.org/10.1016/j.ijproman.2014.06.006

Part IV
Environmental Security Pathways

Chapter 11
Climate and Land Use Impacts on Ecosystem Services Supporting Food Security in Siem Reap Province, Cambodia

Fahad Ahmed and Ho Huu Loc

11.1 Introduction

Water resource availability is recognized as crucial for agriculture, national economies, and sustainable development in the Lower Mekong River Basin (LMRB) to secure food productivity and livelihoods for more than 80% of its population (MRC 2011). Demand for agricultural products from the LMRB is expected to increase 20–50% in the next 30 years to support world population growth (MRC 2015). Considering climate change in LMRB some areas would be affected by altered rainfall patterns, particularly during the dry season from November through May (Linke 2014). Water shortages in the dry season have been observed and will be the major constraints for food production. Climate change affects food security and the nutritional well-being of populations worldwide. By 2050, an estimated 183 million additional people will be at risk of hunger linked directly to climate change (Shukla et al. 2019).

The rate of deterioration of the water quality of freshwater resources, for example, lakes, ponds, rivers, etc., has become a global problem (Mushtaq and Pandey 2014). There are different kinds of pollutants for decreasing the water quality in freshwater resources, and most pollutants are from non-point sources. Non-point source pollution refers to the pollution caused by different sources. When entering

F. Ahmed
Water Engineering and Management, Asian Institute of Technology,
Khlong Nueng, Pathum Thani, Thailand

H. H. Loc (✉)
Earth Systems and Global Change Group, Wageningen University and Research,
Wageningen, Netherlands
e-mail: josh.ho@wur.nl

A. Sharifi et al. (eds.), *Frameworks and Practices Towards a Sustainable and Peaceful Future*, World Sustainability Series,
https://doi.org/10.1007/978-981-95-2310-8_11

streams or rivers, the water quality will be affected, causing harm to plants, aquatic life, and livestock, which will ultimately affect food production.

Ecosystem services (ESs) are defined as the direct and indirect benefits people obtain from ecological systems (Sukhdev 2008). Water-related ecosystem services (WESs), mainly including water yield, soil retention, water purification, climate regulation, and biodiversity, are the products of interactions between water ecosystems and their surrounding terrestrial ecosystems (Chen et al. 2018; Schmalz et al. 2016). Water yield can act on irrigation water supply (Ghimire and Johnston 2019), while soil retention is beneficial for maintaining a healthy agricultural environment (Shi et al. 2017) and both are vital for food production. Inevitably, with human activities and economic development, the land cover on the Earth's surface has changed greatly, and the ability for WESs provision has been weakened (Hao et al. 2019). This has resulted in impaired water purification (Gounand et al. 2018). In addition, increasing some supply services may cause a decline in other regulating services (Bai et al. 2019). Therefore, it is necessary to better understand the mechanisms of WESs change and the driving factors.

Many factors influence WESs, most importantly, land use (Liu et al. 2019) and climate change. The rapid change in land use, associated with growing climate change, has dramatically influenced ecosystem structure and services. Recent research has shown that human-made changes in land use and land cover such as deforestation, cultivated land expansion, and urbanization have irreversible impacts on WESs (Aliani et al. 2019; Bounoua et al. 2018)

Increasing population growth directly and indirectly contributes to Land Use and Land Cover (LULC) changes, especially from the perspective of demand for built-up area, agricultural activities, and water resources. LULC changes in a watershed will affect water quality, leading to increased surface runoff, reduced groundwater discharge, and transfer of pollutants (Butt et al. 2015; Wright and Wimberly 2013). Several studies have shown that the variety of land uses plays an essential role in the water quality of freshwater resources. Significantly, urban, and built-up and cultivated areas influence the water quality within the catchment (Hua 2017; Huang et al. 2013). LULC changes in the watershed area for urbanization and deforestation will continuously have negative impacts on water quality and quantity and indirectly affect the nature of a watershed ecosystem.

Water yield is one of the most vital services provided by ecosystems and essential for human irrigation, food production. It also affects ecological functions such as biomass, carbon cycle, and sediment output (Yang et al. 2021). As an essential ES, water regulation is affected by the combined effects of land use and climate change (Sun et al. 2015). Climate change can influence water yield by changing the temperature and precipitation in a catchment. Land use changes can also affect water yield by altering the underlying surface of the river basin (Song and Deng 2017).

On other hand, previous studies have demonstrated that climate change and LULC have serious effects on sediment and nutrient retention (Su and Fu 2013). Land managers can obtain explicit information to reduce damage and costs by learning from the changes in nutrient retention resulting from climate and land use change (Uwimana et al. 2018). Climate change deeply affects the intensity and spatial pattern of soil erosion and nutrient delivery (Bangash et al. 2013). For example, the average monthly nutrient load in South Australia decreases due to a decrease in average precipitation under a future climate (Shrestha et al. 2017). However, land use change influences sediment and nutrient retention capacity. Also, the heavy use of chemical fertilizers and pesticides for agricultural production results in significant deterioration of water quality in most catchments.

These drastic changes in climate and land cover and land use are likely to have deep impacts on soil erosion, nutrient delivery capacity, and water yield, which is crucially important for sustainable food production. Although many scholars have examined the impact climate change and LULC on ecosystem services in many regions of the world (Sample et al. 2016; Schirpke et al. 2017), the difference in the impact of climate changes and LULC on ecosystem services have attracted little attention. Thus, it is necessary to explore the effects of LULC and climate change on variations in nutrient delivery and water yield.

In recent years, with the development of Geographic Information System (GIS) and Remote Sensing (RS), models such as Soil and Water Assessment Tool (SWAT) (Santhi et al. 2001), Hydrological Simulation Program Fortran (HSPF) (Mishra et al. 2009), and Integrated Valuation of Ecosystem Services and Tradeoffs (InVEST) (Hoyer and Chang 2014) have been developed and used to evaluate ecosystem services.

The InVEST model is one of the most widely used tools due to its simplicity, efficiency, and strong spatial representation, particularly for quantifying ecosystem services in areas with fewer data along with that InVEST also requires moderate data inputs (Bagstad et al. 2013) and is globally applicable (Yang et al. 2021). InVEST has been used to evaluate impacts of land use and climate change in the LMB, including Thailand (Trisurat et al. 2016; Trisurat 2013) and Vietnam (Tue et al. 2014).

Trisurat et al. (2018) used the InVEST model to forecast water yield in Lower Mekong River Basin (LMRB), while the Nutrient Delivery Ratio (NDR) module in the model was specifically used for the simulation of nitrogen and phosphorus exports and has been applied in many regions of the world (Hoyer and Chang 2014; Keeler et al. 2012; Keller et al. 2015; Leh et al. 2013). Moreover, previous studies have successfully evaluated the water yield of different basin scales in different regions of the world, including Iran, India, Europe, and China, using the InVEST model (Aghsaei et al. 2020; Boithias et al. 2014; Goyal and Khan 2017; Hu et al. 2020; Song et al. 2015).

While previous studies have applied the InVEST model in various regions, this study is among the first to integrate CMIP6 climate projections with ecosystem service modeling (AWY and NDR) in Siem Reap province. It uniquely isolates the differential impacts of climate versus land use change on water yield and nutrient export under multiple future pathways, providing region-specific insights to inform food security strategies. This study was conducted in Siem Reap province of Cambodia and aims to evaluate the climate and land use change impacts on the water-related ecosystem services that support food security. The specific objectives of the present study were thus to (1) project the future climate scenarios using CMIP6; (2) develop baseline ecosystem services models using InVEST; (3) assess the ecosystem services under the past and present (2018 and 2022) climate and land use conditions; (4) assess the ecosystem services under future scenarios including near future (2023–2050), mid future (2051–2080), and far future (2081–2100). The results of the study may provide scientific reference for quantitative assessment, effective management, and sustainable development of water resources for food security in Siem Reap province, Cambodia.

11.2 Study Area and Data Source

Siem Reap province is one of the 10 largest provinces in Cambodia. Siem Reap is located in a tropical area from 14.8 °N to 9.9 °N, and from 102.2 °E to 107.9 °E and has total area of around 10,299 km^2 (Fig. 11.1). The elevation ranges from 6 m above sea level at Tonle Sap Lake (the largest freshwater lake in Southeast Asia) to 469 m on Phnom Kulen Mountain National Park. The landscape is a mosaic of dryland, rangelands and edaphic forests, rice fields, shifting cultivation, and urban areas. The wet period is from May to November, whereas December to April is characterized by dry period. Most of the rainfall occurs during the wet season with an annual precipitation range from 1050 to 1800 mm. Siem Reap province in

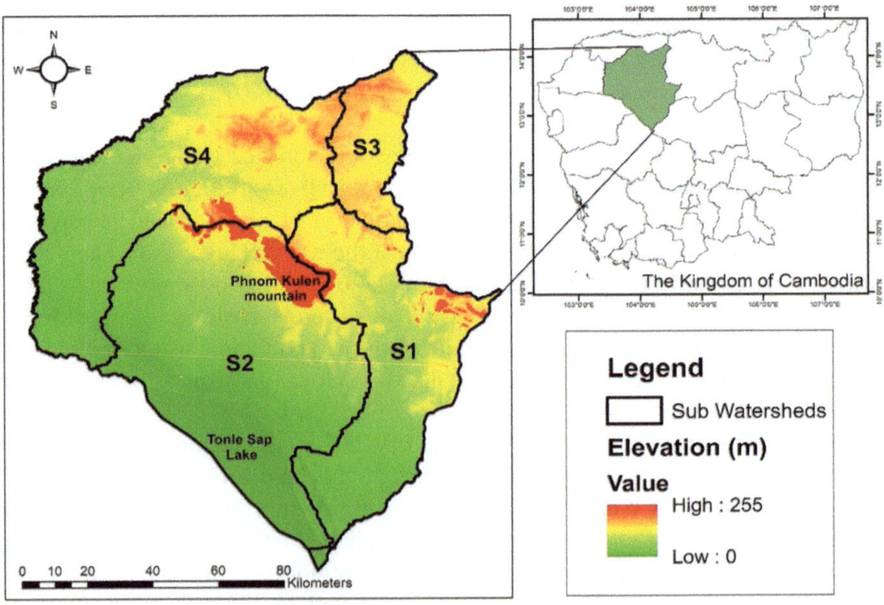

Fig. 11.1 Siem Reap province along with its four sub-watersheds and digital elevation model (DEM)

Cambodia was chosen for this study because it is an important part of Lower Mekong River Basin (LMRB). 45% of the Cambodian population constantly faces food insecurity, and the country ranks at 17.1 on the Global Hunger Index and has a score of 49.4 on Global Food Security Index 2019.

Table 11.1 presents the data, which is divided into three categories, including climate, geographic information system, and biophysical data. The first category includes the data about climate in which observed precipitation was obtained from Aphrodite for the duration of 1980 to 2014, whereas for future projections three GCMs were obtained having the data from 1980–2100. The second category has data from GIS which includes DEM, evapotranspiration, LULC maps, watershed shapefiles, plant available water content (PAWC), root restricting layer depth, etc. The third category includes the biophysical tables which were developed after extensive literature review.

11.3 Methodology

The study follows the overall methodology which is depicted in Fig. 11.2. First the future climate scenarios were projected using Coupled Model Intercomparison Project Phase 6 (CMIP6). Then baseline ecosystem models including annual water yield (AWY) and nutrient delivery ratio (NDR) were developed using InVEST. The models were then utilized to assess the past (2018) and present (2022) climate and land use conditions. And then finally, ecosystem services models were assessed under future climatic scenarios including near future (2023–2050), mid future (2051–2080), and far future (2081–2100).

11.3.1 Climate Change Projection

The selection of Global Climate Models (GCMs) was based on their ability to replicate historical precipitation patterns observed in the study area. Observed daily precipitation data from 1980 to 2014 were obtained from the Aphrodite dataset, and the compatibility of CMIP6 GCM outputs with this observed data was assessed. Three GCMs including NorESM2-MM, EC-EARTH3-CC, and EC-EARTH3 were selected based on their statistical performance using key evaluation metrics, coefficient of determination (R^2), root mean square error (RMSE), and standard deviation (STDV). Models with R^2 values greater than 0.80, low RMSE, and STDV closest to observed values were considered most suitable. Both visual inspection and statistical analysis were used to validate the consistency of these models with local precipitation trends.

To account for systematic biases and spatial resolution limitations inherent in GCM outputs particularly the discretization and averaging within coarse-resolution grid cells, bias correction was applied using the quantile mapping (QM) technique.

Table 11.1 Data and sources

S. No.	Data		Source	Spatial/temporal resolution	Time period	Related model
1	Climate	Observed precipitation	Aphrodite	0.5°/daily	1980–2014	NDR, AWY
		GCM models				
2		EC-Earth3	European consortium of national meteorological services	100 km/daily	1980–2014 2015–2100	NDR, AWY
3		EC-Earth3-CC	European consortium of national meteorological services	100 km/daily	1980–2014 2015–2100	NDR, AWY
4		NorESM2-MM	The Norwegian Earth System Model	100 km/daily	1980–2014 2015–2100	NDR, AWY
5	Geographic information system	Digital elevation model (DEM), SRTM	Earth Explorer, USGS	30 m	2018, 2022	NDR
6		Land use and land cover map	ESRI, Sentinel-2	10 m		NDR, AWY
7		Depth to root restricting layer	ISRIC, SoilGrids	250 m		AWY
8		Plant available water content (PAWC)	ISRIC, SoilGrids	250 m		AWY
9		Reference evapotranspiration	NASA, Global Land Data Assimilation System Version 2 (GLDAS-2)	0.25 × 0.25°/monthly		AWY
10		Watersheds	HydroSHEDS	15 arc s		NDR, AWY
11	Biophysical data	Root_depth	InVEST user's guide and Literature—Vatanakreasey et al. (2015), Jiraporn Kulsoontornrat (2021)	–		NDR, AWY
12		Kc				
13		load_n				
14		load_p				
15		eff_n				
16		eff_P				
17		crit_len_n				
18		crit_len_P				

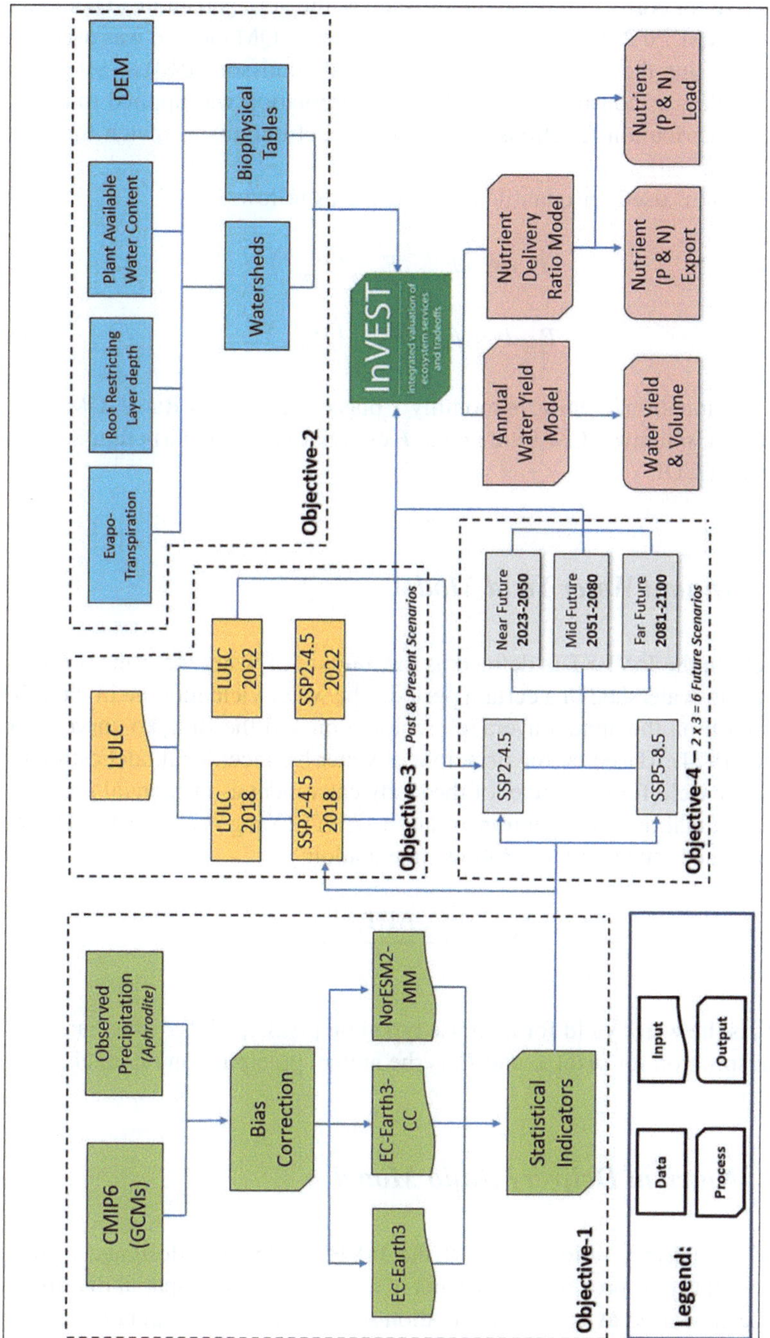

Fig. 11.2 Overall research methodology

QM is widely used to reduce biases in daily precipitation estimates and is particularly effective for correcting peak values, such as the 90th percentile (Themeßl et al. 2012; N'Tcha M'Po 2016). In this study, an empirical QM method was used, implemented in R using the qmap package (Lukas Gudmundsson 2014). A 99-percentile reference table was constructed, and linear interpolation was applied to adjust the cumulative distribution functions (CDFs) of model outputs to match those of the observed data.

The equations used for quantile mapping are as follows:

$$P_{\text{his}}d* = F^{-1}_{\text{obs},m}\, F_{\text{his},m}\left(P_{\text{his},m}\right) \tag{11.1}$$

$$P_{\text{sim}}d* = F^{-1}_{\text{obs},m}\, F_{\text{sim},m}\left(P_{\text{sim},m}\right) \tag{11.2}$$

where *P* precipitation, *d* daily, *m* monthly * bias corrected, *his* Raw GCM data, *obs* observed data, *sim* raw GCM future data, *F* cumulative distribution function (CDF), F^{-1} inverse of CDF.

11.3.2 Annual Water Yield Model

Water yield in the InVEST is defined as the amount of water that runs off the landscape to a sub-watershed in a certain period. The water yield module in the InVEST model is built on the annual average precipitation and the Budyko curve (Budyko and Miller 1974). Based on the principle of water balance, the model estimates the annual water yield for each pixel of the study catchment as the annual precipitation minus the annual actual evapotranspiration (AET) (Yang et al. 2019). The model algorithm is generated with the following equation:

$$Y_{xj} = \left[1 - \frac{\text{AET}_{xj}}{P_x}\right] \times P_x \tag{11.3}$$

where Y_{xj} is the water yield for land use type j on pixel *x*, AET_{xj} is the annual actual evapotranspiration for pixel *x*, and P_x is the annual precipitation on pixel *x*.

11.3.3 Nutrient Delivery Ratio Model

The nutrient delivery ratio (NDR) of the InVEST model is designed to map the source and transport process of nutrients in a river basin. The spatial distribution of nutrients can be used to assess the retention service of natural vegetation. Retention services play an important role in water quality. The model is mainly used for the simulation of nitrogen and phosphorus. The NDR model uses the mass conservation method to simulate the transfer of nutrients in space. Unlike other complex nutrient models, the model does not depict the details of the nutrient cycle and simulates the long-term stable movement of nutrients. The NDR model estimates the total removal amount of pollutants of nitrogen and phosphorus in the runoff by vegetation and soil and the final export amount of total nitrogen and total phosphorus of the grid to reflect its contribution to water purification (Leh et al. 2013; Redhead et al. 2018). The model uses the following equations for calculation:

$$X_{\text{expton}} = \sum_{i} X_{\text{expi}} \tag{11.4}$$

$$X_{\text{expi}} = \text{load}_{\text{surf},i} \times NDR_{\text{surf},i} + \text{load}_{\text{subs},i} \times \text{NDR}_{\text{subs},i} \tag{11.5}$$

where X_{expton} is the total export amount of nutrients in the river basin (kg year^{-1}), and X_{expi} is the export amount of each grid of nutrients (kg year^{-1}). The $\text{load}_{\text{surf,i}}$ is the surface nutrient load (kg ha^{-1} year^{-1}), $\text{NDR}_{\text{suf,i}}$ is the surface nutrient transfer rate, $\text{load}_{\text{subs,i}}$ is the subsurface nutrient load (kg ha^{-1} year^{-1}), and $\text{NDR}_{\text{subs,i}}$ is the subsurface nutrient transfer rate.

11.3.4 InVEST Model Assumptions

The InVEST AWY and NDR models are based on several key assumptions that affect the interpretation of results. For the AWY model, it is assumed that water yield is a function of long-term average precipitation and potential evapotranspiration, represented using the Budyko curve. This model does not consider sub-annual dynamics (e.g., seasonal runoff), groundwater flow, or surface water routing. Water yield is calculated as a static annual value, and spatial variability within a land cover type is not captured in detail.

The NDR model assumes that nutrient export depends primarily on land use specific loading rates and that nutrient transport follows a steady-state, mass-conservative process. It does not account for in-stream processes such as nutrient uptake, decay, or sedimentation. Vegetative retention and soil filtration are assumed to be spatially uniform within each LULC class, and the model does not simulate event-based pollutant dynamics.

These simplifications allow the InVEST models to function with relatively modest data requirements and make them appropriate for regional-scale assessments in data-scarce regions like Siem Reap. However, they also limit temporal resolution and omit detailed hydrological and biochemical interactions. Therefore, model outputs should be interpreted as indicative of spatial and relative trends, rather than precise forecasts.

11.3.5 Land Use and Land Cover

The LULC data used in this study were directly obtained from Esri's pre-classified Sentinel-2 land cover product, which provides annual global LULC maps at 10-m resolution. The dataset was clipped to the boundaries of the study watershed using a mask function in ArcGIS Pro. The original LULC classes were reclassified into broader categories required for use in the InVEST models such as water, crop land, forests, built-up areas, flooded vegetation, bare-land, and rangeland, based on their suitability and representation in the biophysical table. This ensured consistency across model inputs and compatibility with both the annual water yield (AWY) and nutrient delivery ratio (NDR) modules. The LULC in the study area consists of. The LULC maps were obtained for the years 2018 and 2022.

11.3.6 Climate Change and LULC Change Scenarios

Scenarios for the past, present, and future were developed to evaluate the impacts of climate change and LULC on ecosystem services. First, the past and present scenarios were developed for the years 2018 and 2022, respectively. The relevant land use maps for the years 2018 and 2022 were obtained from Esri and extracted using GIS and classification was done as mentioned above, then precipitation data (SSP2-4.5 scenario) were obtained from the best GCM along with other model parameters for AWY and NDR models, which includes evapotranspiration for the years 2018 and 2022, DEM, plant available water content (PAWC), root restricting layer depth, biophysical tables, watershed shapefiles, etc. The outputs were obtained from InVEST for two scenarios.

The future scenarios were divided into three parts, namely near future (NF: 2023–2050), mid future (MF: 2051–2080), and far future (FF: 2081–2100). All the required data were obtained as mentioned above including precipitation data from the best GCM for two pathways, namely SSP2-4.5 and SSP5-8.5. The outputs from the InVEST model were obtained in the form of annual water yield and nutrient delivery including nitrogen and phosphorus loads and exports for 6 future scenarios.

11.3.7 Alignment with Study Objectives

The methods adopted in this study were designed to directly support the research objectives. The use of CMIP6 GCM projections under SSP2-4.5 and SSP5-8.5 enabled us to project future climate conditions (Objective 1). The InVEST annual water yield (AWY) and nutrient delivery ratio (NDR) models were used to simulate baseline ecosystem services (Objective 2). Historical and present-day scenarios (2018 and 2022) were assessed using observed climate and LULC data to evaluate changes in ecosystem services over time (Objective 3). Future scenario modeling was conducted for three future time periods (NF, MF, FF) under different climate pathways to assess future impacts (Objective 4). Together, these methodological steps are tightly aligned with the study's aim to assess the biophysical impacts of climate and land use change on water-related ecosystem services supporting food security.

11.4 Results and Discussion

11.4.1 Siem Reap Climate Change Projection

The GCM selection in this study is based on the results of the statistical indicators for each model as shown in Table 11.2, ranked according to the target values in Sect. 11.3.1.

The GCM models are ranked based on performance through statistical indicators, with the best three models underlined. Overall, the best three performing models are NorESM2-MM, EC-EARTH3-CC, and EC-EARTH3 consecutively. According to the statistics, the NorESM2-MM performs exceptionally well. According to the overall trend analysis (Figs. 11.3 and 11.4), future precipitation in the Siem Reap is expected to decrease in NF for the SSP5-8.5 scenario, whereas for MF it will increase and then for FF it will again slightly decrease.

11.4.2 Past and Present Scenarios

For the past and present scenarios, it has been observed that LULC has changed from 2018 to 2022, as the amount of forests, vegetation, and rangeland decreased in 2022, while the amount of cropland and built-up areas increased significantly, as shown in Fig. 11.5. The increase in croplands and built-up areas can cause increase in pollutants while decrease in forests and vegetation will cause a drop in water yield.

The water yield has also decreased in 2022 scenario as depicted in Fig. 11.6 and Tables 11.3 and 11.4. LULC and climate change are the main culprits for this cause. As the amount of precipitation has decreased, and land use and cover have been altered as mentioned above, there was a 52–69% decrease in annual water yield in 2022. Hence, the same trend can continue depending on LULC and climate change.

As depicted in Figs. 11.7 and 11.8, Tables 11.5 and 11.6, the nitrogen and phosphorus exports have increased in 2022 because of the LULC. The natural vegetation and forests are transformed into croplands and built-up areas by deforestation. Hence, due to heavy application of fertilizers and pollution sources from built-up areas the amount of nitrogen and phosphors export have increased by 627 and 186

Table 11.2 Statistical indicators and rank of the bias-corrected model

Precipitation station	Indicator	Value			Rank		
		ECEARTH3	EC-EARTH3-CC	NorESM2-MM	ECEARTH3	EC-EARTH3-CC	NorESM2-MM
0	RMSE	67.965	63.659	61.603	3	2	1
	R^2	0.47	0.50	0.53	3	2	1
	STDV (79.38)	85.85	81.24	81.36	3	1	2
1	RMSE	62.769	62.643	57.189	3	2	1
	R^2	0.56	0.55	0.61	2	3	1
	STDV (82.50)	86.99	85.57	84.62	3	2	1
2	RMSE	73.673	70.490	67.253	3	2	1
	R^2	0.52	0.54	0.57	3	2	1
	STDV (92.23)	97.04	95.19	93.89	3	2	1
3	RMSE	65.845	61.072	62.838	3	1	2
	R^2	0.52	0.57	0.55	3	1	2
	STDV (84.46)	85.41	83.63	84.45	3	2	1
4	RMSE	68.367	62.780	63.931	3	1	2
	R^2	0.50	0.59	0.58	3	1	2
	STDV (88.86)	90.42	89.81	88.89	3	2	1
5	RMSE	71.318	63.879	62.954	3	2	1
	R^2	0.53	0.61	0.61	2	1	1
	STDV (91.34)	93.91	93.70	90.87	3	2	1

6	RMSE	64.087	59.510	60.453	3	1	2
	R^2	0.54	0.59	0.58	3	1	2
	STDV (85.08)	85.24	83.88	84.45	1	3	2
7	RMSE	69.06	64.95	64.72	3	2	1
	R^2	0.56	0.61	0.61	2	1	1
	STDV (93.54)	94.83	94.74	93.50	3	2	1
8	RMSE	69.422	64.118	61.461	3	2	1
	R^2	0.58	0.64	0.66	3	2	1
	STDV (96.60)	98.11	98.21	95.79	2	3	1
				Overall sum	75	48	35
				Rank	3	2	1

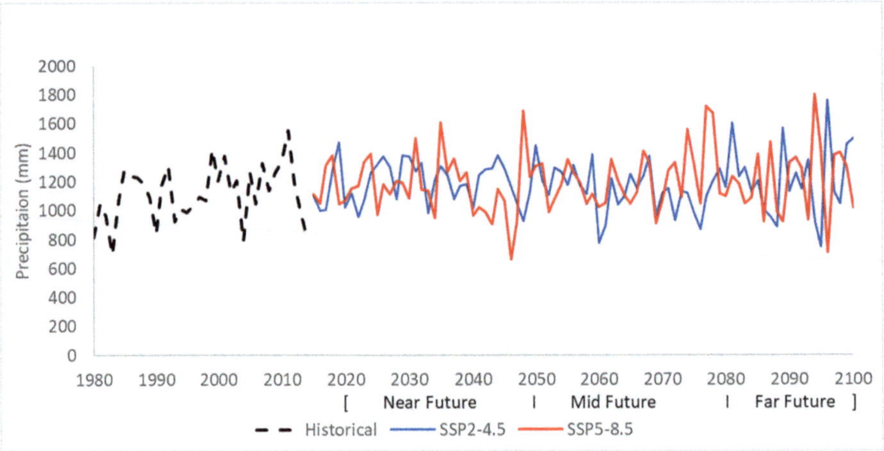

Fig. 11.3 Future climate projections in Siem Reap province

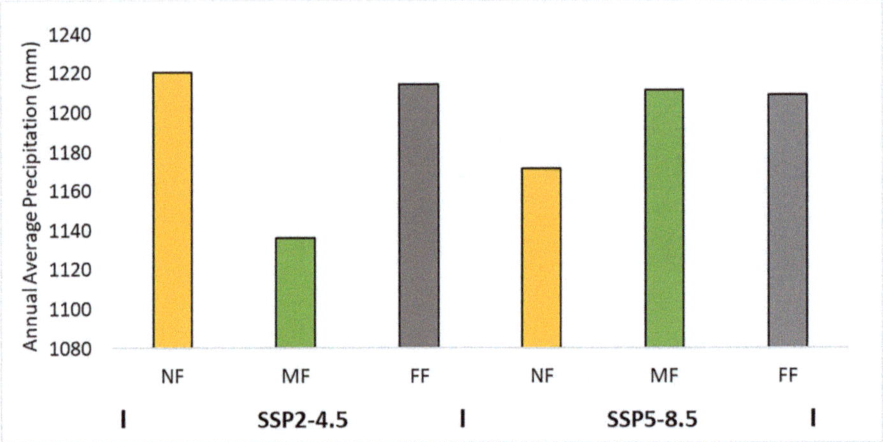

Fig. 11.4 Future precipitation trend for SSP2-4.5 and SSP5-8.5 pathways

tons per year, respectively. It can be observed that LULC is a major driver of nutrient delivery while climate change has an insignificant impact on it.

These findings demonstrate that land use change, particularly the expansion of croplands and built-up areas, is a primary driver of increased nutrient export, while reduced forest and vegetation cover contributes significantly to declining water yield.

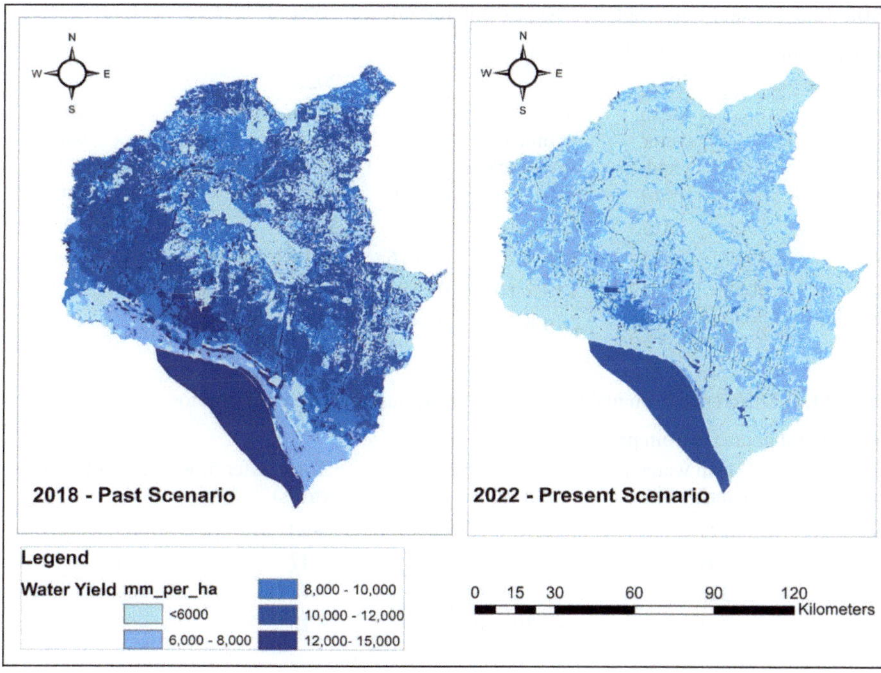

Fig. 11.5 Comparison of LULC between the past (2018) and present (2022) scenarios

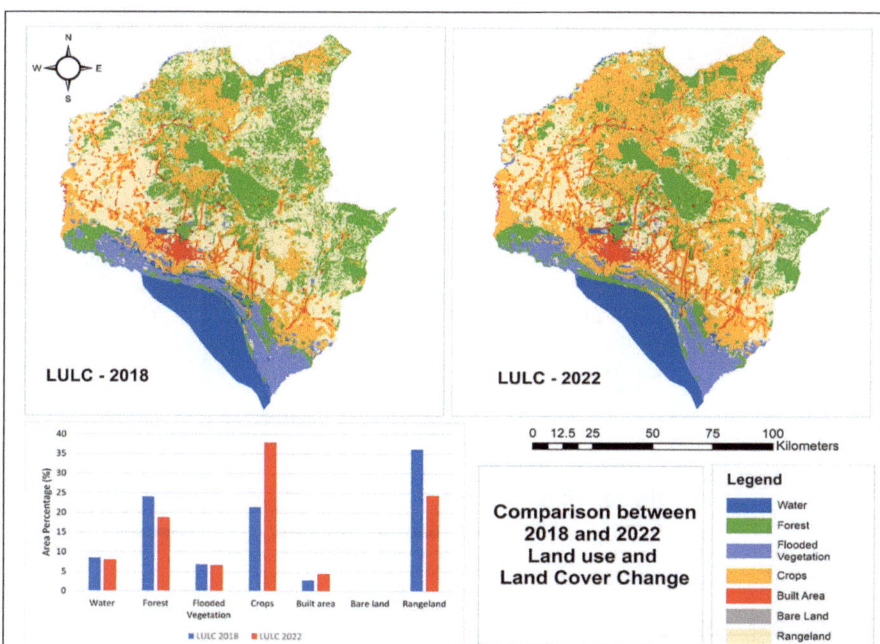

Fig. 11.6 Annual water yield for the past and present scenarios

Table 11.3 Annual water yield for the past and present scenarios

Invest model		Annual water yield					
Land use year		2018			2022		
Sub-watershed	Area (km^2)	Mean water yield (mm)	Total volume of water yield (m^3)	Total water flow (m^3/s)	Mean water yield (mm)	Total volume of water yield ($\times 10^6$ m^3)	Total water flow (m^3/s)
S3	966	790.72	7.64×10^8	24.21	511.38	4.94×10^8	15.66
S2	4917	768.59	3.78×10^9	119.76	469.08	2.31×10^9	73.09
S4	4059	771.17	3.13×10^9	99.18	455.50	1.85×10^9	58.58
S1	2940	735.01	2.16×10^9	68.48	483.55	1.42×10^9	45.05

Table 11.4 Difference in annual water yield for the past and present scenarios

Water yield decrease from past to present				
Sub-watershed	Mean water yield (mm)	Total volume of water yield (m^3)	Total water flow (m^3/s)	Overall decrease (%)
S3	−279	-2.70×10^8	−9	−55
S2	−300	-1.47×10^9	−47	−64
S4	−316	-1.28×10^9	−41	−69
S1	−251	-7.39×10^8	−23	−52

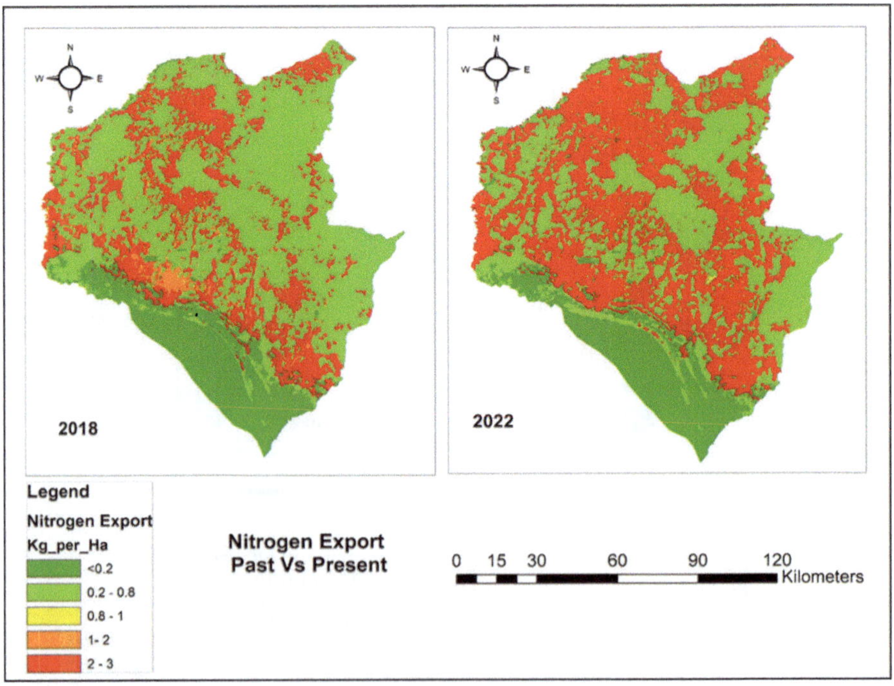

Fig. 11.7 Nitrogen exports for the past and present scenarios

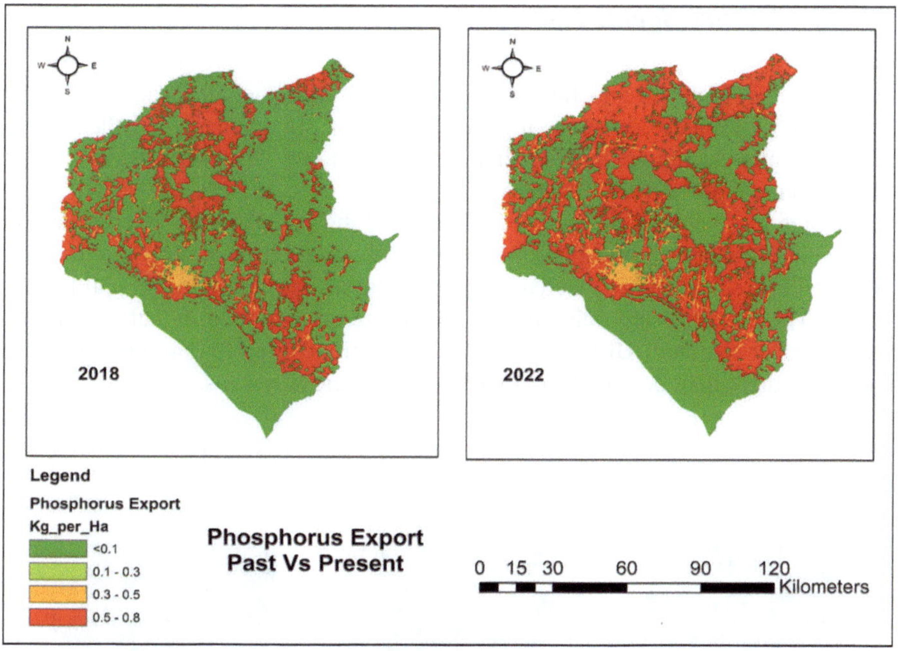

Fig. 11.8 Phosphorus exports for the past and present scenarios

Table 11.5 Nitrogen and phosphorus exports for the past and present scenarios

Invest model		Nutrient delivery ratio							
Land use year		2018				2022			
Sub-watershed	Area (km^2)	*N* load	*N* export	*P* load	*P* export	*N* load	*N* export	*P* load	*P* export
		Tons				Tons			
S3	966	372	60	56	10	660	138	149	33
S2	4917	2024	417	385	88	2579	578	558	134
S4	4059	2009	382	380	78	2809	619	648	150
S1	2940	1081	195	156	34	1650	346	335	78
Total	12883	5486	1054	977	210	7699	1681	1690	395

Table 11.6 Difference in nitrogen and phosphorus exports for the past and present scenarios

Invest Model		Nutrient delivery ratio			
		Nutrients increase from past to present			
		N load	*N* export	*P* load	P Export
Sub-watershed	Area (km^2)	Tons			
S3	966	+288	+78	+93	+23
S2	4917	+556	+161	+173	+46
S4	4059	+800	+237	+268	+72
S1	2940	+569	+151	+179	+44
Total	12883	+2213	+627	+713	+186

11.4.3 Future Scenarios

As depicted in Fig. 11.9 and Table 11.7, the annual water yield has decreased in NF for SSP5-8.5 scenario, whereas it has increased in MF scenario and then, it has insignificantly decreased in FF scenario. Hence, it can be observed that climate change is the major driver of water yield.

Figures 11.10 and 11.11, alongside Table 11.8 show that there is no considerable change in nitrogen and phosphorus exports. The exports have slightly decreased in NF for SSP5-8.5 scenario, whereas they have slightly increased in MF and FF scenarios. These exports are only because of change in precipitation patterns and for future scenarios. The LULC map of 2022 was utilized due to data unavailability, therefore, there may be no impact by LULC on future scenarios.

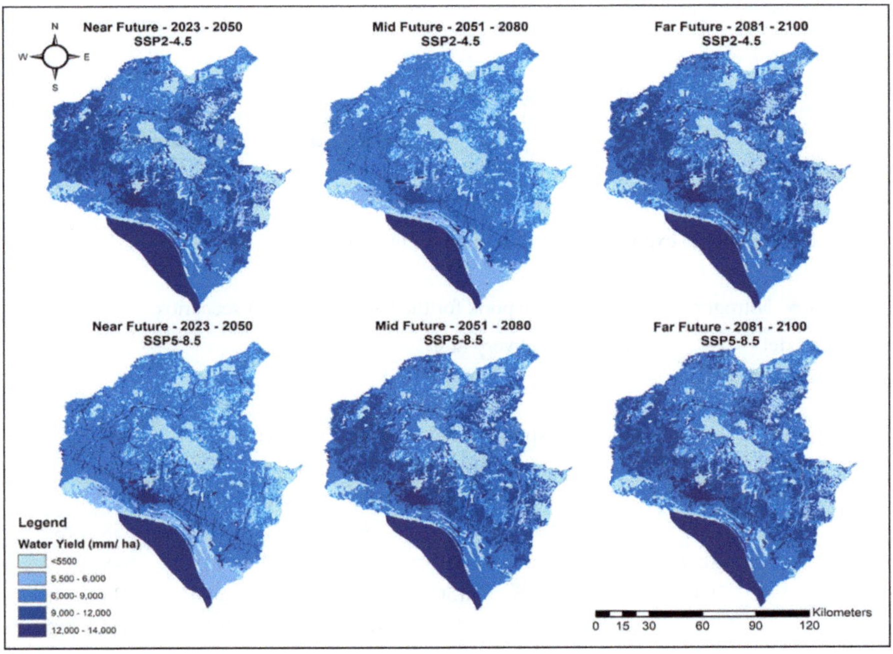

Fig. 11.9 Annual water yield for future scenarios

Table 11.7 Annual water yield difference from SSP2-4.5 to SSP5-8.5 scenario

Overall water yield difference from SSP2 to SSP5 (%)			
Sub-watershed	NF	MF	FF
S3	−7	+11	−1
S2	−6	+11	−1
S4	−6	+12	0
S1	−8	+10	−2

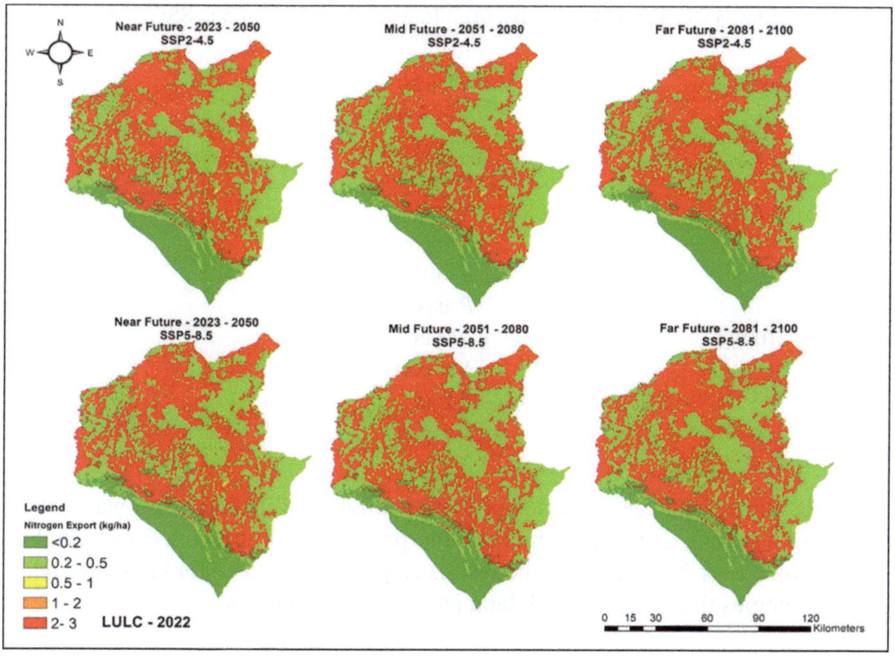

Fig. 11.10 Nitrogen exports for future scenarios

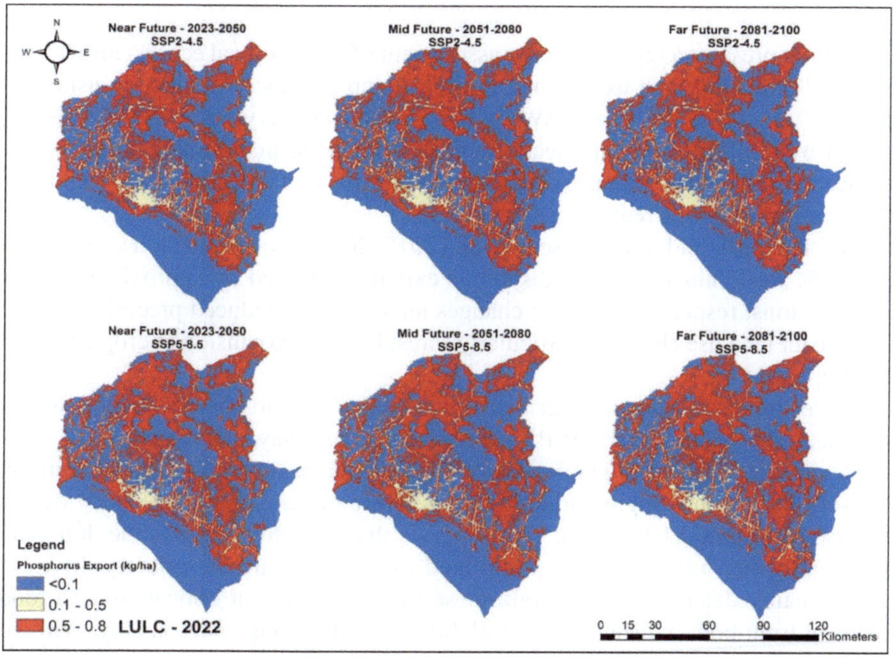

Fig. 11.11 Phosphorus exports for future scenarios

Table 11.8 Nitrogen and phosphorous exports difference from SSP2-4.5 to SSP5-8.5 scenario

Scenario	Tons	
NF (2023–2050)	*N* load	−2.15
	N export	−0.42
	P load	−0.36
	P export	−0.07
MF (2051–2080)	*N* load	+12.2
	N export	+3.0
	P load	+3.2
	P export	+0.8
FF (2081–2100)	*N* load	+10.1
	N export	+2.4
	P load	+2.9
	P export	+0.7

While climate change scenarios primarily influence fluctuations in water yield, the use of a static LULC map limits the interpretation of nutrient delivery trends in future projections. Nevertheless, the results emphasize the need for integrated adaptation strategies that consider both land and climate dynamics.

11.5 Conclusions and Recommendations

This study presents a region-specific assessment of hydrological ecosystem services under climate and land use change in Siem Reap province, Cambodia, using the InVEST annual water yield (AWY) and nutrient delivery ratio (NDR) models, driven by CMIP6 climate projections. It is among the first studies to isolate the respective impacts of climate change and land use dynamics on water yield and nutrient export in this region.

For the past and present scenarios (2018–2022), water yield decreased by 52–69%, while nitrogen and phosphorus exports increased by approximately 627 and 186 tons, respectively. These changes are driven by reduced precipitation and significant land use changes, particularly forest loss and expansion of croplands and built-up areas.

Under future scenarios, water yield shows sensitivity to climate change, with a decrease of 6–8% under the SSP5-8.5 near-future pathway, an increase of 10–12% in the mid future, and a slight decline of 1–2% in the far future. In contrast, nutrient delivery shows only slight variation across future climate pathways, likely due to the use of a static LULC map. This suggests that climate change is the dominant factor affecting water quantity, while land use changes are the primary driver of water quality deterioration. Overall, these findings highlight critical risks to food security from both reduced water availability and declining water quality. Without proactive intervention, these trends may undermine agricultural productivity and local ecosystem resilience.

The climate projections in this study were based on three GCMs, therefore, it is recommended that future studies consider using at least five GCMs to improve the robustness of climate simulations. Additionally, the inclusion of other InVEST models such as crop production, carbon storage and sequestration, and sediment delivery ratio could offer a more comprehensive evaluation of the impacts of climate change and land use change on ecosystem services relevant to food security. Moreover, while this study focused on biophysical drivers, incorporating socio-economic factors such as population growth, land tenure systems, agricultural practices, and local water governance could provide a more holistic understanding of the challenges and potential solutions. Integrating such human dimensions with ecosystem service modeling is essential for developing effective, context-specific adaptation strategies in Siem Reap province.

In addition to technical modeling recommendations, this study suggests several policy and practical actions for local stakeholders to mitigate the impacts of climate and land use change on water-related ecosystem services. For local governments, integrating the findings into land use planning and water resource management strategies particularly in agricultural and peri-urban zones is essential. Policies promoting afforestation, sustainable agriculture, and green infrastructure (e.g., buffer strips, wetlands, and infiltration zones) can help enhance water yield and reduce nutrient runoff. Community-based water conservation programs and improved fertilizer management practices should be encouraged to minimize nutrient loading. Moreover, collaboration among provincial agencies, research institutions, and civil society organizations will be vital to develop adaptive and inclusive strategies that align with both food security and climate resilience goals in Siem Reap province.

11.6 Limitations

This study has a few important limitations. First, due to the unavailability of projected land use data for Siem Reap province, the same LULC map from 2022 was used for all future scenarios. This static assumption does not account for dynamic land use transitions such as urban expansion, agricultural intensification, or deforestation, which may significantly influence water yield and nutrient export in the coming decades. However, this approach enabled the study to isolate and assess the impacts of climate change alone on hydrological ecosystem services.

Second, the analysis relied primarily on global open-access datasets, including the Aphrodite precipitation dataset, CMIP6 climate projections, and Esri's Sentinel-2 land cover product. While these datasets are widely used and valuable in data-scarce regions, the lack of ground-based hydrological and water quality measurements limited model calibration and validation. As a result, the outputs should be interpreted as indicative of spatial and relative trends rather than precise quantitative predictions.

Future research is encouraged to incorporate dynamic LULC projections using land change models or scenario-based planning and to integrate ground-based monitoring data to improve the accuracy and reliability of ecosystem service modeling in Siem Reap province.

Acknowledgments This study was conducted under the project "Safeguarding the Regional Food Security under Climate Change impacts via mainstreaming Nature-based Solutions-centered adaptation strategies (NAFOS)" funded by the Asia-Pacific Network for Global Change Research (APN).

References

Aghsaei, H., Mobarghaee Dinan, N., Moridi, A., Asadolahi, Z., Delavar, M., Fohrer, N. & Wagner, P. D. (2020). Effects of dynamic land use/land cover change on water resources and sediment yield in the Anzali wetland catchment, Gilan, Iran. *Science of The Total Environment*, *712*, 136449. https://doi.org/10.1016/j.scitotenv.2019.136449

Aliani, H., Malmir, M., Sourodi, M. & Kafaky, S. B. (2019). Change detection and prediction of urban land use changes by CA–Markov model (case study: Talesh County). *Environmental Earth Sciences*, *78*(17), 546. https://doi.org/10.1007/s12665-019-8557-9

Bagstad, K. J., Semmens, D. J., Waage, S. & Winthrop, R. (2013). A comparative assessment of decision-support tools for ecosystem services quantification and valuation. *Ecosystem Services*, *5*, 27–39. https://doi.org/10.1016/j.ecoser.2013.07.004

Bai, Y., Ochuodho, T. O. & Yang, J. (2019). Impact of land use and climate change on water-related ecosystem services in Kentucky, USA. *Ecological Indicators*, *102*, 51–64. https://doi.org/10.1016/j.ecolind.2019.01.079

Bangash, R. F., Passuello, A., Sanchez-Canales, M., Terrado, M., López, A., Elorza, F. J., Ziv, G., Acuña, V. & Schuhmacher, M. (2013). Ecosystem services in Mediterranean river basin: Climate change impact on water provisioning and erosion control. *Science of The Total Environment*, *458–460*, 246–255. https://doi.org/10.1016/j.scitotenv.2013.04.025

Boithias, L., Acuña, V., Vergoñós, L., Ziv, G., Marcé, R. & Sabater, S. (2014). Assessment of the water supply: demand ratios in a Mediterranean basin under different global change scenarios and mitigation alternatives. *Science of The Total Environment*, *470–471*, 567–577. https://doi.org/10.1016/j.scitotenv.2013.10.003

Bounoua, L., Nigro, J., Zhang, P., Thome, K. & Lachir, A. (2018). Mapping urbanization in the United States from 2001 to 2011. *Applied Geography*, *90*, 123–133. https://doi.org/10.1016/j.apgeog.2017.12.002

Butt, A., Shabbir, R., Ahmad, S. S. & Aziz, N. (2015). Land use change mapping and analysis using Remote Sensing and GIS: A case study of Simly watershed, Islamabad, Pakistan. *The Egyptian Journal of Remote Sensing and Space Science*, *18*(2), 251–259. https://doi.org/10.1016/j.ejrs.2015.07.003

Chen, J., Cui, T., Wang, H., Liu, G., Gilfedder, M. & Bai, Y. (2018). Spatio-temporal evolution of water-related ecosystem services: Taihu Basin, China. *PeerJ*, *6*, e5041. https://doi.org/10.7717/peerj.5041

Ghimire, S. R. & Johnston, J. M. (2019). Sustainability assessment of agricultural rainwater harvesting: Evaluation of alternative crop types and irrigation practices. *PLOS ONE*, *14*(5), e0216452. https://doi.org/10.1371/journal.pone.0216452

Gounand, I., Little, C. J., Harvey, E. & Altermatt, F. (2018). Cross-ecosystem carbon flows connecting ecosystems worldwide. *Nature Communications*, *9*(1), 4825. https://doi.org/10.1038/s41467-018-07238-2

Goyal, M. K. & Khan, M. (2017). Assessment of spatially explicit annual water-balance model for Sutlej River Basin in eastern Himalayas and Tungabhadra River Basin in peninsular India. *Hydrology Research, 48*(2), 542–558. https://doi.org/10.2166/nh.2016.053

Hao, H., Li, Y., Zhang, H., Zhai, R. & Liu, H. (2019). Spatiotemporal variations of vegetation and its determinants in the National Key Ecological Function Area on Loess Plateau between 2000 and 2015. *Ecology and Evolution, 9*(10), 5810–5820. https://doi.org/10.1002/ece3.5165

Hoyer, R. & Chang, H. (2014). Assessment of freshwater ecosystem services in the Tualatin and Yamhill basins under climate change and urbanization. *Applied Geography, 53*, 402–416. https://doi.org/10.1016/j.apgeog.2014.06.023

Hu, W., Li, G., Gao, Z., Jia, G., Wang, Z. & Li, Y. (2020). Assessment of the impact of the Poplar Ecological Retreat Project on water conservation in the Dongting Lake wetland region using the InVEST model. *Science of The Total Environment, 733*, 139423. https://doi.org/10.1016/j.scitotenv.2020.139423

Hua, A. K. (2017). Land Use Land Cover Changes in Detection of Water Quality: A Study Based on Remote Sensing and Multivariate Statistics. *Journal of Environmental and Public Health, 2017*, 1–12. https://doi.org/10.1155/2017/7515130

Huang, J., Zhan, J., Yan, H., Wu, F. & Deng, X. (2013). Evaluation of the Impacts of Land Use on Water Quality: A Case Study in The Chaohu Lake Basin. *The Scientific World Journal, 2013*, 1–7. https://doi.org/10.1155/2013/329187

Keeler, B. L., Polasky, S., Brauman, K. A., Johnson, K. A., Finlay, J. C., O'Neill, A., Kovacs, K. & Dalzell, B. (2012). Linking water quality and well-being for improved assessment and valuation of ecosystem services. *Proceedings of the National Academy of Sciences, 109*(45), 18619–18624. https://doi.org/10.1073/pnas.1215991109

Keller, A. A., Fournier, E. & Fox, J. (2015). Minimizing impacts of land use change on ecosystem services using multi-criteria heuristic analysis. *Journal of Environmental Management, 156*, 23–30. https://doi.org/10.1016/j.jenvman.2015.03.017

Kulsoontornrat, J. (2021). An optimal scenario of land use and land cover allocation to minimize sediment and nutrient export, *Upper Ing watershed, Phayao province*. Suranaree University of Technology.

Leh, M. D. K., Matlock, M. D., Cummings, E. C. & Nalley, L. L. (2013). Quantifying and mapping multiple ecosystem services change in West Africa. *Agriculture, Ecosystems & Environment, 165*, 6–18. https://doi.org/10.1016/j.agee.2012.12.001

Linke, P. (2014). *On the development of strategies for water and energy management in the context of the water-energy-food nexus* (pp. 196–201). https://doi.org/10.1016/B978-0-444-63433-7.50020-1

Liu, W., Zhan, J., Zhao, F., Yan, H., Zhang, F. & Wei, X. (2019). Impacts of urbanization-induced land-use changes on ecosystem services: A case study of the Pearl River Delta Metropolitan Region, China. *Ecological Indicators, 98*, 228–238. https://doi.org/10.1016/j.ecolind.2018.10.054

Lukas Gudmundsson. (2014). *qmap: Statistical transformations for postprocessing climate model output* (R package version).

M. I. Budyko and David H. Miller. (1974). *Climate and Life* (Vol. 508). Academic Press: New York, NY, USA.

Mishra, A., Kar, S. & Raghuwanshi, N. S. (2009). Modeling Nonpoint Source Pollutant Losses from a Small Watershed Using HSPF Model. *Journal of Environmental Engineering, 135*(2), 92–100. https://doi.org/10.1061/(ASCE)0733-9372(2009)135:2(92)

MRC. (2011). *Assessment of basin-wide development scenarios: main report.* .

MRC. (2015). *Vulnerability report volume 2: basin-wide climate change impact and vulnerability assessment for wetland dependent livelihoods and eco-services.*

Mushtaq, F. & Pandey, A. C. (2014). Assessment of land use/land cover dynamics vis-à-vis hydrometeorological variability in Wular Lake environs Kashmir Valley, India using multitemporal satellite data. *Arabian Journal of Geosciences, 7*(11), 4707–4715. https://doi.org/10.1007/s12517-013-1092-1

N'Tcha M'Po, Y. (2016). Comparison of Daily Precipitation Bias Correction Methods Based on Four Regional Climate Model Outputs in Ouémé Basin, Benin. *Hydrology, 4*(6), 58. https://doi.org/10.11648/j.hyd.20160406.11

Redhead, J. W., May, L., Oliver, T. H., Hamel, P., Sharp, R. & Bullock, J. M. (2018). National scale evaluation of the InVEST nutrient retention model in the United Kingdom. *Science of The Total Environment, 610–611*, 666–677. https://doi.org/10.1016/j.scitotenv.2017.08.092

Sample, J. E., Baber, I. & Badger, R. (2016). A spatially distributed risk screening tool to assess climate and land use change impacts on water-related ecosystem services. *Environmental Modelling & Software, 83*, 12–26. https://doi.org/10.1016/j.envsoft.2016.05.011

Santhi, C., Arnold, J. G., Williams, J. R., Dugas, W. A., Srinivasan, R. & Hauck, L. M. (2001). VALIDATION OF THE SWAT MODEL ON A LARGE RWER BASIN WITH POINT AND NONPOINT SOURCES. *Journal of the American Water Resources Association, 37*(5), 1169–1188. https://doi.org/10.1111/j.1752-1688.2001.tb03630.x

Schirpke, U., Kohler, M., Leitinger, G., Fontana, V., Tasser, E. & Tappeiner, U. (2017). Future impacts of changing land-use and climate on ecosystem services of mountain grassland and their resilience. *Ecosystem Services, 26*, 79–94. https://doi.org/10.1016/j.ecoser.2017.06.008

Schmalz, B., Kruse, M., Kiesel, J., Müller, F. & Fohrer, N. (2016). Water-related ecosystem services in Western Siberian lowland basins—Analysing and mapping spatial and seasonal effects on regulating services based on ecohydrological modelling results. *Ecological Indicators, 71*, 55–65. https://doi.org/10.1016/j.ecolind.2016.06.050

Shi, H., Wang, X., Xu, M., Zhang, H. & Luo, Y. (2017). Characteristics of soil C:N ratio and δ13C in wheat-maize cropping system of the North China Plain and influences of the Yellow River. *Scientific Reports, 7*(1), 16854. https://doi.org/10.1038/s41598-017-17060-3

Shrestha, M. K., Recknagel, F., Frizenschaf, J. & Meyer, W. (2017). Future climate and land uses effects on flow and nutrient loads of a Mediterranean catchment in South Australia. *Science of The Total Environment, 590–591*, 186–193. https://doi.org/10.1016/j.scitotenv.2017.02.197

Shukla P. S. J. B. E. M.-D. V. P. H. R. D. et al. (2019). *Climate Change and Land: an IPCC special report on climate change, desertification, land degradation, sustainable land management, food security, and greenhouse gas fluxes in terrestrial ecosystems.*

Song, W. & Deng, X. (2017). Land-use/land-cover change and ecosystem service provision in China. *Science of The Total Environment, 576*, 705–719. https://doi.org/10.1016/j.scitotenv.2016.07.078

Song, W., Deng, X., Yuan, Y., Wang, Z. & Li, Z. (2015). Impacts of land-use change on valued ecosystem service in rapidly urbanized North China Plain. *Ecological Modelling, 318*, 245–253. https://doi.org/10.1016/j.ecolmodel.2015.01.029

Su, C. & Fu, B. (2013). Evolution of ecosystem services in the Chinese Loess Plateau under climatic and land use changes. *Global and Planetary Change, 101*, 119–128. https://doi.org/10.1016/j.gloplacha.2012.12.014

Sukhdev, P. (2008). *The Economics of Ecosystems & Biodiversity: An Interim Report.*

Sun, S., Sun, G., Caldwell, P., McNulty, S., Cohen, E., Xiao, J. & Zhang, Y. (2015). Drought impacts on ecosystem functions of the U.S. National Forests and Grasslands: Part II assessment results and management implications. *Forest Ecology and Management, 353*, 269–279. https://doi.org/10.1016/j.foreco.2015.04.002

Themeßl, M. J., Gobiet, A., & Heinrich, G. (2012). Empirical-statistical downscaling and error correction of regional climate models and its impact on the climate change signal. *Climatic Change, 112*, 449–468. https://doi.org/10.1007/s10584-011-0224-4

Trisurat Y. (2013). Assessing ecosystem services at the Sakaerat Environmental Research Station, Nakhon Ratchasima Province (in Thai). *Journal of Forest Management, 6*, 1–17.

Trisurat, Y., Aekakkararungroj, A., Ma, H. & Johnston, J. M. (2018). Basin-wide impacts of climate change on ecosystem services in the Lower Mekong Basin. *Ecological Research, 33*(1), 73–86. https://doi.org/10.1007/s11284-017-1510-z

Trisurat, Y., Eawpanich, P. & Kalliola, R. (2016). Integrating land use and climate change scenarios and models into assessment of forested watershed services in Southern Thailand. *Environmental Research*, *147*, 611–620. https://doi.org/10.1016/j.envres.2016.02.019

Tue, N. T., Dung, L. V., Nhuan, M. T. & Omori, K. (2014). Carbon storage of a tropical mangrove forest in Mui Ca Mau National Park, Vietnam. *CATENA*, *121*, 119–126. https://doi.org/10.1016/j.catena.2014.05.008

Uwimana, A., van Dam, A. A., Gettel, G. M. & Irvine, K. (2018). Effects of agricultural land use on sediment and nutrient retention in valley-bottom wetlands of Migina catchment, southern Rwanda. *Journal of Environmental Management*, *219*, 103–114. https://doi.org/10.1016/j.jenvman.2018.04.094

Vatanakreasey, H., Kimseng, C., Sovann, P., Pheaktra, S. C., & Chhengngunn, A. (2015). *Applied invest tools for quantifying and mapping ecosystem services in Mondulkiri province*, Cambodia.

Wright, C. K. & Wimberly, M. C. (2013). Recent land use change in the Western Corn Belt threatens grasslands and wetlands. *Proceedings of the National Academy of Sciences*, *110*(10), 4134–4139. https://doi.org/10.1073/pnas.1215404110

Yang, D., Liu, W., Tang, L., Chen, L., Li, X. & Xu, X. (2019). Estimation of water provision service for monsoon catchments of South China: Applicability of the InVEST model. *Landscape and Urban Planning*, *182*, 133–143. https://doi.org/10.1016/j.landurbplan.2018.10.011

Yang, J., Xie, B., Zhang, D. & Tao, W. (2021). Climate and land use change impacts on water yield ecosystem service in the Yellow River Basin, China. *Environmental Earth Sciences*, *80*(3), 72. https://doi.org/10.1007/s12665-020-09277-9

[illegible]

Chapter 12
Microplastics Ingestion by Fish Species in Thailand's Aquatic Ecosystems

Khouloud Jaffel, Khawla Jaffel, Cristian Joana, and Arjay Ronald Hosmillo

12.1 Introduction

Marine plastic debris is no longer a distant threat; it has become an urgent global concern that severely impacts aquatic ecosystems and marine biodiversity. The gravity of this issue was first acknowledged at the 1984 International Workshop on the Fate and Impact of Marine Debris, where early research from the 1970s and 1980s examined the harmful interactions between plastics and marine life (Shomura and Yoshida 1985). In 1997, Laist documented the devastating consequences of marine plastic pollution, reporting cases of ingestion and entanglement affecting at least 267 marine species (Laist 1997). Alarmingly, this problem has only intensified since the post-World War II era, as the global production and mismanagement of durable plastic materials have led to widespread and persistent contamination of the world's oceans. By 2015, the number of affected species had risen to 557, including

K. Jaffel (✉)
School of Environment, Resources and Development (SERD), Asian Institute of Technology (AIT), Pathum Thani, Thailand

K. Jaffel
National Institute of Chemical Physics and Biophysics, Tallinn, Estonia

The Compact Muon Solenoid (CMS) experiment, The European Organization for Nuclear Research (CERN), Geneva, Switzerland

C. Joana
Institute of Theoretical Physics (ITP), Chinese Academy of Science (CAS), Beijing, China

A. R. Hosmillo
Analytics and Artificial Intelligence Association of the Philippines (AAP), Metro Manila, Philippines

Philippine Institute of Chemical Engineers (PIChE), Metro Manila, Philippines

A. Sharifi et al. (eds.), *Frameworks and Practices Towards a Sustainable and Peaceful Future*, World Sustainability Series,
https://doi.org/10.1007/978-981-95-2310-8_12

previously undocumented fish and invertebrates (Kühn et al. 2015), and more recent estimates suggest that over 700 species, including seabirds, fish, turtles, and marine mammals, are now known to ingest plastic debris (Kühn and van Franeker 2020).

Fish, in particular, have emerged as one of the most affected taxa, due to their ecological diversity, widespread distribution within aquatic environments, and essential roles in aquatic food webs. Notably, between 1972 and 2019, plastic ingestion was documented in 386 marine and estuarine species, including 210 species of commercial importance (Savoca et al. 2021), underscoring the potential ecological and economic implications of plastic pollution. Microplastics defined as particles smaller than 5 mm are now recognized as one of the most concerning forms of contamination, due to their small size and environmental persistence, and ability to be ingested across trophic levels. These particles are found throughout marine environments, from surface waters to the deep sea (Katija et al. 2017), making them widely accessible to fish via direct ingestion or through trophic transfer. By 2025, an estimated 250 million metric tons of plastic will enter the ocean from land-based sources (Jambeck et al. 2015), underscoring the growing ecological and economic threats posed by microplastic pollution in aquatic food systems.

Although the first documented cases of plastic ingestion by fish date back to the late 1960s (Kenyon and Kridler 1969) and early 1970s (Carpenter et al. 1972; Kartar et al. 1976), research on this issue remained limited until the past decade (Markic et al. 2020). In recent years, however, there has been a notable increase in studies conducted across various aquatic environments, including the world's oceans and seas (Davison and Asch 2011; Lusher et al. 2016), coastal areas (Sainio et al. 2021), estuaries (Pegado et al. 2018), freshwater systems such as lakes and rivers (Campbell et al. 2017; Khan and Setu 2022), and even under controlled laboratory conditions (Rochman et al. 2013). These investigations have documented plastic ingestion across a wide range of fish species, spanning diverse geographic regions and trophic guilds. While some studies have reported a lower incidence of plastic ingestion among aquaculture fish compared to wild fish (Cheung et al. 2018; Ibrahim et al. 2017), cases of plastic ingestion in farmed fish have nonetheless been observed, indicating that aquaculture systems are not entirely exempt from this form of contamination.

Plastic ingestion can occur through two primary pathways: direct (primary) ingestion, where plastics particles are mistaken for prey or consumed accidentally, and indirect (secondary) ingestion, which results from the consumption of contaminated prey (Ryan 2016). Secondary ingestion, also referred to as the trophic transfer of microplastics, occurs through the consumption of contaminated organisms across various levels of the food web (Au et al. 2017). Empirical studies across marine and freshwater ecosystems have demonstrated the trophic transfer of microplastics and nanoplastics among various aquatic organisms. Evidence spans multiple trophic levels, including transfer from blue mussels (*Mytilus edulis*) to crabs (Carcinus maenas) (Farrell and Nelson 2013), from mesozooplankton to macrozooplankton within planktonic food webs (Setälä et al. 2014), and from zooplankton (*Daphnia magna*) to fish (*Pimephales promelas*) in freshwater environments (Elizalde-Velázquez et al. 2020). Higher trophic-level organisms are also affected; for instance, microplastics have been detected in grey seals (*Halichoerus grypus*) as a result of consuming contaminated Atlantic mackerel (*Scomber scombrus*) (Nelms et al. 2018). Similarly, nanoplastic transfer has been observed across trophic links from algae to zooplankton and ultimately to fish in freshwater systems (Cedervall et al. 2012).

Despite the growing global evidence, studies on microplastic ingestion by fish in Thailand remain limited. Given the country's diverse aquatic environments and the increasing concern over the ingestion and trophic transfer of microplastics within the marine food web and its potential impact on human health, it is crucial to understand the extent of plastic contamination in local fish species.

This study aims to systematically review and synthesize existing literature on plastic ingestion by fish in Thailand aquatic ecosystems to identify key ecological and methodological factors influencing reported ingestion rates. Specifically, it addresses three research objectives: (i) to assess the prevalence of microplastic ingestion among fish species and examine its variation based on ecological traits such as trophic level and habitat type; (ii) to evaluate the influence of methodological approaches on reported ingestion rates and detection efficiency; (iii) and to explore the broader ecological and socio-economic implications of plastic ingestion for fisheries management, conservation strategies, and public health.

12.2 Methodology

12.2.1 Literature Search and Study Selection

This study employed a systematic literature review to synthesize existing evidence on plastic ingestion by fish in Thailand, following the Preferred Reporting Items for Systematic Review and Meta-Analyses (PRISMA) guidelines (Moher et al. 2010). A total of 18 peer-reviewed articles published between 2018 and 2024 were identified and analyzed. These compromised eight studies on marine fish, six on freshwater fish, and four on estuarine fish. Early studies from 2018 focused solely on marine ecosystems, whereas research from 2020 onward expanded to include freshwater and, subsequently, estuarine environments. Only two studies investigated the relationship between fish feeding habits and plastic ingestion (Hajisamae et al. 2022; Soe et al. 2022).

The literature search was conducted using two major bibliographic databases, ScienceDirect and Scopus, to search for relevant studies on plastic ingestion by fish in Thailand. The search strategy involved employing Boolean combinations of keywords such as fish (ingestion, consumption, contamination, occurrence, prevalence, or abundance), (plastic or microplastic), and (Thailand or Gulf of Thailand). This search returned 267 records, including 251 from ScienceDirect and 16 from Scopus. After removing 2 duplicates using Zotero, 265 unique records remained. Title, abstract, and keyword screening reduced this number to 41 studies addressing microplastic ingestion in fish within Thai aquatic ecosystems. Following full-text screening, 23 studies were excluded due to inappropriate methodology or insufficient data (Fig. 12.1).

Eighteen studies were selected for final inclusion based on the following criteria: (1) they investigated plastic or microplastic ingestion in fish species from Thailand; (2) they provided both quantitative and qualitative information on sampling methods and analytical procedures used for microplastic isolation; and (3) they reported measurable outcomes such as the frequency of microplastic occurrence, sample size, and species identification. Studies were excluded if they focused on organisms other than fish (e.g., invertebrates), lacked primary data or direct analysis of plastic ingestion, or did not meet the methodological and reporting standards required for synthesis (Hermsen et al. 2018; Lusher et al. 2017).

12.2.2 Data Collection

Both qualitative and quantitative data were systematically extracted from the selected studies. The variables and definitions used are summarized in Table 12.1. Data collation and preliminary analysis were conducted using Microsoft Excel. Additional biological and ecological attributes of the examined fish species such as

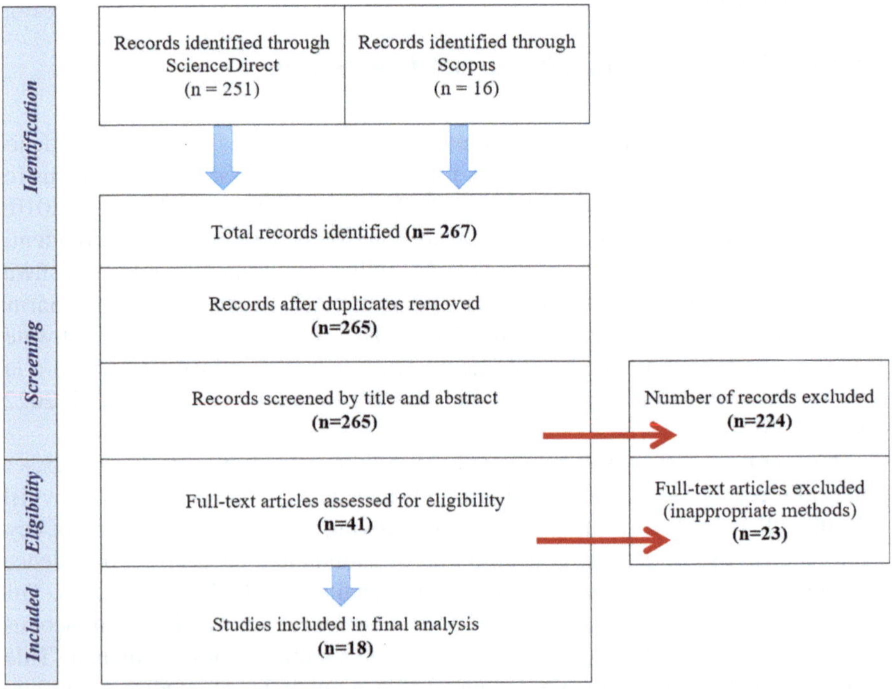

Fig. 12.1 PRISMA flow diagram illustrating the selection process of 18 peer-reviewed studies on plastic ingestion by fish in Thailand

Table 12.1 Overview of quantitative and qualitative variables extracted from 18 peer-reviewed studies on plastic ingestion in fish from Thailand

Variables	Description
Publication year	Year of publication of the study (2018–2024)
Search engine	Database used for literature retrieval: ScienceDirect and Scopus
Environment	Habitat type: freshwater, estuarine, or marine
Sampling method	Techniques employed for fish collection
Order	Taxonomic classification of fish at the order level
Family	Taxonomic classification of fish at the family level
Species	Scientific name of fish species
Common name	Widely recognized name of each fish species
Trophic level	Position of the fish species in the food web, ranging from 2 to 4.5
Habitats	Ecological classification: benthopelagic, demersal, pelagic, pelagic-neritic, or reef-associated
Sample size (*N*)	Total number of fish specimens examined for microplastic ingestion
FO %	Frequency of occurrence of microplastics in the examined samples
IUCN conservation status	Classification according to IUCN red list: vulnerable (VU), near threatened (NT), least concern (LC), data deficient (DD), or not evaluated (NE)
Anatomical compartments	Body parts analyzed for microplastic ingestion, including: whole-body samples; gastrointestinal tract (GIT); GIT and gills; GIT and tissue; gills, stomach, and tissue: stomach only; stomach and intestines; and tissue alone
Sample preparation	Methods used for microplastics extraction *Chemical Digestion*: 10% KOH; 10% KOH + 30% H_2O_2; 2M NaOH + 10M NaOH; 30% H_2O_2; NaOH 1 M and SDS 0.5% w/v *Visual identification (no digestion)*: Petri dish sorting; Sedgwick-Rafter chamber (1 mL)

taxonomic classification of fish (at the order and family levels), habitat type, trophic level, conservation status, and fisheries significance were obtained from authoritative sources, including FishBase (Froese and Pauly 2025), the World Register of Marine Species (WoRMS Editorial Board 2025), and the IUCN Red List of Threatened Species (IUCN 2025).

12.2.3 Meta-Analytical Approach

This study assesses plastic ingestion by fish species in Thailand's aquatic ecosystems using frequency of occurrence (FO%), also referred to as the ingestion rate.

Given that some fish species were investigated in multiple studies, we used each assessment, a single species examined in a single study as an independent data point. For each assessment, two key metrics were recorded

- Sample Size (N): The number of fish specimens examined.
- FO%: The percentage of those specimens in which microplastics were detected.

While some studies provided clear and complete data with sample size, number of fish with ingested microplastics, and FO%, others presented only partial data or reported FO% in narrative text. In some cases (e.g., Kasamesiri et al. 2021), FO% was presented only through bar charts, without any accompanying numerical values. In such instances, values were estimated using WebPlotDigitizer to extract data from visual plots. Assessments lacking either N or FO% were excluded from quantitative synthesis but remained for qualitative discussion.

To account for variability in sample sizes, a weighted averages FO% was calculated in two steps using the following formulas:

1. Species level: For species represented by multiple assessments, a species level FO% was obtained by weighting each study by its sample size

$$\text{Weighted Average FO\%}_{\text{species}} = \frac{\left(N_i \times \text{FO\%}_i\right)}{\Sigma N_i}$$

   where N_i sample size in assessment i (a single study for a specific species), $\text{FO\%}_i$ frequency of occurrence in assessment i, $\sum N_i$ total sample size across all assessments for that species.
2. Group level: To enable broader ecological and methodological comparisons, species-level FO% values were further aggregated across various categorical groupings, including taxonomic order, habitat type, trophic level, IUCN conservation status, human use, MP extraction methods, and anatomical compartments examined. Within each group, a weighted average FO% was calculated using the same weighting principle, where each species' FO% was weighted by its total sample size:

$$\text{Weighted Average FO\%}_{\text{group}} = \frac{\Sigma\left(N_j \times \text{ Weighted Average FO\%}_{\text{species},j}\right)}{\Sigma N_j}$$

   where N_j total sample size for species j (already aggregated from multiple assessments), weighted average; $\text{FO\%}_{\text{species,j}}$ weighted average FO for species j; $\sum N_j$ total sample size across all species within a group.

This weighting approach ensures that assessments with larger sample sizes contribute more substantially to the overall ingestion rate estimate, thereby reducing bias from small-sample studies and enhancing the reliability of the calculated central tendency.

12.3 Results

12.3.1 Sampling Framework

12.3.1.1 Sampling Methods

The reviewed studies employed a variety of sampling approaches, reflecting diverse research objectives, target species, and geographical settings. Fish specimens were sourced from local wet markets or fishing piers, particularly when studies focused on commercially important species.

Different fishing techniques were used across other studies, including gill nets, trawling, fish traps, and mackerel net. These methods were often employed by local fishermen using small-scale traditional vessels.

In terms of temporal scope, most studies conducted sampling over short durations or a few months, which limits the ability to capture potential seasonal fluctuations in microplastic ingestion rates, particularly those influenced by monsoon-driven runoff, fishing intensity, or changes in fish feeding behavior across different times of the year.

Only two studies (Hajisamae et al. 2022; Soe et al. 2022) conducted sampling over periods exceeding 12 months, offering potentially more representative and comprehensive data.

12.3.1.2 Sample Size and FO% Patterns

Sampling effort varied considerably across the reviewed studies, highlighting significant methodological heterogeneity in both species' coverage and sample size. The number of fish species analyzed per study ranged from a single species (e.g. (Jitkaew et al. 2024) to as many as 34 species (Soe et al. 2022). Across all studies, 125 distinct fish species were assessed for microplastic ingestion. However, repeated investigations of 39 species across different studies and regions resulted in a total of 176 species-specific assessments. Sample sizes of these species-specific assessments ranged widely from $N = 1$ to $N = 522$, with a mean of 75.15 ± 109.58, reflecting differing research objectives. For instance, Azad et al. (2018b) analyzed only three specimens of short mackerel (*Rastrelliger brachysoma*), while Soe et al. (2022) examined 522 specimens of the same species. One assessment of Osteogeneiosus militaris (Pradit et al. 2023) was excluded from statistical analyses due to the absence of species-level aggregation, as FO% values were reported separately for individual anatomical compartments (gills, stomach, and tissue). Another assessment of Arius maculatus was conducted by (Jitkaew et al. 2024) and (Pradit et al. 2021); however, only the data reported by (Jitkaew et al. 2024) were included in the analyses, as (Pradit et al. 2021) did not report the number of fish with ingested microplastics or FO% values. Nonetheless, both studies were retained for qualitative analysis.

Across the 175 species-specific assessments, a total of 5877 fish specimens were examined. Ingestion levels were categorized into four categories based on weighted average FO% (Fig. 12.2): no ingestion (0%), low (<10%), moderate (10–50%), and high (>50%). High ingestion was reported in 67 assessments (50 species, $N = 733$) with a weighted average FO% of 81.04%. Moderate ingestion was observed in 52 assessments (34 species, $N = 1162$) with a weighted average FO% of 18.67%. Twenty assessments (20 species, $N = 543$) reported no ingestion, while 36 assessments (21 species, $N = 3439$) were classified as low ingestion category, yielding the lowest weighted average FO% (5.15%) despite comprising the largest total sample size.

To examine the relationship between sample size and ingestion frequency, a scatter plot analysis was conducted (Fig. 12.3), revealing a weak negative linear correlation ($y = -0.0011x + 0.3514$; $R^2 = 0.1722$). This trend was statistically validated by Spearman's rank correlation test (rs = −0.415; $N = 61$; $T = 3.503$; $df = 59$; $p = 0.001 < 0.05$). These findings suggest that higher ingestion rates are more frequently reported in assessments with smaller sample sizes, whereas larger samples are generally associated with lower estimates.

These disparities in sampling effort and species representation underscore key limitations in data comparability and potential bias that may influence the interpretation of ingestion frequency patterns. Therefore, methodological variability, particularly in sample size and species selection, should be acknowledged when synthesizing results across studies.

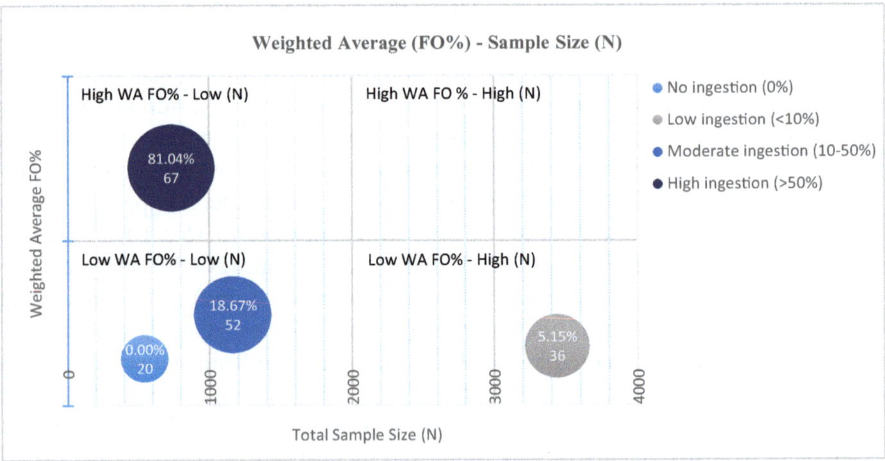

Fig. 12.2 Bubble chart showing ingestion level categories based on sample size (N) and weighted average FO%

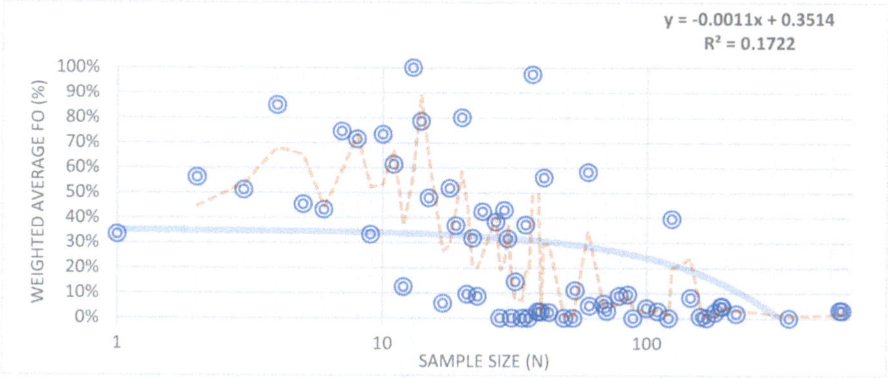

Fig. 12.3 Scatter plot depicting the relationship between sample size (N) and weighted average FO% of microplastic ingestion

Fig. 12.4 Taxonomic distribution of fish species examined for microplastic ingestion

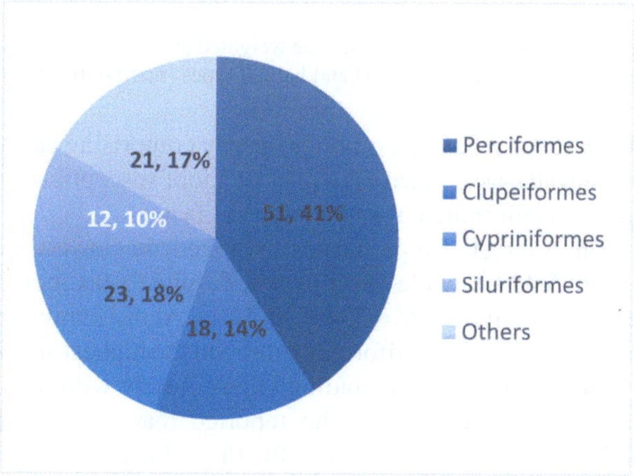

12.3.2 *Taxonomic Classification of Species*

Microplastic ingestion was assessed in a total of 125 fish species, comprising 77 marine, 47 freshwater, and 1 estuarine species, representing 15 taxonomic orders and 47 families. The most represented order was Perciformes, accounting for 51 species (40.80%), followed by Cypriniformes (23 species, 18.40%), Clupeiformes (18 species, 14.40%), and Siluriformes (12 species, 9.60%), with the remaining 11 orders were represented in lower percentages (Fig. 12.4).

Microplastic ingestion was detected in 105 out of the 125 species (84%), with considerable variation in the weighted average FO% across taxonomic orders (Fig. 12.5). Cypriniformes, composed exclusively of freshwater species, exhibited the highest weighted average FO% (N = 346; 77.75%), followed by Siluriformes

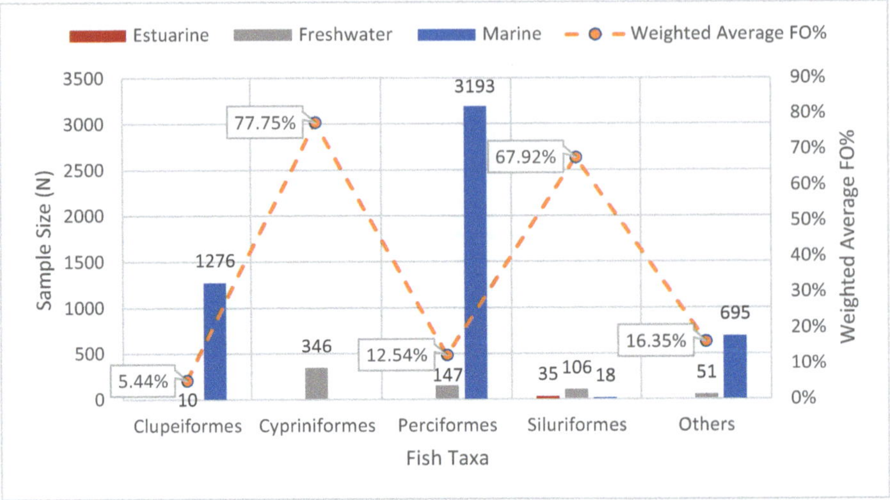

Fig. 12.5 The graph illustrates the weighted average FO% across four major fish orders, with corresponding sample sizes (N) and habitat types (marine, freshwater, estuarine)

(N = 159; 67.92%), a group primarily comprising catfish. These results suggest that freshwater-associated taxa may be more susceptible to microplastic contamination due to environmental and ecological factors.

In contrast, Clupeiformes exhibited the lowest weighted average FO% (N = 1286; 5.44%), while Perciformes, despite its taxonomic richness, showed a moderate ingestion rate (N = 3340; 12.54%). At the family level, Cyprinidae (minnows or carps) within Cypriniformes exhibited the highest species-level contamination, with 16 affected taxa. This finding aligns closely with the observations by Sawangproh and Paejaroen (2025), who reported that Cyprinidae had the most widespread microplastic contamination, affecting 33 taxa.

These patterns likely reflect both ecological and methodological influences. Freshwater ecosystems may act as sinks for microplastics due to their proximity to pollution sources. Additionally, feeding behavior of certain species may further increase ingestion risk (Sutthacheep et al. 2018). Nevertheless, methodological differences among studies such as variations in sample sizes, as illustrated in Fig. 12.5, likely contributed to the observed disparities.

12.3.3 Fisheries and Livelihoods

The occurrence of microplastic ingestion varied notably across different fisheries-use categories. Fish species harvested by subsistence fisheries exhibited the highest weighted average frequency of occurrence (FO%) at 51.85%, followed by commercial (24.84%), minor commercial (18.16%), and highly commercial species, which showed the lowest ingestion rate at 7.30%. Ten species categorized as "of no interest" or "not applicable (n/a)" were excluded from the human uses analysis due to small sample sizes and unclear classification. These findings suggest that fish harvested through traditional or small-scale subsistence fisheries may face higher risks of plastic ingestion, potentially due to their greater exposure to nearshore pollution hotspots. This raises concerns not only about food safety but also about the health and livelihoods of coastal communities who rely heavily on these fisheries for nutrition and income.

12.3.4 Conservation Status

The conservation assessment of 125 fish species based on the IUCN Red List reveals a broad distribution across risk categories. Most species (108 species, 86.40%) were classified as Least Concern (LC), while 4 were Not Evaluated (NE), 7 were Data Deficient (DD), 2 were Near Threatened (NT), and 4 were Vulnerable (VU). This distribution highlights a predominant research focus on species considered to be at lower extinction risk.

Despite this dominance, microplastic ingestion is observed across all conservation categories, with an overall weighted average Frequency of Occurrence (FO%) of 16.81%.

Notably, species categorized as NE exhibited the highest weighted average FO% at 25.00%, despite comprising only 3.20% of the examined species. This suggests a disproportionate exposure to plastic pollution among taxa whose extinction risk has yet to be evaluated.

Among LC species ($N = 3948$), the weighted average FO% was 20.59%, indicating plastic ingestion is widespread even among species presumed to be at lower conservation concern. Similarly, DD species displayed a weighted average FO% of 14.38%, raising concerns about potential underestimation of threats among taxa with insufficient ecological or conservation data. Conversely, VU and NT species exhibited lower ingestion frequencies, with weighted average FO% values of 8.78% and 1.08%, respectively.

The disparity in sampling effort across species is evident. For instance, Rastrelliger brachysoma, a vulnerable species with high commercial and ecological significance in Southeast Asia and Melanesia, accounts for 1220 out of the 1253 VU specimens, which were assessed across four studies, whereas the Harpadon nehereus, a commercially significant and near threatened species, was only represented by only 22 specimens in a single study out of 370 NT. Such inconsistencies can over- or underestimate category-level FO % and obscure actual patterns of ecological risk.

12.3.5 Ecology

12.3.5.1 Habitats

The prevalence of microplastic ingestion among fish species varies across different habitat types. Of the 125 species analyzed, 105 species (84%) exhibited microplastic ingestion, while 20 species (16%) showed no evidence of ingestion. Benthopelagic species had the highest occurrence, with 33 species affected, followed closely by demersal species (30 species) and reef-associated species (25 species). Pelagic-neritic species exhibited a moderate level of plastic ingestion (15 species), whereas strictly pelagic species showed the lowest occurrence (2 species) (Fig. 12.6).

Microplastic ingestion rates varied significantly across habitats. Benthopelagic species exhibited the highest ingestion rates (mean = 71.47%, median = 90.21%), followed by pelagic species (mean = 54.0%, median = 54.0%). Demersal species showed moderate ingestion rates (mean = 36.40%, median = 25.0%), while reef-associated (mean = 26.20%, median = 15.88%) and pelagic-neritic species (mean = 26.96%, median = 2.48%) had the lowest levels of contamination. Benthopelagic fish exhibited an interquartile range (IQR) of 46.46%, with a lower whisker (1.5 × IQR) at −16.16%, whereas reef-associated fish had an IQR of 19.38%, with an upper whisker at 58.44%, indicating potential extreme ingestion cases (Fig. 12.7).

These findings suggest that benthopelagic species are at a higher risk of microplastic ingestion than other habitat groups, likely due to their benthic and near-bottom foraging strategies, which increase their exposure to sinking plastic debris. This aligns with previous research (Keerthika et al. 2023) indicating that benthic species accumulate a higher abundance of microplastics in their gastrointestinal tracts compared to pelagic species, reinforcing the link between habitat-specific feeding strategies and microplastic exposure risk.

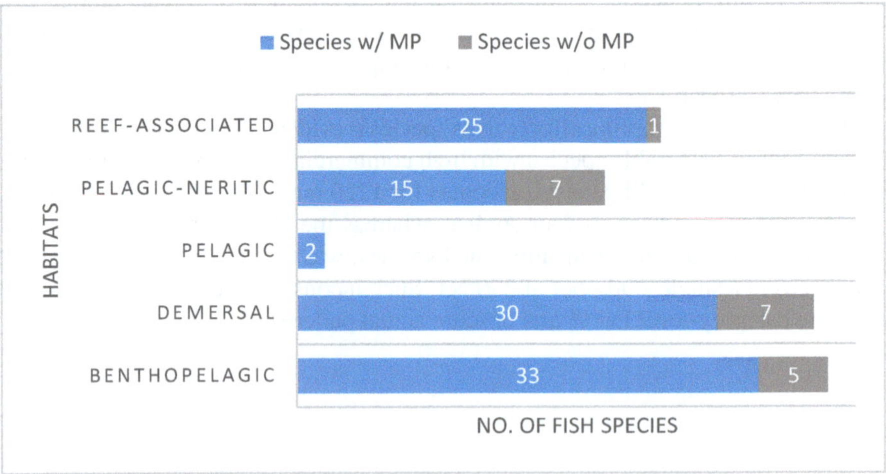

Fig. 12.6 Number of species across different habitat types with and without ingested microplastics

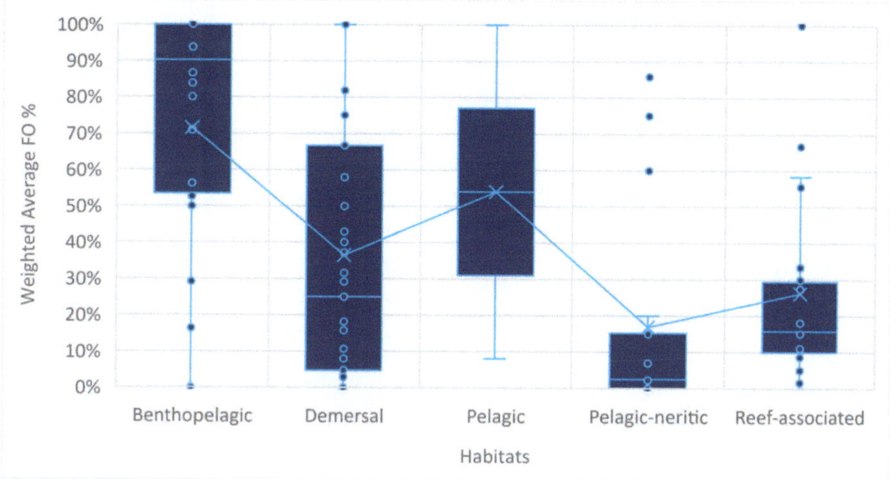

Fig. 12.7 Box plots illustrate plastic ingestion frequency across fish species and habitats. The horizontal line is the median, the bars represent the interquartile ranges (IQR), and the whiskers extend to 1.5 × IQR

Similarly, pelagic species also exhibit relatively high ingestion rates, potentially due to their interaction with buoyant plastic particles suspended in the water column (Rummel et al. 2016). In contrast, the lower ingestion rates in pelagic-neritic and reef-associated species may reflect differences in feeding strategies, habitat characteristics, and localized plastic accumulation patterns.

12.3.5.2 Trophic Level

The weighted-average frequency of occurrence FO% declines markedly with increasing trophic level (Fig. 12.8).

A linear regression analysis describes this negative correlation with the equation $y = -0.2514\ x + 1.2542$, where y represents the weighted average FO% and x denotes the trophic level. The model yielded an R^2 of 0.1775, indicating that trophic level explains approximately 17.75% of the variation in FO%

The negative slope (−0.2514) suggests that lower-trophic-level species are more susceptible to direct plastic ingestion (Walkinshaw et al. 2020), while higher-trophic-level species may ingest plastics indirectly through trophic transfer mechanisms (Kangas et al. 2023; Wang et al. 2021).

This inverse relationship was further supported by a statistically significant Spearman's rank correlation analysis, ($rs = -0.375$; $N = 84$; $T = 3.66$; df = 82; $p = 0.0004 < 0.05$), reinforcing that plastic ingestion declines at higher trophic positions.

A categorical analysis of trophic groups highlights this trend. Herbivorous and detritivorous fish at the lowest trophic levels (2.0–2.5) exhibited the highest weighted average FO% at 67.80%, while top predators (4.0–4.5) showed the lowest weighted

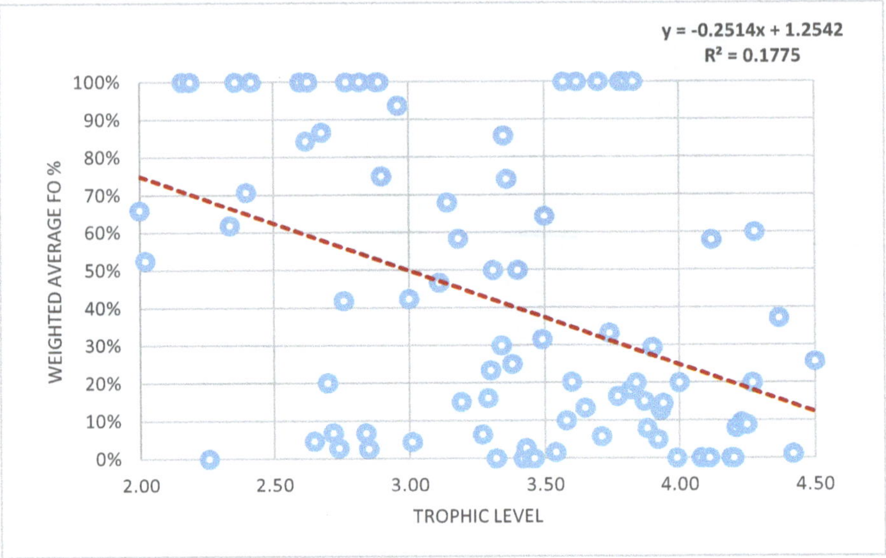

Fig. 12.8 The scatter plot illustrates the correlation between plastic ingestion frequency and the trophic levels of fish species

average FO% at only 7.65%. Intermediate trophic levels displayed variables ingestion rates, with FO% values of 12.79% (2.5–3.0), 23.21% (3.0–3.5), and 19.08% (3.5–4.0).

These results indicate that trophic position is a significant predictor of microplastic ingestion, highlighting a greater direct exposure risk among low-trophic-level species, which underscore the need to consider trophic structure in future risk assessments.

12.3.6 Analytical Methods

12.3.6.1 Sample Preparation

This systematic review reveals that microplastic detection in fish samples is significantly influenced by both the extraction method and the anatomical compartments examined, as summarized in Table 12.2 and Fig. 12.9.

Analytical approaches are categorized into chemical digestion and visual identification without chemical treatment. Chemical digestion breaks down organic matter while preserving plastic particles without damaging plastic polymers (Pfeiffer and Fischer 2020), with common digestion solutions including 30% hydrogen peroxide (H_2O_2), 10% potassium hydroxide (KOH), mixtures of 10% KOH + 30% H_2O_2, sequential treatments of 2M and 10M sodium hydroxide (NaOH), and an alkaline-detergent mixture of NaOH (1M) with sodium dodecyl sulfate (SDS, 0.5%

Table 12.2 Effectiveness of chemical digestion protocols and visual sorting techniques in detecting microplastics within fish anatomical compartments

		Whole-body		GI tract		GI tract and gills		GI tract and tissue		Gills, stomach, and tissue		Stomach		Stomach and intestines		Tissue		Total	
		WA (FO %)	*N*	WA (FO %)	*N*	WA (FO %)	*N*	WA (FO %)	*N*	WA (FO %)	*N*	WA (FO %)	*N*	WA (FO %)	*N*	WA (FO %)	*N*	Weighted average (FO%)	Sample size (*N*)
Chemical digestion	10% KOH	8.00	100					54.17	48	100.00	35							37.70	183
	10% KOH + 30% H_2O_2					13.14	274											13.14	274
	2M NaOH + 10M NaOH			8.33	492													8.33	492
	30% H_2O_2			69.55	266	34.72	72					100.00	30	87.23	274			74.61	642
	1 NaOH + 0.5% SDS (w/v)															46.86	239	46.86	239
No digestion	Sedgwick-Rafter Chamber (1 ml)			3.14	541							2.07	3236					2.22	3777
	Petri dish visual sorting											61.85	270					61.85	270
Total		8.00	100	18.71	1299	17.63	346	54.17	48	100.00	35	7.47	3536	87.23	274	46.86	239	16.81	5877

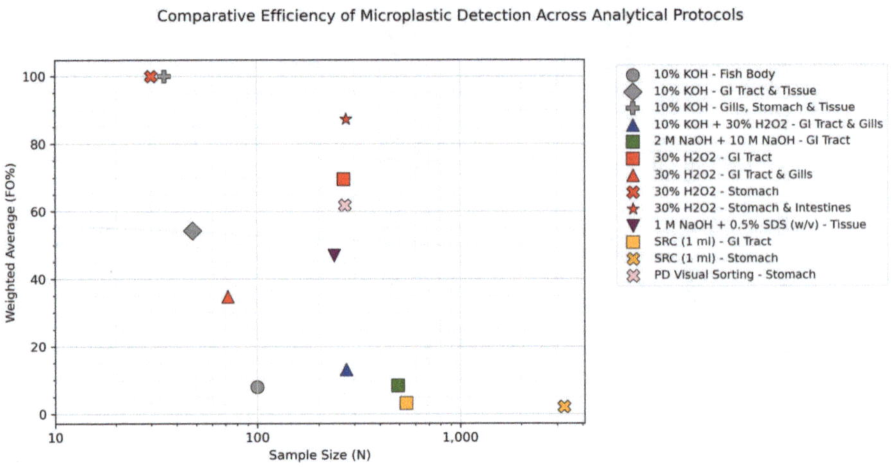

Fig. 12.9 The scatter plots illustrate the comparative efficiency of microplastic detection across analytical protocols

w/v). In contrast, visual identification methods rely on direct examination without chemical treatment, such as petri dish-based visual sorting (Azad et al. 2018a, b) and the Sedgwick-Rafter chamber technique, which involves a standardized 1 mL subsample for microscopic observation (Hajisamae et al. 2022; Soe et al. 2022).

Overall, chemical digestion methods yielded a substantially higher weighted average FO% (40.27%, N = 1830) than visual approaches (6.20%, N = 4047). Although visual methods avoid potential polymer degradation, they may underestimate microplastic abundance due to the absence of biological compartment organic matter digestion and the limited subsample volume, particularly evident with the Sedgwick-Rafter chamber which yielded a weighted average FO% of just 2.22 % (N = 3777). This indicates that chemical digestion enhances microplastic extraction and identification, potentially revealing particles that may be overlooked in direct visual examinations.

The efficiency of microplastic detection also varied among different digestion protocols. Lower FO% values were observed with 10% KOH + 30% H_2O_2 (13.14%, N =274) (Phaksopa et al. 2021), and with sequential digestion using 2M NaOH followed by 10M NaOH (8.33%, N = 492) (Klangnurak and Chunniyom 2020). In contrast, the highest detection rates were associated with 30% hydrogen peroxide (H_2O_2), NaOH (1M) and SDS (0.5% w/v), and 10% KOH, yielding weighted average FO% values of 74.61% (642 samples), 46.86% (239 samples), and 37.70% (183 samples), respectively.

This methodological effect may partly account for the high FO% observed in Cypriniformes (77.75%), a benthopelagic freshwater group, in which all assessments employed 30% H_2O_2. Similarly, Siluriformes, predominately composed of catfish species, exhibited a high FO% of 67.92%, based on assessments that utilized either 30% H_2O_2, 10% KOH, or a mixture of NaOH (1M) + SDS (0.5% w/v) as digestion solutions.

These findings highlight the critical role of digestion protocols and underscore that reported prevalence reflects both biological exposure and the sensitivity of the extraction technique; hence, cross-study comparisons must account for methodological heterogeneity.

12.3.6.2 Anatomical Compartments

In addition to extraction techniques, the frequency of microplastic occurrence (FO%) also varies significantly across anatomical compartments, reflecting differences in particle retention and detection efficiency.

Whole-body analyses using 10% KOH and 0.7 μm filtration yielded the lowest FO% (8.00%, $N = 100$), likely due to dilution effects from averaging microplastic content across the entire organism (Pradit et al. 2022).

Similarly, stomach analyses showed low FO% (7.47%, $N = 3536$), which may be attributed to the predominant use of visual identification methods without prior chemical digestion, limiting particle recovery (Azad et al. 2018a, b; Soe et al. 2022).

In contrast, gastrointestinal tract (GI) samples, often processed with digestion protocols such as 2M + 10M NaOH or H_2O_2, and filtered at 20 μm, reported higher FO% (18.71%, $N = 1299$) (Klangnurak and Chunniyom 2020; Kittichai et al. 2023; Seetapan and Prommi 2023). Comparable detection rates (17.63%, $N = 346$) were also observed in GI tract and gills digested using either 30% H_2O_2 and filtered at 0.45 μm (Labcom et al. 2024; Phaksopa et al. 2021) or 10% KOH + 30% H_2O_2 and filtered at 20 μm (Phaksopa et al. 2021).

Substantially higher FO% values were found in tissue samples digested with NaOH (1M) and SDS (0.5% w/v) and filtered at 20 μm (46.86%, $N = 239$) (Srisiri et al. 2024), and even higher rates were reported in GI tract and tissue analyses in Cyprinidae freshwater fish 54.17% ($N = 48$) using 10% KOH and 0.45 μm filters (Yasaka et al. 2022).

The highest FO% values were associated with multi-compartment assessments. Notably, stomach and intestine samples from 18 freshwater species yielded 87.23% ($N = 274$) using H_2O_2 with 1.2 μm filtration (Kasamesiri and Thaimuangphol 2020; Kasamesiri et al. 2021), while the maximum FO% (100%, $N = 35$) was observed in demersal estuarine fish when gills, stomach, and tissue were simultaneously analyzed using 10% KOH (Jitkaew et al. 2024).

These findings indicate that multi-compartment analysis and the use of chemical digestion method enhance microplastic detection, providing a more comprehensive representation of ingestion patterns and potential bioaccumulation. Therefore, standardizing the selection of anatomical compartments and digestion protocols is essential to improve data comparability and ecological interpretation in microplastic contamination research.

12.4 Discussions and Conclusions

The findings of this systematic meta-analysis reveal that microplastic ingestion is widespread across Thailand's aquatic ecosystems, with 84% of the 125 investigated fish species testing positive for ingestion, and a weighted average frequency of occurrence of 16.81%. The analysis demonstrates that ingestion risk varies significantly by habitat, trophic level, and analytical method. Benthopelagic and pelagic fish species showed notably higher ingestion rates, particularly those at lower trophic levels (2.0–2.5), such as herbivorous and detritivorous species, which exhibited a weighted average frequency of occurrence (FO%) of 67.80%. These species are often targeted in small-scale or subsistence fisheries, underscoring the elevated exposure risk faced by vulnerable communities that rely on these fish for food security.

Freshwater species, especially Cypriniformes also exhibited high ingestion levels at 77.75%. This highlights the critical importance of addressing plastic pollution in inland water bodies, particularly in regions near urban centers and agricultural zones where plastic leakage is common.

Methodological differences, especially in digestion protocols, emerged as a major factor influencing reported ingestion rates. Studies using chemical digestion methods, especially 30% hydrogen peroxide (H_2O_2), reported substantially higher FO% at weighted average FO% 74.61%.

The implications of these findings are wide-reaching. For fisheries managers, the results provide evidence to prioritize monitoring and management of species at greater ingestion risk, especially those harvested in subsistence contexts. Conservation authorities should integrate ingestion risk into habitat protection and pollution control measures, while public health agencies may need to reassess food safety protocols for fish consumed locally.

By highlighting how plastic pollution disproportionately affects low-income communities and ecologically sensitive habitats, this study contributes to the broader discourse on peace and sustainability. Protecting aquatic food sources from microplastic contamination supports not only ecosystem health but also food equity and social justice.

12.5 Recommendations

Based on the findings of this study, we propose a set of recommendations to improve the methodological rigor and ecological relevance of future research on microplastic ingestion in fish.

First, incorporating geographic coordinates during sample collection is essential for enabling spatial analyses of contamination patterns, identification of pollution hotspots, and examination of regional trends. This spatial dimension enhances reproducibility, supports cross-study comparisons, and informs location-specific conservation strategies. In parallel, the investigation of temporal trends is recommended to detect changes in plastic ingestion over time and to identify emerging risks in response to shifting pollution dynamics in aquatic ecosystems.

Second, improving taxonomic resolution is critical. Studies should avoid generalized classifications such as genus-level (e.g., *Nemipterus* spp., as in Srisiri et al. (2024)), or family level (e.g., *Engraulidae*, as in Pradit et al. (2022)) identifications, as these may obscure species-specific ingestion patterns and hinder accurate ecological risk assessments. Species-level identification enables more targeted management and conservation responses. Also, there is a need to report habitat classifications of sampled species in accordance with their ecological niches. Our meta-analysis highlighted inconsistencies and ambiguities in habitat labeling; to address this, we used standardized classifications from FishBase. However, we emphasize the importance of collecting accurate, site-specific habitat information to enhance ecological precision and support robust cross-study comparisons of habitat-specific ingestion patterns.

Third, future studies should ensure statistically adequate sample sizes to reduce sampling bias and improve the precision of ingestion frequency estimates. Sampling efforts must also be expanded to include underrepresented and conservation-priority taxa such as species listed as vulnerable (VU), near threatened (NT), data deficient (DD), and not evaluated (NE). Additionally, incorporating sex as a biological variable by separately analyzing male and female specimens, as seen in Azad et al. (2018a, b), can provide more accurate insights into ingestion patterns, especially in relation to reproductive vulnerability and population dynamics.

Fourth, analytical methods must be standardized to improve data quality. The adoption of validated chemical digestion protocols is essential; among the reviewed studies, 30% hydrogen peroxide (H_2O_2) proved to be the most effective across various tissue types (gastrointestinal tract, stomach, intestines, gills) for maximizing microplastic recovery while minimizing polymer degradation. Furthermore, while the use of Sedgwick-Rafter chambers (1 mL subsample) has provided a standardized framework for observation, as seen in Hajisamae et al. (2022) and Soe et al. (2022), this method may underestimate the variability of microplastic presence due to the limited sample volume. Researchers should consider analyzing larger or multiple subsamples and report the total volume and subsampling strategy to improve detection accuracy. Importantly, studies should provide numerical data rather than relying solely on visual approximations from graphs, as seen in Kasamesiri et al. (2021), as this compromises data precision and limits meta-analytic integration.

Finally, we urge for greater consistency in reporting methodologies. Standardized documentation of sampling, processing, digestion, and quantification protocols will significantly enhance the reproducibility, comparability, and reliability of future research. Establishing such methodological harmonization is crucial to support comprehensive ecological risk assessments, guide conservation strategies, and inform evidence-based policy interventions at national and global levels.

References

Au, S., Lee, C., Weinstein, J., van den Hurk, P., & Klaine, S. (2017). Trophic transfer of microplastics in aquatic ecosystems: Identifying critical research needs. *Integrated Environmental Assessment and Management, 13*(3), 505-509. https://doi.org/10.1002/ieam.1907

Azad, S., Towatana, P., Pradit, S., Patricia, B., & Hue, H. (2018a). Ingestion of microplastics by some commercial fishes in the lower Gulf of Thailand: A preliminary approach to ocean conservation. *International Journal of Agricultural Technology, 14*, 1017-1032. Retrieved from http://www.ijat-aatsea.com

Azad, S., Towatana, P., Pradit, S., Patricia, B., Hue, H., & Jualaong, S. (2018b). First evidence of existence of microplastics in stomach of some commercial fishes in the lower Gulf of Thailand. *Applied Ecology and Environmental Research, 16*(6), 7345-7360. https://doi.org/10.15666/aeer/1606_73457360

Campbell, S., Williamson, P., & Hall, B. (2017). Microplastics in the gastrointestinal tracts of fish and the water from an urban prairie creek. *FACETS, 2*(1), 395-409. https://doi.org/10.1139/facets-2017-0008

Carpenter, E., Anderson, S., Harvey, G., Miklas, H., & Peck, B. (1972). Polystyrene Spherules in Coastal Waters. *Science, 178*(4062), 749-750. https://doi.org/10.1126/science.178.4062.749

Cedervall, T., Hansson, L.-A., Lard, M., Frohm, B., & Linse, S. (2012). Food Chain Transport of Nanoparticles Affects Behaviour and Fat Metabolism in Fish. *PLoS ONE, 7*(2), e32254. https://doi.org/10.1371/journal.pone.0032254

Cheung, L., Lui, C., & Fok, L. (2018). Microplastic Contamination of Wild and Captive Flathead Grey Mullet (Mugil cephalus). *International Journal of Environmental Research and Public Health, 15*(4), 597. https://doi.org/10.3390/ijerph15040597

Davison, P., & Asch, R. (2011). Plastic ingestion by mesopelagic fishes in the North Pacific Subtropical Gyre. *Marine Ecology Progress Series, 432*, 173-180. https://doi.org/10.3354/meps09142

Elizalde-Velázquez, A., Carcano, A., Crago, J., Green, M., Shah, S., & Cañas-Carrell, J. (2020). Translocation, trophic transfer, accumulation and depuration of polystyrene microplastics in Daphnia magna and Pimephales promelas. *Environmental Pollution, 259*, 113937. https://doi.org/10.1016/j.envpol.2020.113937

Farrell, P., & Nelson, K. (2013). Trophic level transfer of microplastic: Mytilus edulis (L.) to Carcinus maenas (L.). *Environmental Pollution, 177*, 1-3. https://doi.org/10.1016/j.envpol.2013.01.046

Froese, R., & Pauly, D. (2025). *Fish Base. World Wide Web Electronic Publication*. Retrieved from www.fishbase.org

Hajisamae, S., Soe, K., Pradit, S., Chaiyvareesajja, J., & Fazrul, H. (2022). Feeding habits and microplastic ingestion of short mackerel, Rastrelliger brachysoma, in a tropical estuarine environment. *Environmental Biology of Fishes, 105*(2), 289-302. https://doi.org/10.1007/s10641-022-01221-z

Hermsen, E., Mintenig, S., Besseling, E., & Koelmans, A. (2018). Quality Criteria for the Analysis of Microplastic in Biota Samples: A Critical Review. *Environmental Science & Technology, 52*(18), 10230-10240. https://doi.org/10.1021/acs.est.8b01611

Ibrahim, Y., Rathnam, R., Anuar, S., & Khalik, W. (2017). Isolation and Characterisation of Microplastic Abundance in Lates calcarifer from Setiu Wetlands, Malaysia. *Malaysian Journal of Analytical Science, 21*(5), 1054–1064. https://doi.org/10.17576/mjas-2017-2105-07

IUCN, (2025) *The IUCN Red List of Threatened Species*, Version 2025-1, http://www.iucnredlist.org.

Jambeck, J., Geyer, R., Wilcox, C., Siegler, T., Perryman, M., Andrady, A., Law, K. (2015). Plastic waste inputs from land into the ocean. *Science, 347*(6223), 768-771.

Jitkaew, P., Pradit, S., Noppradit, P., Sengloyluan, K., Yucharoen, M., Suwanno, P.,Murugiah, K. (2024). Microplastics in estuarine fish (Arius maculatus) from Songkhla Lagoon, Thailand. *Regional Studies in Marine Science, 69*, 103342. https://doi.org/10.1016/j.rsma.2023.103342

Kangas, A., Setälä, O., Kauppi, L., & Lehtiniemi, M. (2023). Trophic transfer increases the exposure to microplastics in littoral predators. *Marine Pollution Bulletin, 196*, 115553. https://doi.org/10.1016/j.marpolbul.2023.115553

Kartar, S., Abou-Seedo, F., & Sainsbury, M. (1976). Polystyrene spherules in the Severn Estuary — A progress report. *Marine Pollution Bulletin, 7*(3), 52. https://doi.org/10.1016/0025-326X(76)90092-8

Kasamesiri, P., & Thaimuangphol, W. (2020). Microplastics ingestion by freshwater fish in the Chi River, Thailand. *International Journal of GEOMATE, 18*(67), 114-119. https://doi.org/10.21660/2020.67.9110

Kasamesiri, P., Meksumpun, C., Meksumpun, S., & Ruengsorn, C. (2021). Assessment on microplastics contamination in freshwater fish: A case study of the Ubolratana Reservoir, Thailand. *International Journal of GEOMATE, 20*(77), 62-68. https://doi.org/10.21660/2020.77.6108

Katija, K., Choy, C., Sherlock, R., Sherman, A., & Robison, B. (2017). From the surface to the seafloor: How giant larvaceans transport microplastics into the deep sea. *Science Advances, 3*(8). https://doi.org/10.1126/sciadv.1700715

Keerthika, K., Padmavathy, P., Rani, V., Jeyashakila, R., Aanand, S., Kutty, R., Subash, P. (2023). Microplastics accumulation in pelagic and benthic species along the Thoothukudi coast, South Tamil Nadu, India. *Marine Pollution Bulletin, 189*, 114735. https://doi.org/10.1016/j.marpolbul.2023.114735.

Kenyon, K., & Kridler, E. (1969). Laysan Albatrosses Swallow Indigestible Matter. *The Auk, 86*(2), 339-343. https://doi.org/10.2307/4083505

Khan, H., & Setu, S. (2022). Microplastic Ingestion by Fishes from Jamuna River, Bangladesh. *Environment and Natural Resources Journal, 20*(2), 1-11. https://doi.org/10.32526/ennrj/20/202100164

Kittichai, C., Krittawit, S.-U., Paridchaya Intep, P., & Saritporn, V. (2023). First Evidence of Microplastics Ingestion by Freshwater Fishes from Nong Luang Wetland (Chiang Rai, Thailand). *Sains Malaysiana, 52*(9), 2471-2483. https://doi.org/10.17576/jsm-2023-5209-01

Klangnurak, W., & Chunniyom, S. (2020). Screening for microplastics in marine fish of Thailand: the accumulation of microplastics in the gastrointestinal tract of different foraging preferences. *Environmental Science and Pollution Research, 27*, 27161–27168. https://doi.org/10.1007/s11356-020-09147-8

Kühn, S., & van Franeker, J. (2020). Quantitative overview of marine debris ingested by marine megafauna. *Marine Pollution Bulletin, 151*, 110858. https://doi.org/10.1016/j.marpolbul.2019.110858

Kühn, S., Bravo Rebolledo, E. L., & van Franeker, J. A. (2015). Deleterious Effects of Litter on Marine Life. In M. Bergmann, L. Gutow, & M. Klages, *Marine Anthropogenic Litter* (pp. 75-116). Springer, Cham. https://doi.org/10.1007/978-3-319-16510-3_4

Labcom, C., Inmuong, U., Sornlorm, K., Grudpan, C., Thongdamrongtham, S., Chaikhan, S., & Chakhamrun, N. (2024). The accumulation of microplastics in fish freshwater in the Mun river, Thailand. *International Journal of Public Health Asia Pacific*, 12-24. https://doi.org/10.62992/123kyn29

Laist, D. (1997). Impacts of Marine Debris: Entanglement of Marine Life in Marine Debris Including a Comprehensive List of Species with Entanglement and Ingestion Records. In *Marine Debris* (pp. 99-139). Springer, New York, NY. https://doi.org/10.1007/978-1-4613-8486-1_10

Lusher, A., O'Donnell, C., Officer, R., & O'Connor, I. (2016). Microplastic interactions with North Atlantic mesopelagic fish. *ICES Journal of Marine Science, 73*(4), 1214-1225. https://doi.org/10.1093/icesjms/fsv241

Lusher, A., Welden, N., Sobral, P., & Cole, M. (2017). Sampling, isolating and identifying microplastics ingested by fish and invertebrates. *Analytical Methods, 9*(9), 1346-1360. https://doi.org/10.1039/C6AY02415G

Markic, A., Gaertner, J.-C., Gaertner-Mazouni, N., & Koelmans, A. (2020). Plastic ingestion by marine fish in the wild. *Critical Reviews in Environmental Science and Technology, 50*(7), 657-697. https://doi.org/10.1080/10643389.2019.1631990

Moher, D., Liberati, A., Tetzlaff, J., & Altman, D. (2010). Preferred reporting items for systematic reviews and meta-analyses: The PRISMA statement. *International Journal of Surgery, 8*(5), 336-341. https://doi.org/10.1016/j.ijsu.2010.02.007

Nelms, S., Galloway, T., Godley, B., Jarvis, D., & Lindeque, P. (2018). Investigating microplastic trophic transfer in marine top predators. *Environmental Pollution, 238*, 999-1007. https://doi.org/10.1016/j.envpol.2018.02.016

Pegado, T., Schmid, K., Winemiller, K., Chelazzi, D., Cincinelli, A., Dei, L., & Giarrizzo, T. (2018). First evidence of microplastic ingestion by fishes from the Amazon River estuary. *Marine Pollution Bulletin, 133*, 814-821. https://doi.org/10.1016/j.marpolbul.2018.06.035

Pfeiffer, F., & Fischer, E. (2020). Various Digestion Protocols Within Microplastic Sample Processing—Evaluating the Resistance of Different Synthetic Polymers and the Efficiency of Biogenic Organic Matter Destruction. *Frontiers in Environmental Science, 8*. https://doi.org/10.3389/fenvs.2020.572424

Phaksopa, J., Sukhsangchan, R., Keawsang, R., Tanapivattanakul, K., Thamrongnawasawat, T., Worachananant, S., & Sreesamran, P. (2021). Presence and Characterization of Microplastics in Coastal Fish around the Eastern Coast of Thailand. *Sustainability, 13*, 1-12. https://doi.org/10.3390/su132313110

Pradit, S., Noppradit, P., Goh, B., Sornplang, K., Ong, M., & Towatana, P. (2021). Occurrence of microplastics and trace metals in fish and shrimp from songkhla lake, thailand during the covid-19 pandemic. *Applied Ecology and Environmental Research, 19*(2), 1085-1106. https://doi.org/10.15666/aeer/1902_10851106

Pradit, S., Noppradit, P., Sengloyluan, K., Nitiratsuwan, T., & Chuan, O. (2022). Low occurrence of microplastic contamination in anchovies, a transboundary species, in Thai waters. *ScienceAsia, 48*(4), 459. https://doi.org/10.2306/scienceasia1513-1874.2022.069

Pradit, S., Noppradit, P., Jitkaew, P., Sengloyluan, K., Yucharoen, M., Suwanno, P., Nitiratsuwan, T. (2023). Microplastic Accumulation in Catfish and Its Effects on Fish Eggs from Songkhla Lagoon, Thailand. *Journal of Marine Science and Engineering, 11*(4), 723. https://doi.org/10.3390/jmse11040723

Rochman, C., Hoh, E., Kurobe, T., & Teh, S. (2013, 11 21). Ingested plastic transfers hazardous chemicals to fish and induces hepatic stress. *Scientific Reports, 3*(1), 3263. https://doi.org/10.1038/srep03263

Rummel, C., Löder, M., Fricke, N., Lang, T., Griebeler, E.-M., Janke, M., & Gerdts, G. (2016). Plastic ingestion by pelagic and demersal fish from the North Sea and Baltic Sea. *Marine Pollution Bulletin, 102*(1), 134-141. https://doi.org/10.1016/j.marpolbul.2015.11.043

Ryan, P.G. (2016). Ingestion of Plastics by Marine Organisms. In: Takada, H., Karapanagioti, H.K. (eds) *Hazardous Chemicals Associated with Plastics in the Marine Environment. The Handbook of Environmental Chemistry, vol 78*. Springer, Cham. https://doi.org/10.1007/698_2016_21

Sainio, E., Lehtiniemi, M., & Setälä, O. (2021). Microplastic ingestion by small coastal fish in the northern Baltic Sea, Finland. *Marine Pollution Bulletin, 172*, 112814. https://doi.org/10.1016/j.marpolbul.2021.112814

Savoca, M., McInturf, A., & Hazen, E. (2021). Plastic ingestion by marine fish is widespread and increasing. *Global Change Biology, 27*(10), 2188-2199. https://doi.org/10.1111/gcb.15533

Seetapan, K., & Prommi, T. (2023). Microplastics in commercial fish digestive tracts from freshwater habitats in Northern Thailand. *Ecologica Montenegrina*, 48-65. https://doi.org/10.37828/em.2023.68.6

Setälä, O., Fleming-Lehtinen, V., & Lehtiniemi, M. (2014). Ingestion and transfer of microplastics in the planktonic food web. *Environmental Pollution, 185*, 77-83. https://doi.org/10.1016/j.envpol.2013.10.013

Shomura, R., & Yoshida, H. (1985). Proceedings of the Workshop on the Fate and Impact of Marine Debris, 27-29 November 1984, Honolulu, Hawaii. *Workshop on the Fate and Impact of Marine Debris (1984 : Honolulu, Hawaii).*

Soe, K., Hajisamae, S., Sompongchaiyakul, P., Towatana, P., & Pradit, S. (2022). Feeding Habits and the Occurrence of Anthropogenic Debris in the Stomach Content of Marine Fish from Pattani Bay, Gulf of Thailand. *Biology, 11*(2), 331. https://doi.org/10.3390/biology11020331

Srisiri, S., Haetrakul, T., Dunbar, S., & Chansue, N. (2024). Microplastic contamination in edible marine fishes from the upper Gulf of Thailand. *Marine Pollution Bulletin, 198*, 115785. https://doi.org/10.1016/j.marpolbul.2023.115785

Sutthacheep, M., Yeemin, T., Sangiamdee, D., Musig, W., Awakairt, S., Rangseethampanya, P., & Rongprakhon, S. (2018). Microplastics in the digestive tracts of reef fish in the Eastern Gulf of Thailand. *Ramkhamhaeng International Journal of Science and Technology, 1*, 9-14.

Sawangproh, W., & Paejaroen, P. (2025). Thailand's animals in peril: a systematic review of microplastic contamination and its ecological consequences. *Environmental Sciences Europe, 37*(1), 14. https://doi.org/10.1186/s12302-024-01043-z

Walkinshaw, C., Lindeque, P., Thompson, R., Tolhurst, T., & Cole, M. (2020). Microplastics and seafood: lower trophic organisms at highest risk of contamination. *Ecotoxicology and Environmental Safety, 190*, 110066. https://doi.org/10.1016/j.ecoenv.2019.110066

Wang, F., Wu, H., Wu, W., Wang, L., Liu, J., An, L., & Xu, Q. (2021). Microplastic characteristics in organisms of different trophic levels from Liaohe Estuary, China. *Science of The Total Environment, 789*, 148027. https://doi.org/10.1016/j.scitotenv.2021.148027

WoRMS Editorial Board. (2025). World Register of Marine Species. Retrieved from https://www.marinespecies.org at VLIZ

Yasaka, S., Pitaksanurat, S., Laohasiriwong, W., Neeratanaphan, L., Jungoth, R., Donprajum, T., & Taweetanawanit, P. (2022). Bioaccumulation of Microplastics in Fish and Snails in the Nam Pong River, Khon Kaen, Thailand. *EnvironmentAsia, 15*(1), 81-93. https://doi.org/10.14456/ea.2022.8

[illegible]

Part V
Contested Pathways and Critical Perspectives

Chapter 13
Power Struggle and Conflict Over Urban Space in Nepal: Implications for Sustainable Peace

Akash Pandey

13.1 Introduction

Power struggles and conflict over urban space have remained a key feature of Nepal's informal economy. While authorities crack down on street vendors, viewing them as illegitimate, street vendors in turn resist through individual and collective resistance strategies. This repeated cycle of contestation continues to shape the power dimension of urban space in Kathmandu and imperils the livelihood of 30,000 street vendors (Adhikari 2011).

The 2015 Constitution of Nepal introduced a federal governance system with provisions for independent and autonomous local governments. It restructured the governance system into three levels, including a federal government, seven provincial governments, and 753 local bodies. This governance structure vests local governments with significant authority, which they lacked previously in the centralized system. Local bodies now hold 22 independent powers and 15 concurrent powers shared with federal and provincial governments. Adoption of federalism paved the way for the first local government election in 2017, allowing citizens to exercise their voting rights, after 20 years of political turmoil (Democracy Resource Center 2017). Afterward, in May 2022, citizens again participated in the local-level elections. Of the 753 local units, 35,221 representatives submitted their candidacies, with most representing political parties (Democracy Resource Center 2022). At the

A. Pandey (✉)
Central Department of Conflict, Peace and Development Studies, Tribhuvan University, Kirtipur, Nepal

Graduate School of Humanities & Social Sciences, Hiroshima University, Higashihiroshima, Japan
e-mail: akash.787533@dcpds.tu.edu.np; m250308@hiroshima-u.ac.jp

A. Sharifi et al. (eds.), *Frameworks and Practices Towards a Sustainable and Peaceful Future*, World Sustainability Series,
https://doi.org/10.1007/978-981-95-2310-8_13

same time, the rise of independents was also observed, which later turned out to be a landslide for traditional parties as independents won over several constituencies.

The rise of independents stemmed from citizen dissatisfaction with traditional party leaders. Initially viewed as an alternative to party leaders, the activities of independents, particularly the work of Kathmandu Metropolitan City (KMC), have come under withering criticism. The administration, under the newly elected independent leadership, has been demolishing street-side businesses. Street vendors are not new to this issue, but the election of a new independent leader instilled a renewed sense of hope. However, the struggle persists. Left without any choice, vendors operate under a cloud of fear and insecurity and have adopted different resistance strategies. Kathmandu Metropolitan City, on the other hand, continues to evict vendors. Both actors continue to struggle over urban space in Kathmandu, Nepal.

This chapter aims to unravel the power struggle and conflict over urban space in Kathmandu, Nepal. Earlier works on Nepal's street vendors have explored livelihood (Timalsina 2012), socio-economic aspects (Adhikari et al. 2021), and the spatial location pattern of the petty street vendors in the Kathmandu Valley (Sharma and Pradhan 2017). Regional and international scholars have primarily centered their focus on the conflict and resistance strategies of vendors. In Colombia, Falla and Valencia (2019) explored the subtle resistance tactics of vendors, which they termed "quite encroachment." In Chile, Stillerman (2006) studied the spatial and political approaches adopted by vendors to access space. Hanser (2016) in China and Boonjubun (2017) in Thailand examined the conflict between vendors and authorities induced by restrictive policies. Others have delved into the violence faced by vendors, such as evictions in Mumbai (Saha 2010), harassment in Bangladesh (Etzold 2015a), and forceful displacement in Zimbabwe (Yatmo 2008).

Together, these literatures are important in enhancing the understanding of urban space conflict. Compared to Nepalese scholarship, international contributions have situated the urban space issue in the broader discourse of peace and conflict research. In Nepal, conflict understanding is mainly limited to Maoist Insurgency (Upreti 2006). Persisted for over four decades, the everyday micro-level urban space conflict between vendors and Metropolitan City has received minimal scholarly attention. Recognizing this critical gap, this chapter aims to contribute toward enhancing the existing knowledge base of Nepal's peace and conflict discourse by discussing urban space conflict through the lens of power, peace, and conflict. This study elucidates why there is a contestation over space between vendors and Kathmandu Metropolitan City (KMC). Using Weberian and Foucauldian concepts of power, it examines how spatial authority is constructed and negotiated. Then, using Galtung's violence triangle, it analyzes the implications induced by the constant power struggle for sustainable peace. It further provides recommendations for stakeholders on creating an inclusive and humane space as well as directions for future research. This chapter is organized into six sections. It begins with an introduction followed by methodology, theoretical framework, findings, discussion, and conclusion.

13.2 Methodology

This study is based on a qualitative research design. It employed in-depth interviews and key informant interviews to collect data. Twenty in-depth interviews were conducted with street vendors. Key informant interviews were conducted with representatives of Kathmandu Metropolitan City and the trade unions of street vendors. Trade unions interviewed in this study include the Nepal Trade Union Congress (NTUC), All Nepal Trade Union Federation (ANTUF, Maoist Centre), and the Nepal Street Vendors Union (NEST), associated with General Federation of Nepalese Trade Unions (GEFONT). Researcher also participated in the monthly meetings of trade unions and observed the movements led by the Nepal Street Vendors Union. Purposive sampling method was used to select the sample. While street vendors were selected based on their vending nature, experience, location, and gender, trade union and Metropolitan representatives were selected based on leadership and grassroots-level experiences of working and dealing with street vendors. Both mobile vendors (who do not have fixed locations) and street vendors (who have fixed locations) were included. To get access to vendors, snowball sampling method was used. Because of the fear and insecurity created by the eviction activities of Metropolitan City, vendors hesitated to participate in the study. Through trade unions, contacts were established with vendors.

Field work was conducted from March 2024 to May 2024, in different episodes. Two different clusters were identified in Kathmandu Metropolitan City. Considering the limitations of earlier studies that relied on a single area, such as Ratnapark (Bhattarai and Pathak 2020) and NewBuspark (Adhikari 2018), this study utilized a clustered method to capture more distinct data. The first cluster includes Pepsi-Cola, Jadibuti, and Koteshwor areas situated in ward number 32 of Kathmandu. The second cluster comprises Sundhara-Ratnapark-Asan-Thamel area. A standard time framework was designed to interview vendors in both clusters. Vendors were interviewed during morning, day, evening, and night periods. For instance, during morning and evening, when the Metropolitan Police are inactive, vendors operating in the first cluster were approached. Inactive here means off duty. Vendors in the second cluster were approached during day and night periods. Specifically, vendors of Sundhara and Ratnapark were interviewed during the day, as it is when vendors actively seize the opportunity. Vendors of Asan and Thamel were interviewed during the night because vendors have shifted their vending time due to the insecure day environment.

Altogether, twenty samples were interviewed, ten each from both clusters. Previously conducted qualitative studies on Nepal's street vendors have an average sample size of 30 (Timalsina 2012), 11 (Shrestha 2019), and 72 (Gautam and Salike 2021). On the other hand, quantitative studies have studied 403 samples (Adhikari et al. 2021) and 450 (Bhattarai and Pathak 2020). There are varied sample sizes considered by previous studies. According to Hennink and Kaiser (2022), 9–17 interviews are adequate for reaching saturation. Taking these into account, and other factors like time and resources, this study studied 20 samples.

Interviews were conducted in Nepali, spanning around 25–30 minutes. To further understand the issue in depth, follow-up visits were made. Before data collection, ethical approval was received from the Central Department of Conflict, Peace and Development Studies, Tribhuvan University. Informed consent was taken to record the overall conversation. Collected data were translated into English and imported into NVivo for thematic analysis. A preliminary analysis was first conducted to ensure the adequacy of the collected data. In this context, researcher briefly read, analyzed, and summarized the data. Ensuring that the data were sufficient to answer study objectives, researcher then began the coding process. Both inductive and deductive methods were used for coding. Since preliminary codes were already developed based on research questions and a theoretical framework, I used a deductive approach for coding, while also remaining open for codes emerging from the data. After the initial coding, redundant codes were removed. The remaining codes were grouped into categories. Afterward, patterns and relationships between categories were examined, and themes were developed. This systematic process of analysis helped maintain analytic rigor.

13.3 Theoretical Framework

Theoretical foundation of this study is rooted in the concepts of Max Weber, Michel Foucault, and Johan Galtung. Firstly, Weber's concept of "Monopoly over the legitimate use of power" is used to analyze how Kathmandu Metropolitan City maintains its legitimate control over urban space. In his contribution entitled "Politics as a Vocation," Weber (1946) theorized that the state has a monopoly over the legitimate use of physical force in the enforcement of its order. While the Weberian concept of legitimate monopoly is particularly articulated in the context of the state, this study explored the legitimate monopoly maintained by local government, specifically the Kathmandu Metropolitan City. The 2015 Constitution of Nepal established an independent and autonomous local government with the provision for 22 independent and 15 concurrent powers shared with the federal and provincial governments. These constitutional provisions grant Kathmandu Metropolitan City the power to govern and implement decisions over several matters, including urban space management. Taking this into account, the Weberian power concept is relevant to examine how authority legitimizes its interventions taken for urban space management.

Secondly, Michel Foucault's concept of power and resistance is used to examine vendors resistance. Foucault (1978) contends that power is everywhere. Unlike thinkers who view power as repressive, Foucault (1978, p. 94) argues that power is productive. Not only does power operate at the macro level, but also at the micro level. This suggests that power is not only exercised by authorities but also by grassroots actors like street vendors. Previous studies have shown that street vendors employ different strategies to confront state-restrictive regulations (Adama 2020; Lata et al. 2021). Such acts of resistance were not limited to a certain time; vendors have constantly employed them to challenge the authority repression, which continues in the present context of Kathmandu, Nepal. Then, using Galtung's (2009) Attitude, Behavior and Contradiction (ABC) triangle and violence triangle, the implications of this constant power struggle for sustainable peace are examined.

13.4 Findings and Discussion

13.4.1 Vendors Struggle for Urban Space

13.4.1.1 Gender, Informality, and Street Vending

This study explored the underlying causes behind street vendors everyday struggle for urban space. In this section, I show how family issues like violence push individuals to engage in vending. Then, I present the structural causes of vending, including unemployment, labor migration, disaster, and civil war.

Evidence from this research shows that family issues like violence and separation cause individuals to seek livelihood alternatives like vending. These categories of vendors include single women, people with disabilities, and elderly citizens. In Nepal, the responsibility to care for elderly citizens is often taken as a burden by family members (Yadav et al. 2018). Elderly citizens interviewed in this study recounted similar experiences. An elderly woman stated:

> I have been engaged in street vending for 16 years. My life has been full of hardships. I lost my left leg at an early age. My husband passed away soon after I got married. I used to live with my son, but after he got married, everything changed. I was often mistreated and was taken as a burden by my son and daughter-in-law. After having such experiences, I left them and came to the street. Slowly and gradually, I began to sell goods for my survival. Now, I have nowhere to go but to rely on the street.

Similarly, women vendors shared how separation compelled them to pursue vending as a means of survival. Most of the interviewed vendors were single women. A women vendor who has been engaged in street vending for 17 years recounted her struggles:

> I primarily came to the street when my husband left me and married another woman. At that time, my son was only three years old, and on top of that, I already had two daughters. I was alone and had no source of income. So, to raise my children, support their education, and run our livelihood, I saw street vending as an alternative.

What is evident from these individual accounts is the broader context of women's participation in the informal sector. As per the Nepal Labor Force Survey 2017/18, 62.2% of the labor force is engaged in the informal sector, with women constituting the majority (Central Bureau of Statistics 2019, p. 25). 90.5% of women in Nepal are engaged in informal work (Central Bureau of Statistics 2019, p. 29). Trade union estimates that around half of vendors in Kathmandu, i.e., 50–60%, constitute women. Clearly, these figures demonstrate the gendered dimension of informal economy and signify how street vending emerges as a crucial livelihood option for women in vulnerable situation.

13.4.1.2 Internal Migration, Labor Migration, and Street Vending

Closely linked to the issue of street vending is the internal migration. Internal migrants residing in big cities like Kathmandu for study and work purposes have relied on street vending. Recent data suggest a 5% increase in urban population from 22.31% in 2011 to 27.07% in 2021 (National Statistics Office 2024, p. 32). The present population of Kathmandu stands at 1,994,555 (National Statistics Office 2024, p. 39). The 2021 Census identifies Kathmandu, Bhaktapur, and Lalitpur, with the highest in-migration rates, with people mainly migrating for employment, study, and training purposes (National Statistics Office 2025). Since socio-economic opportunities primarily exist in urban cities like Kathmandu, the capital offers an attractive hubs for both formal and informal economic activities. However, internal migrant faces significant barriers in entering formal sectors because of lack of networking, skills, and qualifications. As a result, street vending becomes an essential livelihood alternative to support the expensive urban life, which is cost-friendly and easy to operate.

Additionally, the trend of labor migration is responsible for the influx of internal migrants. During the period 2019/20 and 2021/22, the government issued over 1.1 million labor approvals for foreign employment (Ministry of Labour, Employment and Social Security 2022). With breadwinners migrating abroad, the left-behinds have internally migrated to urban areas like Kathmandu for children's education and a better life. Left behind are particularly women who take on the responsibility of rearing children. The income of migrants is utilized to support left-behind livelihoods. Women vendors interviewed in this study reported the inadequate income of their spouses to support everyday lives. In some cases, the prolonged absence of labor migrants has imperiled the livelihood of left-behinds. In such a situation, women have felt the necessity to engage in vending. Adhikari and Hobley (2015) also note that women engage in informal and home-based enterprises in response to the irregular flow of remittance. While Adhikari (2011) estimates around 30,000 street vendors working in the Kathmandu Valley, more recent estimations of trade unions report around 40–50 thousand vendors, predominantly composed of internal migrants. Thus, these discussions demonstrate that street vending is not merely driven by economic choice but a survival strategy in the face of internal migration and labor migration.

13.4.1.3 Disaster, Informality, and Street Vending

Along with migration factors, natural disaster is another major driver of internal migration and informality in Nepal. Natural disaster displaces thousands of families every year. Between 2011 and 2021, 3.4 million people were displaced, with the

2015 Gorkha earthquake causing massive displacement of 2.6 million alone (Internal Displacement Monitoring Center 2022). Both the Metropolitan officials and trade union representatives reported the surge in the number of street vendors in the aftermath of the 2015 earthquake. The 7.5 magnitude earthquake rendered 2.8 million homeless, with about 864,000 of them already living below the poverty line (World Health Organization. Regional Office for South-East Asia 2016, p. 7). Gurung (2018) confirms that young people displaced by earthquake internally migrated to revive their lost livelihood. On top of being disproportionately affected by natural disaster (Kim 2012), the marginalized people also lack the capacity to cope with and recover from losses (Hallegatte et al. 2020). Following the earthquake in Nepal, studies have documented the use of remittances and loans as coping mechanisms (Raut 2021). Not everyone, however, possesses equal coping abilities. Therefore, migration to urban centers like Kathmandu becomes a necessity to restore the broken livelihood, where informal work like vending remains a key survival strategy.

13.4.1.4 Maoist Insurgency and Street Vending

The involvement of ex-Maoist combatants and the Internally Displaced Persons (IDPs), affected by the decade-long Maoist insurgency, was also noted. An estimated 200,000 people were internally displaced by the conflict (Internal Displacement Monitoring Center and Norwegian Refugee Council 2006). Among the total 17,052 ex-combatants listed by the Special Committee Secretariat, 1,422 were integrated into the Nepal Army,15,624, opted for voluntary retirement, and 6 selected the rehabilitation package (Bhandari 2015). Both the IDPs and the ex-rebels faced livelihood challenges in the post-conflict period. Despite some repatriation efforts of non-government organizations (Internal Displacement Monitoring Center and Norwegian Refugee Council 2008), IDPs did not return to their native places because of insecurity, livelihood obstacles, poor shelter, and inadequate support mechanisms (Gurung 2022). Voluntary retired ex-insurgents settled near cantonments or migrated to cities like Kathmandu, Lalitpur, Bhaktapur, Biratnagar, Pokhara and Nepalgunj (Subedi 2014). Although the cash packages helped conclude the Disarmament, Demobilization and Re-integration (DDR) process, limited entrepreneurial skills, misuse of cash, and social backlash impeded the effective re-integration process. In addition to ex-combatants and IDPs, the conflict also left the child soldiers in a precarious situation. In view of this, it is worth noting that the Nepalese civil war not only reshaped the country's political trajectories but also induced livelihood insecurity among those who were victimized and those who fought for the change.

Looking at all these aspects, it is evident that street vending in Kathmandu is primarily driven by the force of necessity. Although different situations have led them to vending, the most common theme observed is "survival." This aligns with the findings of Bhowmik (2005) and Bromley (2000). They contend that street vending is a survival strategy for the urban poor. Unlike other studies that attribute street vending to personal and structural factors, this study also found that the Nepalese civil war induced livelihood insecurity among those who were victimized and who rebelled for change, compelling them into street vending. That said, street vending emerges as a post-war coping mechanism that enables individuals to reclaim and restore agency and stability.

13.4.2 Legitimate Monopoly over Urban Space

In this section, I discuss how the Kathmandu Metropolitan City has maintained legitimate control over urban space. I explored this question by adopting Max Weber's concept of monopoly over the legitimate use of force. Particularly, I critically delve into the legal rational framework, charismatic authority, and discourses used by KMC to legitimize actions and maintain control over space. In examining this, I also elucidate why the Metropolitan City intends to control urban space.

13.4.2.1 Legal Rationality and the Modernist Vision

The control over urban space in Kathmandu is primarily maintained through the legal framework. The Local Governance Operation Act (LGOA), 2017, and the Solid Waste Management Act, 2011, are two legal frameworks brought into effect. Referring to Sections 11 and 97 of the Local Governance Act, 2017, and Section 39 of the Solid Waste Management Act, 2011, the Metropolitan City declared a ban on street vending on January 10, 2023 (Kathmandu Metropolitan City 2023). While the former allows the Metropolitan City to mobilize the Metropolitan Police to prevent public property encroachment, the latter prohibits any activity polluting the public space.

Scholars have framed the use of legal instruments as a form of disciplinary technology (Adama 2021), where the law serves as a tool to marginalize certain populations in the name of order. Nahar Lata (2021) describes such regulations as exclusionary practices adopted to advance the neoliberal vision of modernist cities. In Kathmandu, vendors are not simply viewed as unlawful actors for legal reasons, but because they pose a danger to the aesthetics of a modernist city. Metropolitan Police representative pointed out the problems created by vendors, such as traffic congestion, road accidents, pollution, and disruption in the flow of pedestrian movement. Within the broader vision of making a clean, green, and pollution-free city, Metropolitan representatives argued that vendors have remained a prime challenge.

The above perspectives suggest that authorities perceive street vendors as harmful to the urban order (Yatmo 2008). A number of studies have reported the negative consequences of vending. According to Bhattarai and Pathak (2020), street vendors are the source of environmental pollution and crowded roadsides in Kathmandu. The public health hazards emanating from vending have also been noted. Since vendors operate in unclean environments, without proper hygiene, their foods increase the risk of microbial contamination (Rane 2011). When the number of vendors exceeds the critical threshold, authorities may find it problematic (Bayat 1997). That is why vendors are perceived as offensive and illegitimate (Rajagopal 2001) and legal actions are viewed as indispensable in maintaining the urban order (Etzold 2015a). Within this context, the KMC has deployed Metropolitan Police to oversee the public property management. Rule 6(1) of the Municipal Police Regulation, 2015 (2072) empowers municipal police to "remove obstacles created by any construction or business in the mobility of the public" (Kathmandu Metropolitan City 2015). Based on this legal framework, the KMC conducts and justifies evictions. Although trade unions criticize evictions as a demonstration of power, the Metropolitan City justifies the use of force to prevent public space encroachment.

Considering this, a critical contradiction has emerged in the claim made by the authorities. KMC perceives vendors as a source of pollution and road accidents. Empirical evidence, however, indicates households as prime producers of waste (Central Bureau of Statistics 2021) and poor road conditions and carelessness as major causes of road accidents (Manandhar 2022). Interms of waste, rapid urbanization and population growth have quantified waste production in Kathmandu (Sharma et al. 2018). Theses discussions are opposite to what KMC argued and shows how authorities deploy narratives and justify the use of force to oust vendors from the urban core. Critical to emphasize that vendors and trade unions acknowledge the disruption in pedestrian movements. Without alternatives, vendors have no choice but to operate in the street. This fundamental contradiction between the modernist vision and the survival of the vendors has resulted in conflict. The legal frameworks are thus introduced to justify the intervention that supports the idea of creating a clean and green city.

13.4.2.2 Charismatic Authority

According to Weber (1978), charismatic authority is considered to be individuals with exceptional leadership qualities whose legitimacy rests on followers' subjective loyalty and belief. The charisma defined by Weber echoes the context of Kathmandu Metropolitan City leadership. Since the last local government election, the newly elected leader has garnered huge support from citizens. Whether it be the demolition of buildings or the illegality of street vending, the public has hailed it. Traditional authorities' inability to meet the expectations of citizens has triggered public discontentment. Satisfaction, transparency, and accountability determine citizen trust (Molnar 2019), which Nepal's traditional authority does not fulfill. In such a context, charismatic authority is seen as an alternative (House and Howell 1992). The determination shown by the KMC leader through several initiatives has furthered people's confidence and support for the authority and its work. However, it is crucial to note which group and nature of people have supported the KMC action.

Field data and observation show that people operating businesses within the Metropolitan City view street vendors as a problem, as vendors offer cheaper goods in comparison. Culturally sensitive locals also perceive street vendors as a problem to their culture and heritage. These groups have supported Metropolitan-led eviction activities. Meanwhile, civil society has expressed serious reservations towards the KMC ban on street vending.

It is interesting to observe the influence of charismatic authority among street vendors themselves. Compared to traditional authorities, street vendors view the present Metropolitan leader as an effective alternative. Such perception stems from the inability of the traditional leaders to meet not only the needs of vendors but also the broader socio-economic goals of society. Notably, some vendors have left the vending profession and engage in the work provided by the Metropolitan City. Regardless of this, the background of the vendor, who left the profession, should be critically examined, including their status as residents and the nature of work,whether permanent or temporary. Given these findings, what can be identified is that along with the general public, charismatic authority is also supported by the vendors, who ironically resist the same authority.

13.4.2.3 Discourse and Unaccountability

The unaccountability of KMC has further exacerbated the issue of street vendors. Field interviews reveal that the Kathmandu Metropolitan City has consistently ignored the concerns of street vendors. Despite several attempts for co-operation, KMC has often remained unresponsive to the grievances of vendors. Instead, KMC has deployed narratives of unplanned cities, framing vendors as chaotic and illegal users of public space. These narratives are primarily promoted through the media. Kathmandu Metropolitan City operates its online media, called Metro News, and some online reporters have supported the dissemination of such narratives. Persistent unaccountability of authorities is not unique to the present administration, as vendors have endured such challenges in earlier administration, too. It is because authorities are not interested in exploring reliable alternatives for street vendor management (Anjaria 2010). And the strategic framing of vendors as urban problems is a way to promote the elite-led transformation (Crossa 2016; Tucker 2017). This understanding aligns with the views expressed by vendors in this study. They view the move of the Metropolitan to give rise to elite-centered cities while deterring their image publicly. A common theme that was mostly repeated in each interview was about the class issue: Is this city only for the rich? Don't we have any place in this city? This indicates urban space is a class issue.

Taken together, what emerges from these findings is that the KMC assertion over urban space is primarily legitimized through the interplay of legal rational and charismatic authority. The local government laws and regulations and the deployment of the Metropolitan Police collectively serve to institutionalize KMC authority and justify the eviction of street vendors. Weber's (1978) conception of charismatic authority, based on exceptional leadership qualities, is evident in this study. Exemplary work led by the Metropolitan leader has widely appealed to the public, helping the authority to garner public support, including from the street vendors.

Although Weber (1978, p. 954) contended that charismatic authority lacks the institutional form and operates without rational rules, the KMC case suggests that charismatic authority relies on legal rational frameworks to reinforce its action. In fact, the basis of modern charismatic authority is the legal framework. Yet, the independent nature of leadership, with no accountability mechanisms, unlike traditional party leaders, suggests a potential danger of erosion of institutional norms. This is what Weber also warned about charismatic authority (Weber 1978, p. 242). Overall, this intersection of authority and the approaches to maintain a legitimate monopoly over urban space has significance for understanding the power and politics of urban governance.

13.4.3 Resistance of Street Vendors

13.4.3.1 Individual Resistance

In response to the KMC attempt to maintain legitimate control over urban space, vendors have adopted individual and collective resistance strategies. In this section, I analyze vendors resistance using Michel Foucault's concept of power and resistance. I first discuss the individual strategies followed by collective resistance efforts led by trade unions and civil society.

Individual resistance shows routinized vending practices adopted by vendors. For example, vendors vend during morning and evening, when the Metropolitan Police are inactive, while during day, they move to areas with minimal surveillance such as public events and sports games. Due to the stringent monitoring of Metropolitan City, vendors operating in prime areas have shifted their operation to nighttime, such as in touristic areas like Thamel where the night market operates. By operating at various times, vendors do not merely vend but strategically vend.

In their quest for everyday survival, vendors have reduced the quantity of goods to simplify their operations, to avoid evictions. This approach is specifically performed by mobile vendors, often known as hawkers, who do not have a fixed location. Street vendors have also adopted this practice. Additionally, vendors utilize bags to conceal products and warn each other via mobile phones. This is to note that vendors engaged in trade unions have local subcommittees in distinct locations of Kathmandu. Through this organized network, they inform each other about the police patrol. This phenomenon was also evident in Peru, where vendors utilized social networks to inform each other about evictions, which Steel (2012) termed the "social security network." Those without any formal association have maintained an informal network. Unlike formal networks, this informal network consists of several informal workers including vendors, beggars, and sex workers. Together, these diverse informal workers have built an alliance to resist authorities in covert ways.

Some have effectively utilized informal networks by making use of private spaces. Street vendors have rented private spaces with whom they have a good relationship. Mobile vendors visit door to door to deliver goods for individuals they interact with daily. The role of non-state actors was also documented in Georgia (Polese et al. 2016) and Mexico (Crossa 2016), where networking became a key source of power for vendors, which is evident in this study as well. The strategic use of spatial tactics was also noted. To illustrate, vendors operate in bridges that demarcate the district borders, and when officials spot them, they simply cross the border. This highlights that vendors are aware of the jurisdiction of the authorities.

Nonetheless, everyday micro-level resistance of vendors illustrates that resistance manifests not only through outright rebellion, but also through subtle, informal, and mundane practices that silently negotiate power without confrontation, which Scott (1985) conceptualize as weapons of the weak. Yet, some vendors directly confronted authorities. They do not comply with KMC restrictions, and non-compliance itself is vendors coping mechanism to express dissatisfaction with the political and economic system. Due to the lack of formal employment opportunities, vendors perceive their use of space as legitimate.

13.4.3.2 Collective Resistance

The collective resistance of vendors is led by trade unions, civil society, and independent activists. In this section, I discuss the collective efforts of trade unions and then detail the support expressed by broader civil society.

13.4.3.2.1 Trade Union Movements and Grassroots Practices

Trade unions of street vendors have played a major role in organizing collective movements. These unions have constantly called on authorities by facilitating peaceful movements, negotiations, and political lobbying. The role of trade unions has also been documented in several other geographical settings. In Bogotá, Colombia, trade unions helped vendors to assert their right to space (Munoz 2018). In Ghana, the Ghana Trade Union Congress and its associated network, the StreetNet Ghana Alliance, exhibited an apparent advocacy role for street vendor rights (Gillespie 2017). GEFONT, ANTUF, and NTUC are the major trade unions studied in this research. These trade unions operate under the Joint Trade Union Center (JTUCC), an umbrella organization of trade unions in Nepal. Each trade union has its affiliate organization of street vendors. Specifically, GEFONT has maintained an organized entity entitled "*Nepal Street Vendors Union*." The Nepal Street Vendors Union is affiliated with the StreetNet International organization and has a total of twelve street vendor subcommittees, which include one provincial committee and 11 Yikai (local) committees. In these subcommittees, street vendors are assigned specific roles and responsibilities. Members are responsible for membership renewals, vendor mobilizations, and holding monthly meetings. Each vendor is allowed to share their opinions during meetings. By discussing their collective issues in such grassroots forums, vendors encourage each other and strengthen their collective stance.

13.4.3.2.2 Political Approach

Since the trade union examined in this research serves as a sister organization of one of Nepal's major political parties, they have leveraged political networks to obtain vendor access to space. GEFONT is associated with the Communist Party of Nepal (Unified Marxist-Leninist), the Nepal Trade Union Congress with the Nepali Congress, and the All Nepal Trade Union Federation with the Communist Party of Nepal (Maoist Centre). These unions have constantly engaged in lobbying efforts with ministries, high-level authorities, and their corresponding political parties. While trade unions are affiliated with political parties, street vendors are affiliated with both trade unions and respective political parties. They engage in unions for political protection, and for that, they pay fees and support the party's ideology. Similar cases were found in Mexico, where vendors utilized political connections to assert their right to public spaces (Cross 1998). According to Donovan (2008) and Milgram (2011), vendors use their votes as a negotiation tool to claim their rights. Kathmandu's street vendors are also aware about their voting rights and the role their votes played in helping the leaders to reach power. However, they express dissatisfaction because of the dishonesty of the representative they voted for. Whether vendors are satisfied or dissatisfied, if they are not affiliated with trade unions, they remain vulnerable.

13.4.3.2.3 Legal Approach

Despite several movements and the use of political networks, the independent nature of Metropolitan leaders has thwarted vendors resistance. In past administrations, the situation was comparatively flexible. Vendors were allowed to operate during specific morning and evening hours. They also recognize that the present leadership functions independently, unlike the previous administration, which was affiliated with a political party and was accountable to the party. This indicates that street vending is flexible under leadership affiliated with political parties but difficult during independent leadership. Considering this, trade unions have opted for legal measures. The Nepal Street Vendors Union has filed a case against the KMC for their harsh treatment of street vendors. Vendors have primarily resorted to legal measures because they are aware of the fundamental rights guaranteed by the Constitution. The 2015 Constitution of Nepal has guaranteed thirty-one fundamental rights. Trade union representatives criticize KMC for infringing upon vendors right to work and to live with dignity. The Supreme Court has directed the KMC and the relevant entities of the Nepal Government to implement alternative measures for street vendor management. The court has given a 15-day deadline to formulate a plan and submit it to the apex court via the office of the Attorney General (Rai 2024b). The reliance on legal avenues demonstrate that vendors rely on the same approach as authorities used to control the urban space.

13.4.3.2.4 Civil Society and Independent Youth Activists

Since KMC started the eviction of street vendors, civil society representatives and youth activists have expressed their solidarity with the vendors. Through peaceful rallies, demonstrations, and open letters, civil society has criticized the use of coercive measures and demanded the proper management of street vendors. Across different geographical regions, including India and Indonesia, civil society has played a major role in advocating the rights of vendors (Gibbings 2013; Kumari 2015). Besides, there are national and international organizations that work for the rights of vendors, such as WIEGO (Women in Informal Employment: Globalizing and Organizing) and StreetNet International.

In Kathmandu, apart from the vigilant oversight of civil society, an individual-led movement was observed. Two consecutive movements were initiated by an individual activist demanding an alternative arrangement for street vendors. In the first movement, the activist called on KMC through a standing demonstration for a straight 199 hours. Eventually, the peaceful protest concluded with the four-point agreement with the KMC. When the KMC failed to implement the agreement, the activist embarked on a 9-day foot walk from the birthplace of Buddha (Lumbini) to the Prime Minister's residence, Kathmandu. While the support from broader civil society and individuals showcases their concern for the welfare of vulnerable sections of society, their engagement is often occasional in nature.

Collectively, the individual and collective resistance of vendors supports the idea that power is not only imposed from above, but it is exercised and contested at many different levels, which corresponds with Foucault's dictum that power is everywhere and it comes from everywhere (Foucault 1978, p. 93). Along with mass demonstrations, vendors negotiate repression with micro-level, subtle, and dispersed practices that resonate with what Scott (1985) views as weapons of the weak. These micro-level resistances imply that power is dynamic, and in constant flux. By resisting, vendors are not merely opposing the authority but forming a network of power from below (Foucault 1978, p. 94) that constantly reshapes the urban order and power dynamics. Grounded in shared experiences and everyday negotiations, the overt response of vendors highlights the relational dimension of power circulating in daily practices that sustain both domination and resistance.

13.4.4 Implications for Sustainable Peace

Empirical studies have documented multiple forms of violence faced by street vendors. This includes both direct and indirect violence. In several major cities of India, including Mumbai, Delhi, and Bangalore, vendors goods were evicted by the municipal authorities (Gehlot et al. 2023; Saha 2010; Williams and Gurtoo 2012). In Dhaka, Bangladesh, vendors were physically harassed by authorities (Etzold 2015b). In Zimbabwe, the unrest fueled by state-led cleanup operations resulted in the arrest of 20,000 vendors (Yatmo 2008, p. 13). Similarly, street vendors in

Kathmandu have experienced direct, structural, and cultural violence. Direct violence, legitimized by legal rational framework, remains overt in the form of physical attacks and evictions. Systemic exclusion of vendors from urban space manifests structural violence, and the hegemonic discourses on street vendors, such as chaotic, have reinforced cultural violence, thereby justifying direct and structural violence. Table 13.1 shows this interrelated violence and the implications perpetuated as a result.

Exposure to direct, structural, and cultural violence has created an insecure environment for vendors. Reports suggest the displacement of 10,000 street vendors in Kathmandu (Rai 2024a). According to trade union representatives, vendors have either migrated abroad or internally to safer cities in search of safer livelihoods. The uncertainty has further compelled 89% of vendors to rely on loans (Ghimire 2023), and in extreme cases, some have engaged in survival activities like sex work (Bishwakarma 2024).

As per the internal records provided by the KMC, the Metropolitan Police seized over 251 carts and cycles from August to November 2023. All vendors interviewed in this study, at least once, have experienced confiscation. Anything once taken away is not returned, and it requires a certain level of investment to restart the business. Starting over for vendors is an uphill battle. The eviction activities are justified through legal frameworks and the narratives that frame vendors as urban nuisances. This is not only limited to Nepal but also in different Global South countries like Vietnam, Mexico, and South Africa, where vendors are perceived as urban disorder (Donovan 2008; Drummond 2000). These discourses reinforce cultural violence that normalizes direct and structural violence, and in the case of Kathmandu, the nature of charismatic authority has further legitimized the situation. Every day, evictions reveal that vendors are not governed by law but rather through keeping them under a state of uncertainty (Anjaria 2006). An uncertainty that creates a sense of fear and insecurity (Khamrang et al. 2022). That is why vendors argue that their right to work and the right to live with dignity have been violated. Importantly, vendors consist of single women, elderly and senior citizens, people with disabilities, conflict victims, ex-Maoist combatants, natural disaster victims, and rural migrants, among others. Having already been overlooked by the state and society, the Metropolitan crackdown may further intensify their existing socio-economic woes.

Table 13.1 Direct, structural, and cultural violence faced by street vendors

Type of violence	By/through	Examples	Implications
Direct Violence	Metropolitan Police	Use of force; evictions; verbal abuse; detentions	Constant fear and insecurity
Structural violence	Ban on street vending	Street vending as an illegal activity	Displacement; livelihood insecurity
Cultural violence	Narratives	Vendors as a threat to urban aesthetics; violators of public order; chaotic and polluters	Perpetuation of direct and structural violence

13.4.4.1 Why the Power Struggle Over Space Continues?

Using Galtung's (2009) Attitude, Behavior, and Contradiction (ABC) model, I analyze and report why the power struggle over urban space is likely to continue. The conflict between Kathmandu Metropolitan City and vendors stems from the contradictions over the use of urban *space*. While KMC aims to maintain urban order, vendors are driven by survival. In the pursuit of a modernized vision, KMC views vendors as a prime challenge, adopting a coercive approach, legitimized by legal frameworks, to sideline vendors from the urban core. But vendors view KMC actions as oppressive and resist through their individual and collective strategies.

This everyday conflict and power struggle reflects a broader tension between neoliberal urban governance and the survival of the urban poor. The neoliberal logic prioritizes aesthetics and economic rationality at an exclusion of disadvantaged groups. Through such aesthetic discourse, the exclusion of the poor is justified in the urban planning process (Mitchell 2003). The KMC ban on street vendors in Kathmandu reflects such neoliberal logic. Despite attempts to criminalize, vendors, however, do not go away. They negotiate and re-negotiate in the same space that excludes them. Through everyday resistance, vendors come up with their own way of survival.

Survival is what keeps vendors resilient against neoliberal logic. As Burton (1990) points out, the conflict takes a protracted form when basic needs are unmet or denied. This is because needs are non-negotiable. Although vendors' resistance is constantly undermined by the Metropolitan city, the broader systemic conditions that give rise to urban informality are poverty, unemployment, and uneven development. This is why neoliberal urban regimes like Kathmandu will remain a hub for opportunities, where vending remains an accessible—albeit informal survival strategy. Yet modernist discourse, on the other hand, will continue to frame vendors as urban nuisances, triggering a vicious cycle of conflict and resistance that not only perpetuates direct, structural, and cultural violence but also obscures the conditions for positive peace. The prospects for sustainable peace, thus, remain challenging. Figure 13.1 illustrates this dynamic.

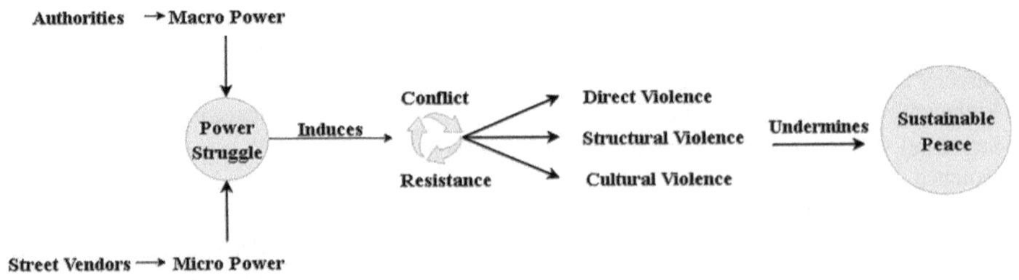

Fig. 13.1 Conceptual diagram showing how power struggle undermines sustainable peace

13.5 Conclusion

This chapter provides an analysis of the everyday conflict between street vendors and Kathmandu Metropolitan City over urban space in Nepal. Firstly, this chapter discusses why vendors struggle and claim their right over urban space. Findings show that a range of structural factors, such as family separation, violence, natural disaster, internal migration, labor migration, and civil war have compelled individuals to pursue vending as a livelihood strategy. In the aftermath of such situations, informal work like vending has become a critical livelihood strategy. In short, vendors are driven by the force of necessity.

Secondly, drawing on Weber's concept of monopoly over the legitimate use of force, this study illustrates how KMC maintains its legitimate control over urban space. The attempt to control urban space is guided by the neoliberalist vision of modernist cities. In other words, aesthetic governmentality is what KMC envisions. In the pursuit of developing a green and clean city, vendors have stood out as a major problem. In line with this, the legal rational frameworks are brought into effect to ban vending and to justify the eviction activities led by the Metropolitan Police. Notably, public support garnered by charismatic leadership has further legitimized the KMC action. Interestingly, even vendors, who are ironically resisting the same authority, are influenced. Exceptional leadership qualities of a charismatic leader, as discussed by Weber (1978), appeal to the general public. The danger, however, lies in the erosion of institutional norms (Weber 1978, p. 242). Metropolitan leadership, operating independently of institutional accountability, raises concern about the potential erosion of institutional norms. Yet, unlike the Weberian understanding of charismatic authority that does not correspond to institutional form, the KMC case suggests that the modern charismatic leader relies on a legal rational framework and discourses to legitimize their action. The interplay between legal rationality and charisma thus frames exclusionary practices as necessary steps toward urban modernization.

Thirdly, by adopting Foucault's concept of power and resistance, this study demonstrates that power not only operates at the macro level but also at the micro level (Foucault 1978). Vendors resist the repression both individually and collectively. By resisting, street vendors do more than oppose authority; they assert agency and generate a network of power from below (Foucault 1978, p. 94) that constantly reshapes the urban power dynamics. Vendors resistance illustrates that power manifests not only in grandiose movements but in subtle and mundane practices that silently negotiate power without confrontation. This supports the view that power is not static and is in constant flux (Foucault 1978, p. 92). Vendors, therefore, negotiate and re-negotiate in the same space that excludes them. They do not reject the idea of modernity but claim space in it. In saying so, I argue that space itself is a power for vendors.

As a result of the constant power struggle, vendors have been exposed to direct, structural, and cultural violence. Vendors, however, do not go away, as they are driven by necessity. When the issue is rooted in survival, the conflict is difficult to resolve. Without considering the root causes of street vending, such as poverty, the pursuit of opportunities, in urban cities like Kathmandu, is likely to continue, as vending remains an accessible livelihood alternative. Consequently, vendors will continue to be framed as an impediment to urban modernity. The vicious cycle of conflict and resistance is therefore likely to persist, which in turn has implications for sustainable peace.

But an inclusive approach and co-operation may lead to a win-win situation. In this context, it is important to analyze the role of street vendors in urban peacebuilding. Multiple studies have documented the socio-economic significance of vendors. Economically, vending serves as a crucial safety net for the urban poor. For instance, in Kenya, 70% of the urban population is engaged in street vending (Mitullah 2003). The income generated through vending helps meet the essential basic needs of the family (Lyons et al. 2014). By catering to different segments of society, vendors facilitate interaction among diverse backgrounds, thereby creating a lively public space (Zaman and Ahmed 2023). Due to their regular presence in the street, they serve as vigilant watchdogs of society (Deore and Lathia 2019). More than a livelihood strategy, vending is an assertion to dignified life and identity.

The significance of street vending is multifold. Recognizing the role of street vendors in urban peacebuilding can pave the way for a more equitable urban imaginary and move beyond the vicious cycle of conflict and resistance. Since the issue of vending is not a new phenomenon, the KMC alone cannot be questioned. The broader stakeholders, including the federal government, political parties, and political leaders, do have an equal role. Indeed, the use of muscular measures by the KMC raises critical questions on legitimacy because citizens themselves are the prime source of legitimacy. But violence faced by vendors is not new, as it has persisted for more than four decades, even during the government led by political parties with whom vendors are associated. This critical turn raises question on trade unions: why do vendors continue to suffer despite having parties in the mainstream government? These dynamics suggest a broader power politics over urban space.

This study enriches the peace and conflict scholarship by discussing everyday micro level conflict between authorities and vendors and its implication for sustainable peace. By highlighting the everyday struggles of vendors, it calls on policymakers, urban planners, and relevant stakeholders to re-think inclusive and humane urban spaces where everyone including vendors can thrive. While this research is among the first to discuss power, space, and conflict in Nepal, I suggest future research delve into understanding the role of street vendors in urban peacebuilding.

Acknowledgment I would like to express my sincere gratitude to the Central Department of Conflict, Peace and Development Studies, Tribhuvan University, for providing the opportunity to conduct this study. I also thank Prof. Chiranjibi Bhandari and Prof. Dev Raj Dahal for their supervision and valuable feedback. Equally grateful to Prof. Pitambar Bhandari, Prof. Rajib Timilsina, Prof. Mibit Dahal, and Prof. Bijaya Dahal for their insightful lectures on peace, research, development, and migration studies, which significantly enhanced my capacity to undertake this research. Lastly, I am thankful to all vendors, representatives of trade unions, and Kathmandu Metropolitan City for giving their valuable time and attention to my study.

References

Adama, O. (2020). Abuja is not for the poor: Street vending and the politics of public space. *Geoforum, 109*, 14–23. https://doi.org/10.1016/j.geoforum.2019.12.012

Adama, O. (2021). Criminalizing Informal Workers: The Case of Street Vendors in Abuja, Nigeria. *Journal of Asian and African Studies, 56*(3), 533–548. https://doi.org/10.1177/0021909620930740

Adhikari, D. B. (2011). Income generation in informal sector: A case study of the street vendors of Kathmandu Metropolitan City. *Economic Journal of Development Issues*, 1–14. https://doi.org/10.3126/ejdi.v13i0.7193

Adhikari, D. B. (2018). Informal Street Food Trade: A Source of Income Generation in Urban Nepal. *Economic Journal of Development Issues*, 1–17. https://doi.org/10.3126/ejdi.v23i1-2.19062

Adhikari, D. B., Gupta, P., Devkota, N., Paudel, U. R., & Parajuli, S. (2021). Challenges in Transformation of Informal Business Sector towards Formal Business Sector in Nepal: Evidence from Descriptive Cross-Sectional Study. *Asian Journal of Agricultural Extension, Economics & Sociology*, 95–106. https://doi.org/10.9734/ajaees/2021/v39i230534

Adhikari, J., & Hobley, M. (2015). "Everyone is leaving. Who Will Sow Our Fields?" The Livelihood Effects on Women of Male Migration from Khotang and Udaypur Districts, Nepal, to the Gulf Countries and Malaysia. *HIMALAYA, the Journal of the Association for Nepal and Himalayan Studies, 35*(1). https://digitalcommons.macalester.edu/himalaya/vol35/iss1/7

Anjaria, J. S. (2006). Street Hawkers and Public Space in Mumbai. *Economic and Political Weekly, 41*(21), 2140–2146.

Anjaria, J. S. (2010). *The Politics of Illegality: Mumbai Hawkers, Public Space and the Everyday Life of the Law.* (1st ed.). Routledge.

Bayat, A. (1997). Un-civil society: The politics of the "informal people." *Third World Quarterly, 18*(1), 53–72. https://doi.org/10.1080/01436599715055

Bhandari, C. (2015). The Reintegration of Maoist Ex-Combatants in Nepal. *Economic and Political Weekly, 50*, 63–68.

Bhattarai, K., & Pathak, B. (2020). Street Vending, Income Generation and Poverty Implication: The Case of the Kathmandu Valley, Nepal. *Quest Journal of Management and Social Sciences, 2*(2), 194–205. https://doi.org/10.3126/qjmss.v2i2.33269

Bhowmik, S. K. (2005). Street Vendors in Asia: A Review. *Economic and Political Weekly*. http://www.wiego.org/sites/default/files/migrated/publications/files/Bhowmik-Street-Vendors-Asia.pdf

Bishwakarma, D. (2024, July 7). "बालेन सटिी" ले जन्माएका नयाँ यौनकर्मी (Balen City' creates new sex workers). *Himal Khabar*. https://www.himalkhabar.com/news/142202. Accessed 5 February 2024

Boonjubun, C. (2017). Conflicts over streets: The eviction of Bangkok street vendors. *Cities, 70*, 22–31. https://doi.org/10.1016/j.cities.2017.06.007

Bromley, R. (2000). Street vending and public policy: a global review. *International Journal of Sociology and Social Policy, 20*(1/2), 1–28. https://doi.org/10.1108/01443330010789052

Burton, J. (1990). *Conflict: Human needs theory*. Springer.

Central Bureau of Statistics. (2019). *Report on the Nepal Labour Force Survey 2017/2018*. Kathmandu: Central Bureau of Statistics. https://nepalindata.com/media/resources/items/20/bNLFS-III_Final-Report.pdf. Accessed 24 May 2025

Central Bureau of Statistics. (2021). *Waste Management Baseline Survey of Nepal 2020*. Government of Nepal, National Planning Commission. https://unstats.un.org/unsd/envstats/Censuses%20and%20Surveys/Waste-Management-Baseline-Survey-of-Nepal-2020.pdf. Accessed 6 April 2024

Cross, J. C. (1998). Co-optation, Competition, and Resistance: State and Street Vendors in Mexico City. *Latin American Perspectives, 25*(2), 41–61. https://doi.org/10.1177/0094582X9802500203

Crossa, V. (2016). Reading for difference on the street: De-homogenising street vending in Mexico City. *Urban Studies, 53*(2), 287–301. https://doi.org/10.1177/0042098014563471

Democracy Resource Center. (2017). *Nepal's Local Elections 2017: Final Observation Report.* https://www.democracyresource.org/wp-content/uploads/2017/11/DRCN_Local-Election-Final-Observation-Report_Eng.pdf

Democracy Resource Center. (2022). *Observation of Nepal Local Level Election 2022*. https://www.democracyresource.org/wp-content/uploads/2022/07/Observation-of-Nepal-Local-Election-Aug-2022_web_English_2022.pdf. Accessed 20 May 2024

Deore, P., & Lathia, S. (2019). Streets as Public Spaces: Lessons from Street Vending in Ahmedabad, India. *Urban Planning*, *4*(2), 138–153. https://doi.org/10.17645/up.v4i2.2058

Donovan, M. G. (2008). Informal Cities and the Contestation of Public Space: The Case of Bogotá's Street Vendors, 1988—2003. *Urban Studies*, *45*(1), 29–51. https://doi.org/10.1177/0042098007085100

Drummond, L. B. W. (2000). Street Scenes: Practices of Public and Private Space in Urban Vietnam. *Urban Studies*, *37*(12), 2377–2391. https://doi.org/10.1080/00420980020002850

Etzold, B. (2015a). *Street vending in the neoliberal city: a global perspective on the practices and policies of a marginalized economy*. (K. Graaff & N. Ha, Eds.). New York ; Oxford: Berghahn Books.

Etzold, B. (2015b). Selling in Insecurity – Living with Violence: Eviction Drives against Street Food Vendors in Dhaka and the Informal Politics of Exploitation (pp. 164–190). https://doi.org/10.1515/9781782388357-010

Falla, A. M. V., & Valencia, S. C. (2019). Beyond state regulation of informality: understanding access to public space by street vendors in Bogotá. *International Development Planning Review*, *41*(1), 85–105. https://doi.org/10.3828/idpr.2019.3

Foucault, M. (1978). *The history of sexuality* (1st American ed.). New York: Pantheon Books.

Galtung, J. (2009). *Theories of Conflict: Definitions, Dimensions, Negations*. Transcend University Press.

Gautam, A., & Salike, I. P. (2021). Space Utilization by Street Vendors in World Heritage Site: A case of Patan Durbar Square.

Gehlot, R., Kapoor, S., Kaur Butalia, G., & Jain, V. (2023). Invisible Entrepreneurs: Unpacking the Challenges of Street Vendors in the Delhi Metropolitan Area. *International Research Journal of Multidisciplinary Scope*, *04*(03), 01–05. https://doi.org/10.47857/irjms.2023.v04i03.096

Ghimire, P. (2023, August 19). MPs take on mayor shahs action on street vendors. *The Annapurna Express*. https://theannapurnaexpress.com/story/45465/. Accessed 4 February 2024

Gibbings, S. (2013). Unseen Powers and Democratic Detectives: Street Vendors in an Indonesian City. *City & Society*, *25*(2), 235–259. https://doi.org/10.1111/ciso.12018

Gillespie, T. (2017). From quiet to bold encroachment: contesting dispossession in Accra's informal sector. *Urban Geography*, *38*(7), 974–992. https://doi.org/10.1080/02723638.2016.1191792

Gurung, G. (2018). 2015 Earthquake as Driver of Labour Migration. In *South Asia Watch on Trade, Economics and Environment*. https://sawtee.org/publications/Chap-4.pdf. Accessed 28 May 2025

Gurung, S. (2022). Rehabilitation of Internally Displaced Persons and the Support of Nepal Peace Trust Fund in Gorkha District. *Journal of Political Science*, 26–39. https://doi.org/10.3126/jps.v22i1.43037

Hallegatte, S., Vogt-Schilb, A., Rozenberg, J., Bangalore, M., & Beaudet, C. (2020). From Poverty to Disaster and Back: a Review of the Literature. *Economics of Disasters and Climate Change*, *4*(1), 223–247. https://doi.org/10.1007/s41885-020-00060-5

Hanser, A. (2016). Street Politics: Street Vendors and Urban Governance in China. *The China Quarterly*, *226*, 363–382. https://doi.org/10.1017/S0305741016000278

Hennink, M., & Kaiser, B. N. (2022). Sample sizes for saturation in qualitative research: A systematic review of empirical tests. *Social Science & Medicine*, *292*, 114523. https://doi.org/10.1016/j.socscimed.2021.114523

House, R. J., & Howell, J. M. (1992). Personality and charismatic leadership. *The Leadership Quarterly*, *3*(2), 81–108. https://doi.org/10.1016/1048-9843(92)90028-e

Internal Displacement Monitoring Center. (2022). *Disaster displacement: Nepal country briefing*. Geneva: Internal Displacement Monitoring Centre. https://doi.org/10.55363/idmc.fvtm8338

Internal Displacement Monitoring Center, & Norwegian Refugee Council. (2006). *Nepal: IDP return still a trickle despite ceasefire*. https://www.raonline.ch/pages/np/pdf/np_IDMCidp1006.pdf. Accessed 6 April 2024

Internal Displacement Monitoring Center, & Norwegian Refugee Council. (2008). *Nepal: sustainability of IDP returns undermined by lack of assistance*. https://www.raonline.ch/pages/np/pdf/np_IDMCreport0608.pdf. Accessed 4 March 2024

Kathmandu Metropolitan City. (2015). Municipal Police Regulation 2015. https://rb.gy/r47bmm. Accessed 28 May 2024

Kathmandu Metropolitan City. (2023). Notice from Kathmandu Metropolitan City regarding the prohibition of conducting or allowing business activities on sidewalks (footpaths) and other public spaces. https://shorturl.at/AvGdT. Accessed 2 April 2024

Khamrang, L., Datey, A., Kim, S., Dema, T., & Rai, B. (2022). Silent Struggle of the Informal Workers: Everyday Lived Experiences, Challenges and Negotiation of the Women Street Vendors in Thimphu City, Bhutan. In M. I. Hassan, S. Sen Roy, U. Chatterjee, S. Chakraborty, & U. Singh (Eds.), *Social Morphology, Human Welfare, and Sustainability* (pp. 407–430). Cham: Springer International Publishing. https://doi.org/10.1007/978-3-030-96760-4_16

Kim, N. (2012). How much more exposed are the poor to natural disasters? Global and regional measurement. *Disasters*, *36*(2), 195–211. https://doi.org/10.1111/j.1467-7717.2011.01258.x

Kumari, P. (2015). Issues and Challenges for Street Vendors in Delhi. *SSRN Electronic Journal*. https://doi.org/10.2139/ssrn.2779343

Lata, L., Walters, P., & Roitman, S. (2021). A marriage of convenience: Street vendors' everyday accommodation of power in Dhaka, Bangladesh. *Cities*, *84*, 143–150. https://doi.org/10.1016/j.cities.2018.08.002

Lyons, M., Brown, A., & Msoka, C. (2014). Do Micro Enterprises Benefit from the 'Doing Business' Reforms? The Case of Street-Vending in Tanzania. *Urban Studies*, *51*(8), 1593–1612. https://doi.org/10.1177/0042098013497412

Manandhar, R. B. (2022). Factors Affecting Road Traffic Accidents in the Kathmandu valley. *International Research Journal of MMC*, *3*(3), 82–90. https://doi.org/10.3126/irjmmc.v3i3.48639

Milgram, B. L. (2011). Reconfiguring Space, Mobilizing Livelihood Street Vending, Legality, and Work in the Philippines. *Journal of Developing Societies*, *27*(3–4), 261–293. https://doi.org/10.1177/0169796X1102700403

Ministry of Labour, Employment and Social Security. (2022). *Nepal Labour Migration Report 2022*. Kathmandu, Nepal: Government of Nepal. https://giwmscdnone.gov.np/media/app/public/298/posts/1715249334_10.pdf. Accessed 28 June 2024

Mitchell, D. (2003). *The Right to the City: Social Justice and the Fight for Public Space*. Guilford Press. Accessed 22 July 2025

Mitullah, W. V. (2003). Street Vending in African Cities: A Synthesis of Empirical Findings from Kenya, Cote D'Ivoire, Ghana, Zimbabwer, Uganda, and South Africa. *Institute for Development Studies, University of Nairobi*. Accessed 13 May 2024

Molnar, A. (Ed.). (2019). *Strategic Management and Innovative Applications of E-Government*. IGI Global. https://doi.org/10.4018/978-1-5225-6204-7

Munoz, L. (2018). "Recovering" public space and race: Afro-Colombian street vendors in Bogotá, Colombia. *Environment and Planning C: Politics and Space*, *36*(4), 573–588. https://doi.org/10.1177/2399654418761889

Nahar Lata, L. (2021). To Whom Does the City Belong? Obstacles to Right to the City for the Urban Poor in Bangladesh. *Journal of Contemporary Asia*, *51*(4), 638–659. https://doi.org/10.1080/00472336.2020.1791934

National Statistics Office. (2024). *Degree of Urbanization (DEGURBA) in Nepal*. Kathmandu: Government of Nepal, Office of the Prime Minister & Council of Ministers. https://censusnepal.cbs.gov.np/results/downloads/degree-urbanization-report?type=report. Accessed 25 May 2025

National Statistics Office. (2025). *Internal Migration in Nepal*. Kathmandu, Nepal: Government of Nepal, Office of the Prime Minister & Council of Ministers. https://censusnepal.cbs.gov.np/results/files/result-folder/Internal%20Migration%20in%20Nepal%20Report.pdf. Accessed 20 April 2025

Polese, A., Rekhviashvili, L., & Morris, J. (2016). Informal Governance in Urban Spaces: Power, Negotiation and Resistance among Georgian Street Vendors. *Georaphy Research Forum, 36*. https://grf.bgu.ac.il/index.php/GRF/article/view/508. Accessed 7 October 2023

Rai, V. (2024a, March 2). महानगरले गरायो १० हजार फुटपाथ व्यवसायी वस्थिापित (The Metropolis displaced 10,000 sidewalk traders). *ekantipur*. https://ekantipur.com/national/2024/03/02/the-metropolis-displaced-10000-sidewalk-traders-24-39.html. Accessed 12 March 2024

Rai, V. (2024b, May 17). फुटपाथ व्यापारीलाई रोजगारीको कार्ययोजना पेस गर्न सर्वोच्चको आदेश (Supreme Court order to submit action plan for employment of pavement traders). *ekantipur*. https://ekantipur.com/news/2024/05/17/supreme-court-order-to-submit-action-plan-for-employment-of-pavement-traders-37-05.html. Accessed 14 March 2024

Rajagopal, A. (2001). The Violence of Commodity Aesthetics. *Social Text*, *19*(3), 91–113. https://doi.org/10.1215/01642472-19-3_68-91

Rane, S. (2011). Street Vended Food in Developing World: Hazard Analyses. *Indian Journal of Microbiology*, *51*(1), 100–106. https://doi.org/10.1007/s12088-011-0154-x

Raut, N. K. (2021). An assessment of livelihood recovery status of earthquake-affected households in Nepal: A study of coping strategies and their effectiveness. *Progress in Disaster Science*, *9*, 100147. https://doi.org/10.1016/j.pdisas.2021.100147

Saha, D. (2010). Collective Bargaining for Street Vendors in Mumbai: Toward Promotion of Social Dialogue. *Journal of Workplace Rights*, *15*(3–4), 445–460. https://doi.org/10.2190/WR.15.3-4.n

Scott, J. C. (1985). *Weapons of the Weak: Everyday Forms of Peasant Resistance*. Yale University Press.

Sharma, A., Ganguly, R., & Gupta, A. K. (2018). Matrix method for evaluation of existing solid waste management system in Himachal Pradesh, India. *Journal of Material Cycles and Waste Management*, *20*(3), 1813–1831. https://doi.org/10.1007/s10163-018-0703-z

Sharma, P., & Pradhan, P. K. (2017). The Petty Street Vendors and Their Livelihoods of the Kathmandu Valley Cities, Nepal. In A. Li, W. Deng, & W. Zhao (Eds.), *Land Cover Change and Its Eco-environmental Responses in Nepal* (pp. 359–379). Singapore: Springer Singapore. https://doi.org/10.1007/978-981-10-2890-8_16

Shrestha, A. (2019, July 3). *Public Space Management and Urban Commons: Rethinking Street vending in Kathmandu, Nepal.*

Steel, G. (2012). Whose Paradise? Itinerant Street Vendors' Individual and Collective Practices of Political Agency in the Tourist Streets of Cusco, Peru. *International Journal of Urban and Regional Research*, *36*(5), 1007–1021. https://doi.org/10.1111/j.1468-2427.2012.01153.x

Stillerman, J. (2006). The Politics of Space and Culture in Santiago, Chile's Street Markets. *Qualitative Sociology*, *29*, 507–530. https://doi.org/10.1007/s11133-006-9041-x

Subedi, D. B. (2014). Conflict, Combatants, and Cash: Economic Reintegration and Livelihoods of Ex-combatants in Nepal. *World Development*, *59*, 238–250. https://doi.org/10.1016/j.worlddev.2014.01.025

Timalsina, K. P. (2012). Contextualizing Livelihood: Livelihood of street vendors in the urban areas of Kathmandu Metropolitan City, Nepal, *3*(3). https://onlineresearchjournals.com/ijopagg/art/95.pdf

Tucker, J. L. (2017). Affect and the dialectics of uncertainty: Governing a Paraguayan frontier town. *Environment and Planning D: Society and Space*, *35*(4), 733–751. https://doi.org/10.1177/0263775816680821

Upreti, B. C. (2006). The Maoist Insurgency in Nepal: Nature, Growth and Impact. *South Asian Survey*, *13*(1), 35–50. https://doi.org/10.1177/097152310501300103

Weber, M. (1946). Politics As a Vocation. In H. Gerth & W. Mills (Eds.), *From Max Weber: Essays in Sociology* (pp. 77–128). New York: Oxford University Press.

Weber, M. (1978). *Economy and Society: An Outline of Interpretive Sociology*. (G. Roth & C. Wittich, Eds.). Los Angeles: University of California Press.

Williams, C. C., & Gurtoo, A. (2012). Evaluating competing theories of street entrepreneurship: some lessons from a study of street vendors in Bangalore, India. *International Entrepreneurship and Management Journal*, *8*(4), 391–409. https://doi.org/10.1007/s11365-012-0227-2

World Health Organization. Regional Office for South-East Asia. (2016). *Nepal Earthquake 2015: An insight into risks, A vision for resilience*. World Health Organization, Regional Office for South-East Asia. https://iris.who.int/handle/10665/255623. Accessed 4 June 2025

Yadav, U. N., Tamang, M. K., Paudel, G., Kafle, B., Mehta, S., Chandra Sekaran, V., & Gruiskens, J. R. J. H. (2018). The time has come to eliminate the gaps in the under-recognized burden of elder mistreatment: A community-based, cross-sectional study from rural eastern Nepal. *PLOS ONE*, *13*(6), e0198410. https://doi.org/10.1371/journal.pone.0198410

Yatmo, Y. A. (2008). Street Vendors as 'Out of Place' Urban Elements. *Journal of Urban Design*, *13*(3), 387–402. https://doi.org/10.1080/13574800802320889

Zaman, S., & Ahmed, N. N. (2023). The Impact of Street Vendors on Social Sustainability of Public Places: The Case of Khulna City in Bangladesh, *10*(3).

Chapter 14
Does Consumer Animosity Lead to Peace and Sustainability? The Case of Filipino Students as Consumers Amidst the Russia–Ukraine War

Reynaldo Bautista Jr., Luz Suplico Jeong, Takanori Osaki, and Michele Stewart

14.1 Introduction

When Russia invaded Ukraine in 2022, some consumers expressed their disapproval through product boycotts, demonstrating a form of consumer animosity (Akhtar et al. 2024). According to Akhtar et al. (2024), consumer animosity is characterized by antagonistic feelings toward a hostile country based on its past or present actions, which may result in punitive behaviors such as product avoidance. This reaction stems from the desire to penalize the country socially, economically, and politically (Leonidou et al. 2019). Studies have shown that animosity can elicit strong negative emotions among consumers, which in turn may adversely impact brands associated with the offending country.

The salience of consumer animosity has increased in the context of the Russia–Ukraine conflict, highlighting how wartime aggressions influence global consumer perceptions and behaviors toward foreign products (Akhtar et al. 2024; Bahaee and Pisani 2009; Shimp and Sharma 1987). By investigating the factors that contribute to consumer animosity toward products from a perceived aggressor, this research helps expand the limited literature on consumer behavior in developing countries, where political context often intersects with brand perception.

R. Bautista Jr. · L. S. Jeong (✉)
De La Salle University, Manila, Philippines
e-mail: reynaldo.bautista@dlsu.edu.ph; luz.suplico@dlsu.edu.ph

T. Osaki
Kagawa University Graduate School of Management, Takamatsu, Japan
e-mail: osaki.takanori@kagawa-u.ac.jp

M. Stewart
University of Windsor, Windsor, ON, Canada
e-mail: michele.stewart@uwindsor.ca

A. Sharifi et al. (eds.), *Frameworks and Practices Towards a Sustainable and Peaceful Future*, World Sustainability Series,
https://doi.org/10.1007/978-981-95-2310-8_14

This study also explores how digital media intensifies consumer attitudes during international conflicts, contributing to the broader disciplines of international marketing, consumer psychology, and media studies. Understanding how digital media fosters consumer animosity allows marketers, policymakers, and media platforms to develop more responsive strategies that acknowledge consumer sensitivities (Ettenson and Klein 2005; Janiszewski and Laran 2024; Sebastian 2024).

Beyond economic retaliation, consumer animosity can serve as a value-driven form of civic engagement. In geopolitical conflicts such as the Russia–Ukraine war, consumer behavior evolves into symbolic acts of solidarity with victims and resistance against violence. These actions reflect a shift in global consciousness—recognizing ethical consumption as a pathway to peacebuilding and accountability for aggressor states and complicit corporations (Anwar et al. 2018). Therefore, consumer animosity becomes not only a product of conflict but also a potential vehicle for advocating peace and systemic change.

In addition to emphasizing consumer response, this study seeks to identify potential strategies for companies facing animosity. One such approach involves repositioning brands through business model innovation, which can improve consumer perceptions (Areiza-Padilla et al. 2021). Companies may also reduce animosity by engaging in corporate social responsibility and community-building initiatives that foster goodwill and promote harmony. Sustainable practices, likewise, can serve as a differentiator in competitive markets, further reducing hostility and contributing to long-term brand resilience.

14.2 Hypotheses About Consumer Animosity

Perceived intrusiveness is when advertising disrupts an individual's cognitive process. It has been found to negatively affect the consumer's ability to absorb and evaluate information. It refers to the extent to which an advertisement interferes with the consumer's knowing, learning, and understanding (Li et al. 2002). Smink et al. (2020) further describe it as the disruption that triggers irritation or anger. Within the framework of the Uses and Gratifications (U&G) theory, perceived intrusiveness can be considered a negative consequence of media exposure (Ducoffe 1995). When the perceived intrusiveness stems from a country with which the consumer feels political or moral conflict—as in the case of Russian advertising during the Russia–Ukraine war—it may amplify negative feelings toward the originating brand. This leads to the following hypothesis: H1. Perceived intrusiveness positively affects consumer animosity toward Russian brands.

Consumer animosity has been defined as intense resentment toward the products of an offending country, usually triggered by political, military, or economic behavior (Harmeling et al. 2015). When such resentment exists, it can heighten the consumer's emotional vulnerability, potentially evoking fear about threats to political stability, peace, or economic security. According to U&G theory, media that evokes irritation—such as news coverage of war—can result in negative emotional

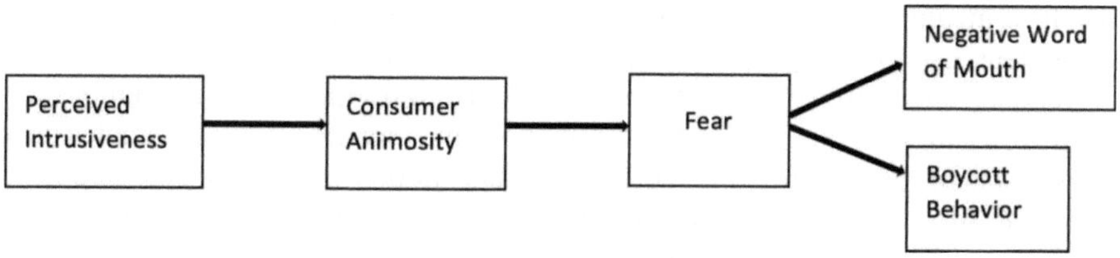

Fig. 14.1 Consumer animosity and consumer behavior

responses, including fear (Ducoffe 1995). Harmeling et al. (2015) explain that fear functions as a retreat emotion, prompting the individual to distance themselves from perceived harm, a point supported by evolutionary psychology (Nesse 1990). Therefore, the next hypothesis is as follows: H2. Consumer animosity is positively related to fear.

As fear increases, consumers may adopt protective behaviors such as distancing themselves from the perceived source of harm. This distancing can manifest in the avoidance of products associated with the offending country (Harmeling et al. 2015). One such form of avoidance is the boycott of products, particularly when consumers believe that continued consumption may lead to personal or collective harm. U&G theory supports this behavior by identifying withdrawal as a typical outcome of negative media engagement (Ducoffe 1995). This behavioral response leads to the next hypothesis: H3. Fear is positively related to product boycott.

In addition to boycotting, fear can also stimulate coping mechanisms such as the dissemination of negative word-of-mouth. Consumers, motivated by a desire to protect others, may engage in warning behaviors by spreading critical information about the offending country's products. Bougie et al. (2003) note that such reactions are rooted in moral emotions, while Siddiqi et al. (2020) suggest that consumers may act prosocially when they perceive a risk to others. U&G theory again aligns with this behavior, as negative outcomes of media engagement often include forms of expression aimed at influencing others (Ducoffe 1995). This leads to the final hypothesis: H4. Fear is positively related to negative word-of-mouth.

Figure 14.1 shows the study's conceptual framework.

14.3 Methodology

The Uses and Gratifications (U&G) theory posits that audiences are not passive recipients of media but active agents who select content based on psychological, emotional, and informational needs (Katz et al. 1973). Media usage is purposeful and motivated, shaped by users' experiences, expectations, and goals. These needs include identity reinforcement, entertainment, integration, and information seeking. Importantly, the theory allows for understanding not only what gratifies users but also how media exposure can lead to negative psychological outcomes such as irritation, fear, and disengagement (Ducoffe 1995; Lim and Ting 2012).

Scholars have applied U&G theory across various digital platforms, demonstrating its robustness in explaining diverse media behaviors. For instance, studies have shown that users engage with the Internet and mobile media for entertainment, escapism, social interaction, and information (Charney and Greenberg 2002; Diddi and LaRose 2006; Ferguson and Perse 2000; Haridakis and Hanson 2009; Kaye 1998; Ko et al. 2005; Papacharissi and Rubin 2000; Stafford and Stafford 2001). Social media, in particular, has been linked to both personal and social integrative needs and tension release (Zolkepli and Kamarulzaman 2011). These motivations shape not just what people consume but how they interpret conflict-related media, which can significantly affect consumer perceptions of foreign brands.

In conflict situations, the same media content that informs can also agitate. Lim and Ting (2012) observed that negative consequences such as web irritation often arise from excessive, intrusive, or emotionally triggering content. Ducoffe (1995) highlighted how intrusive digital advertising can lead to confusion, anxiety, and brand rejection. Harmeling et al. (2015) extended this to consumer animosity, showing that retreat emotions like fear can trigger boycotts and negative word-of-mouth.

To explore these dynamics, the current study employed a quantitative research design grounded in the U&G framework. A cross-sectional survey was conducted online to reach participants from geographically dispersed regions of the Philippines. The online format was particularly appropriate given the focus on digital media use and ensured that respondents were familiar with the digital landscape and likely to be exposed to relevant content regarding the Russia–Ukraine war.

The participants were recruited from universities in Metro Manila, reflecting a population of young, digitally active consumers who are often engaged in global issues and social discourse. Their media habits and exposure to geopolitical narratives made them ideal subjects for examining the emotional and behavioral outcomes of digital media engagement in times of international tension.

The instrument was a structured survey divided into clearly labeled sections to guide respondents through demographic questions and validated measurement scales. Constructs included perceived intrusiveness, consumer animosity, fear, product boycott, and negative word-of-mouth. Each was measured using items from established scales on a five-point Likert scale (5 = Strongly Agree to 1 = Strongly Disagree). The survey was pretested with four respondents, whose feedback improved the survey's flow and readability.

Data analysis was carried out using Structural Equation Modeling (SEM), a statistical technique well-suited for testing complex theoretical models with multiple variables. SEM allowed the researchers to evaluate both direct and indirect relationships between constructs and assess the overall model fit. The method was particularly useful for capturing the nuanced ways in which digital media exposure contributes to consumer attitudes and behaviors in politically charged environments.

In addition to documenting adverse consumer responses, this study also considered how brands might mitigate animosity. Areiza-Padilla et al. (2021) provided an example of a Venezuelan-rooted company, Farmatodo, that faced animosity in Colombia due to political tensions. By upgrading its service experience through sustainable technology, Farmatodo was able to reposition its brand favorably. This suggests that while media-fueled animosity can damage brand perceptions, strategic innovation and socially responsible actions offer a path toward reconciliation and renewed consumer trust.

Ultimately, this chapter illustrates that consumer animosity, while shaped by external events and digital narratives, can also be transformed through intentional and value-aligned brand behavior. Media, far from being a neutral platform, functions as a dynamic space where ethics, commerce, and citizenship intersect.

14.4 Results and Discussion

This chapter presents the findings from the survey data collected from 306 Filipino respondents, offering insights into how digital media exposure during the Russia–Ukraine conflict influences consumer animosity, fear, boycott behavior, and negative word-of-mouth (NWOM). The analysis includes descriptive statistics on respondent demographics, as well as the results of Structural Equation Modeling (SEM) to test the hypothesized relationships between the study's constructs.

The participants in this study were primarily university students from Metro Manila, with representation from both undergraduate and graduate levels. The majority (55.23%) were aged between 19 and 24 years old—classified as Generation Z, known for their technological fluency and heavy reliance on digital media (Ozkan and Solmaz 2017). This age group's digital engagement makes them particularly relevant for analyzing the influence of media content related to geopolitical events.

Table 14.1 presents the demographic profile of respondents across variables such as gender, age, educational attainment, and estimated monthly household income. A nearly even gender split was observed (51.31% male; 48.69% female), while 63.07% of the sample were college students and 36.93% were graduate students. Income data revealed that respondents came from a broad range of economic backgrounds.

Figure 14.2 represents a structural equation model (SEM) path diagram, illustrating the relationships between latent variables and their observed indicators.

Table 14.2 presents the reliability and validity analysis for several constructs, including Boycott, Consumer Animosity (Consu_Animo), Fear, NWOM (Neg_WOM), and Perceived Intrusiveness (Perc_Intrusiv). The constructs were evaluated using Cronbach's alpha, composite reliability (rho_a and rho_c), and average variance extracted (AVE). All metrics indicate that the thresholds for reliability and convergent validity have been met.

Cronbach's alpha measures the internal consistency of the constructs, with values above 0.70 considered acceptable (Nunnally 1978). The alpha values for all

Table 14.1 Respondents' profile

Characteristics	*N* (306)	%
Gender		
Male	157	51.31
Female	149	48.69
Age		
19–24	169	55.23
25–30	112	36.60
31–36	2	0.65
37–42	3	0.98
Over 42	20	6.54
Educational level		
College	193	63.07
Graduate students	113	36.93
Estimated (monthly) household income[a]		
Less than P10,957 ($195.66)	48	15.69
P10,957–P21,194 ($195.66–378.46)	43	14.05
P21,195–P43,828 ($378.48–782.64)	48	15.69
P43,829–P76,669 ($782.66–1369.08)	49	16.01
P76,670–P131,484 ($1369.10–2347.93)	47	15.36
P131,485–P219,140 ($2347.95–3913.21)	27	8.82
P219, 141($3913.23) and up	44	14.38

[a]Exchange rate is $1.00: P56

constructs exceed this threshold, indicating strong reliability. Specifically, Boycott (0.924), Fear (0.922), and NWOM (0.950) demonstrate particularly strong internal consistency, while Consumer Animosity (0.814) and Perceived Intrusiveness (0.815) also meet the criteria for good reliability.

Composite reliability (rho_and rho_c) further supports the reliability of the constructs. According to Hair et al. (2019), values above 0.70 indicate strong reliability, while values above 0.90 are considered excellent. In this study, all constructs exceed the 0.70 benchmark, with NWOM showing especially high reliability (rho_c = 0.967), followed by Boycott (rho_c = 0.952) and Fear (rho_c = 0.951).

Lastly, the average variance extracted (AVE) values for all constructs are above 0.50, which is the standard threshold for adequate convergent validity (Fornell and Larcker 1981). The AVE values for Boycott (0.868), Fear (0.866), and NWOM (0.908) are particularly high, indicating that a substantial portion of the variance is captured by the constructs themselves. Consumer Animosity (0.727) and Perceived Intrusiveness (0.724) also meet the minimum criteria for convergent validity.

Table 14.3 presents the analysis of discriminant validity using the Heterotrait-Monotrait ratio (HTMT), a method that examines whether the constructs are distinct from one another. According to Henseler et al. (2015), discriminant validity is considered adequate when the HTMT values are below the threshold of 0.85. In this analysis, all the constructs—Boycott, Consumer Animosity (Consu_Animosity),

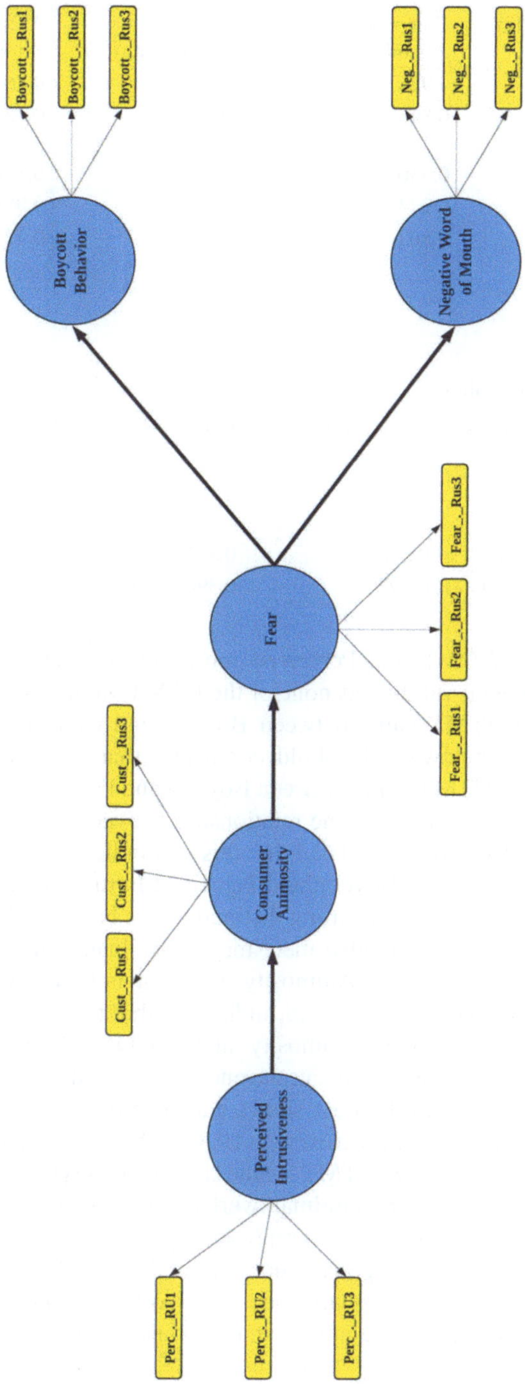

Fig. 14.2 SEM path diagram

Table 14.2 Construct reliability and validity

Constructs	Cronbach's alpha	Composite reliability (rho_a)	Composite reliability (rho_c)	Average variance extracted (AVE)
Boycott	0.924	0.926	0.952	0.868
Consu_ Animo	0.814	0.846	0.889	0.727
Fear	0.922	0.930	0.951	0.866
Neg_WOM	0.950	0.974	0.967	0.908
Perc_ Intrusive	0.815	0.862	0.887	0.724

Table 14.3 Discriminant validity

Constructs	Boycott	Consu_Animosity	Fear	Neg_WOM	Perc_Intrusiveness
Boycott					
Consu_Animosity	0.742				
Fear	0.456	0.768			
Neg_WOM	0615	0.582	0.353		
Perc_Intrusiveness	0.221	0.397	0.390	0.170	

Fear, NWOM (Neg_WOM), and Perceived Intrusiveness (Perc_Intrusiveness)—demonstrate discriminant validity, as none of the HTMT values exceed 0.85.

For instance, the HTMT value between Boycott and Consumer Animosity is 0.742, which falls well below the threshold, confirming that these two constructs are distinct. Similarly, the HTMT value between Boycott and Fear is 0.456, further supporting their discriminant validity. The relationship between Boycott and NWOM shows a moderate HTMT value of 0.615, but this still indicates sufficient discriminant validity. Additionally, the Boycott and Perceived Intrusiveness value is 0.221, indicating a weak relationship and strong distinctiveness between these constructs.

The other pairs of constructs also meet the discriminant validity criteria. The HTMT value between Consumer Animosity and Fear is 0.768, which, although higher, remains below the 0.85 threshold, indicating that they are distinguishable. The values between Consumer Animosity and NWOM (0.582) and between Consumer Animosity and Perceived Intrusiveness (0.397) also confirm that these constructs are distinct. Similarly, Fear shows adequate discriminant validity with NWOM (0.353) and Perceived Intrusiveness (0.390), both of which are below the critical threshold. Lastly, the HTMT value between NWOM and Perceived Intrusiveness is 0.170, suggesting minimal overlap between these constructs.

Table 14.4 shows the diagonal values representing the square root of the AVE for each construct, while the off-diagonal values show the correlations between constructs. For each construct, the square root of its AVE (diagonal value) must be higher than any off-diagonal value in the same row and column.

Table 14.4 Fornell–Larcker criterion

	Boycott	Consu_Animosity	Fear	Neg_WOM	Perc_Intrusiveness
Boycott	0.932				
Consu_Animosity	0.639	0.853			
Fear	0.423	0.681	0.930		
Neg_WOM	0.581	0.504	0.340	0.953	
Perc_Intrusiveness	0.198	0.345	0.354	0.139	0.851

Table 14.5 Direct path coefficients

Paths	Original sample (O)	Sample mean (M)	Standard deviation (STDEV)	T statistics (|O/STDEV|)	*P* values
Consu_Animosity → Fear	0.681	0.682	0.038	17.707	0.000
Fear → Boycott	0.423	0.424	0.069	6.151	0.000
Fear → Neg_WOM	0.340	0.340	0.064	5.300	0.000
Perc_Intrusiveness → Consu_Animosity	0.345	0.351	0.063	5.482	0.000

For the Boycott construct, the square root of its AVE is 0.932, which is greater than its correlations with Consumer Animosity (0.639), Fear (0.423), NWOM (0.581), and Perceived Intrusiveness (0.198). This indicates that the Boycott construct has discriminant validity. Similarly, Consumer Animosity has a square root of AVE of 0.853, exceeding its correlations with other constructs such as Boycott (0.639), Fear (0.681), NWOM (0.504), and Perceived Intrusiveness (0.345). This shows that Consumer Animosity is distinct from other constructs.

For Fear, the square root of its AVE is 0.930, which is greater than its correlations with Boycott (0.423), Consumer Animosity (0.681), NWOM (0.340), and Perceived Intrusiveness (0.354), demonstrating discriminant validity for this construct. NWOM has a square root of AVE of 0.953, which is also higher than its correlations with all other constructs, including Boycott (0.581), Consumer Animosity (0.504), Fear (0.340), and Perceived Intrusiveness (0.139), confirming its discriminant validity. Finally, Perceived Intrusiveness has a square root of its AVE of 0.851, which surpasses its correlations with Boycott (0.198), Consumer Animosity (0.345), Fear (0.354), and NWOM (0.139), ensuring that it is distinct from the other constructs.

Table 14.5 shows the relationships between consumer animosity, fear, boycott behavior, NWOM, and perceived intrusiveness in the context of Filipino perceptions of Russia–Ukraine conflict. Below are the discussions on the relationships between variables.

14.4.1 Consumer Animosity to Fear

The coefficient of 0.681 indicates a very strong and positive relationship between consumer animosity and fear. In the context of the Russia–Ukraine conflict, this suggests that as Filipinos' animosity toward Russia increases, their feelings of fear also significantly rise. The extremely high T-statistic (17.707) and the p-value (0.000) confirm that this relationship is statistically significant and highly robust. This means that the more Filipinos harbor negative emotions toward Russia due to the conflict, the more they experience fear—likely driven by concerns about geopolitical instability, potential economic impacts, or ethical implications. This insight validates the existing literature that consumer animosity can lead to fear (Harmeling et al. 2015). Further, it validates the U&G theory that states that media use can also have negative consequences (Ducoffe 1995).

Understanding the cause of this fear will enable companies to convert fear into more positive positioning and innovations (Areiza-Padilla et al. 2021). According to Areiza-Padilla et al. (2021), companies that try to offer better service and virtual channels can still have positive comments from consumers even if these companies originate from the hostile country. This suggests that companies facing consumer animosity and fear can still engage consumers by innovating on their business models. One possible innovation is to build a sustainable or eco-friendly brand (Scepkova and Zauskova 2024). According to Scepkova and Zauskova (2024), the company can interact with consumers using social media and address their concerns, which may include fear, and can change negative sentiments to more constructive engagement.

This chapter proposes that fear can also catalyze constructive consumer engagement. Fear arising from geopolitical conflicts may initially trigger boycott intentions, but it also prompts consumers to seek transparency, demand ethical accountability, and support brands that align with peace and sustainability values. Companies can channel this consumer fear into dialogue by actively communicating their humanitarian positions, sustainability practices, and support for conflict resolution. For instance, brands facing backlash have launched educational campaigns or partnered with civil society organizations to engage consumers more deeply and constructively (Scepkova and Zauskova 2024). By responding to fear with openness and values-driven communication, brands can convert crisis into an opportunity for peace-driven engagement.

14.4.2 Fear to Boycott Behavior

The coefficient of 0.423 shows a strong and positive relationship between fear and boycott behavior. For Filipinos, this indicates that the fear generated by animosity toward Russia translates into a significant likelihood of boycotting Russian products. The T-statistic (6.151) and the low p-value (0.000) reinforce the statistical significance of this finding. In this context, it can be inferred that as Filipinos' fear concerning the Russia–Ukraine conflict increases, they are more likely to avoid purchasing products associated with Russia as a form of protest or self-preservation. This validates existing studies that show that fear can lead to product boycott (Akhtar et al. 2024; Harmeling et al. 2015). It also validates the U&G theory that states that media use can lead to negative consequences (Ducoffe 1995).

At the consumer level, boycott behavior is a non-violent form of protest against the hostile country as opposed to war (Anwar et al. 2018). It can bring positive changes when companies facing product boycotts can address the consumers' concerns through social actions that can bring the company into a more positive light or through active interaction in social media (Scepkova and Zauskova 2024).

Traditionally seen as a withdrawal of economic support, boycott behavior in this chapter is reinterpreted as a peaceful form of protest. It represents a moral stance wherein consumers disengage from products or brands associated with aggressor states, signaling disapproval through non-violent means. This aligns with transformative approaches to conflict resolution, where civic resistance is channeled through market choices rather than political confrontation. For example, the widespread boycotts of Russian-affiliated brands following the Ukraine invasion reflect consumers' desire to support peace without engaging in hostility (Harmeling et al. 2015). Boycotting thus becomes a civic act, reinforcing peace-oriented consumption and ethical accountability in the global market.

14.4.3 Fear to NWOM

The coefficient of 0.340 suggests a moderately strong and positive relationship between fear and NWOM. This implies that as fear increases among Filipinos regarding the Russia–Ukraine conflict, they are more likely to engage in negative discussions about Russia and its products. The T-statistic (5.300) and p-value (0.000) confirm that this relationship is statistically significant. Essentially, fear driven by animosity leads Filipinos not only to boycott Russian goods but also to speak negatively about Russia within their social circles, potentially influencing others' perceptions. This supports existing literature that shows that consumer animosity can lead to NWOM (Akhtar et al. 2024) and that media use can lead to negative consequences (Ducoffe 1995).

NWOM can take three forms: (1) saying bad things about the hostile country and its products, (2) recommending non-purchases of products from the hostile country, and (3) discrediting products from the hostile country (Grappi et al. 2013). Understanding that NWOM is caused by fear can encourage companies facing animosity to engage in social action that can decrease fear. One such action can be an interactive social media presence where customers' fears are being addressed (Scepkova and Zauskova 2024). This is a step toward decreasing animosity and building good relations with the customers. By addressing and resolving fear and engaging in building good relations with customers, NWOM can be an opportunity to achieve peace and sustainability.

14.4.4 Perceived Intrusiveness to Consumer Animosity

The coefficient of 0.345 indicates a positive relationship between perceived intrusiveness and consumer animosity. This suggests that the more Filipinos perceive the Russia–Ukraine conflict as intrusive—perhaps due to frequent and intense media coverage or perceived relevance to their daily lives—the more likely they are to develop animosity toward Russia. The T-statistic (5.482) and *p*-value (0.000) indicate that this relationship is statistically significant. Perceived intrusiveness in this context might stem from the conflict's visibility in the media, its global political implications, or the moral and ethical considerations it raises among Filipinos. This validates the U&G theory that states that too much pop-up, glaring advertisements and irritating features can drive the consumer to unfavorable behavior (Ducoffe 1995).

Table 14.6 presents several indirect paths that detail the relationships between perceived intrusiveness, consumer animosity, fear, boycott behavior, and NWOM. Below are the discussions of the relationships among the variables.

Table 14.6 Indirect path coefficients

Paths	Original sample (O)	Sample mean (*M*)	Standard deviation (STDEV)	T statistics (|O/STDEV|)	*P* values
Perc_Intrusiveness → Consu_Animosity → Fear → Neg_WOM	0.080	0.082	0.025	3.182	0.001
Consu_Animosity → Fear → Neg_WOM	0.231	0.233	0.050	4.585	0.000
Perc_Intrusiveness → Consu_Animosity → Fear → Boycott	0.099	0.102	0.029	3.477	0.001
Perc_Intrusiveness → Consu_Animosity → Fear	0.235	0.240	0.047	4.952	0.000
Consu_Animosity → Fear → Boycott	0.288	0.291	0.057	5.075	0.000

14.4.5 Perceived Intrusiveness to Consumer Animosity to Fear to NWOM

The path coefficient of 0.080 indicates a significant indirect effect of perceived intrusiveness on negative word-of-mouth (NWOM), mediated through consumer animosity and fear. In the context of the Russia–Ukraine conflict, this suggests that as Filipinos perceive the conflict as more intrusive—possibly due to its extensive media coverage—they develop stronger animosity toward Russia. This animosity then increases their fear, which further leads to a greater likelihood of spreading NWOM about Russia. The T-statistic (3.182) and p-value (0.001) confirm the statistical significance of this indirect pathway. This also supports the U&G theory that states that when media becomes too intrusive, it will lead to negative consequences (Ducoffe 1995).

14.4.6 Consumer Animosity to Fear to NWOM

The coefficient of 0.231 shows a strong and significant indirect effect of consumer animosity on NWOM through fear. Among Filipinos, animosity toward Russia due to the conflict leads to increased fear, which in turn amplifies their likelihood of engaging in negative discussions about Russia. The high T-statistic (4.585) and very low p-value (0.000) indicate this relationship is robust and statistically significant. This supports the U&G theory that media behavior can lead to negative consequences (Ducoffe 1995).

14.4.7 Perceived Intrusiveness to Consumer Animosity to Fear to Boycott

This path (0.099) shows that perceived intrusiveness has a significant indirect effect on boycott behavior through the mediating roles of consumer animosity and fear. For Filipinos, perceiving the Russia–Ukraine conflict as highly intrusive fosters animosity toward Russia, which then increases fear, leading to a higher likelihood of boycotting Russian products. The T-statistic (3.477) and p-value (0.001) confirm this indirect path is statistically significant. Based on the U&G theory, fear can be a negative consequence of media behavior (Ducoffe 1995).

14.4.8 Perceived Intrusiveness to Consumer Animosity to Fear

The coefficient of 0.235 indicates a significant indirect effect of perceived intrusiveness on fear, mediated through consumer animosity. This suggests that when Filipinos perceive the Russia–Ukraine conflict as intrusive, it heightens their animosity toward Russia, which subsequently increases their fear. The T-statistic (4.952) and p-value (0.000) show that this pathway is statistically significant.

14.4.9 Consumer Animosity to Fear to Boycott

This path (0.288) reveals a strong indirect effect of consumer animosity on boycott behavior through fear. Among Filipinos, greater animosity toward Russia due to the conflict leads to increased fear, which in turn significantly drives boycott behavior. The very high T-statistic (5.075) and p-value (0.000) underline the robustness and significance of this indirect pathway.

14.4.10 Filipino Perceptions of the Russia–Ukraine War

In the context of the Russia–Ukraine conflict, the data suggest that Filipino perceptions are significantly influenced by how intrusive they find the conflict. When Filipinos perceive the conflict as highly intrusive, it leads to increased consumer animosity toward Russia. This animosity, in turn, elevates fear, which acts as a powerful driver of both boycott behavior and NWOM.

The strongest indirect effect observed is in the path from consumer animosity to boycott via fear (0.288), indicating that fear is a critical mediator in transforming animosity into tangible consumer actions such as boycotting Russian products. The significant indirect effects also suggest that Filipinos' perceptions of the conflict are not only influenced by direct emotions (like animosity) but are also shaped by the mediating role of fear, which amplifies their likelihood of engaging in negative consumer behaviors against Russia.

For businesses and policymakers, understanding these pathways is crucial in managing and potentially mitigating the negative impacts of consumer animosity in international markets, particularly during sensitive geopolitical conflicts like the one between Russia and Ukraine.

14.4.11 Practical Implications

The findings of this study provide several insights for marketing practitioners, particularly for brands operating in or affiliated with countries involved in geopolitical conflict. As shown in the model, perceived media intrusiveness significantly contributes to consumer animosity, leading to behavioral consequences such as product boycotts and negative word-of-mouth. For brands associated with nations engaged in war or aggression, maintaining a neutral or values-based brand image is crucial. Communication strategies that openly distance the brand from political affiliations or demonstrate alignment with humanitarian efforts may mitigate the impact of negative media exposure.

Marketers must also be sensitive to the emotional tone of media coverage during conflicts. Monitoring shifts in sentiment and identifying early indicators of consumer fear or anger can lead to proactive messaging. Public relations initiatives that express empathy, solidarity with affected populations, or support for peacebuilding may help preserve brand equity during periods of public backlash.

In terms of digital marketing strategy, brands should re-evaluate how intrusive or persistent their content appears, especially in emotionally charged contexts. Aggressive advertising, pop-ups, or emotionally disconnected promotions may worsen animosity. A more effective approach may involve empathetic messaging or campaigns that support dialogue, healing, and collective values. Soft-touch, emotionally resonant content that focuses on unity, sustainability, or community relief efforts can enhance engagement and minimize reputational risk.

Segmentation strategies should also reflect differing levels of sensitivity across markets. As the study suggests, fear and animosity vary by individual exposure to and interpretation of media narratives. Brands can benefit from tailoring messages and engagement efforts according to these regional or psychographic differences, especially in areas where media content about a conflict is highly salient.

Corporate social responsibility (CSR) becomes particularly important in managing conflict-related backlash. Engagement in peace-oriented or humanitarian projects can demonstrate that the brand is not complicit but instead part of a global effort to address suffering or injustice. Brands such as McDonald's and Starbucks, which suspended operations in Russia following the invasion of Ukraine, were perceived more favorably for taking what consumers viewed as principled action (Akhtar et al. 2024).

Ethical consumption also offers a consumer-driven avenue for brands to rebuild trust. Certifications such as Fair Trade or Human Rights Compliance have become tools for ethical signaling in response to global issues, including labor rights and environmental harm. For example, consumer resistance to Xinjiang cotton prompted a shift toward brands that visibly adopted transparency and ethical sourcing standards (Leonidou et al. 2019).

Moreover, brand-led engagement can transform animosity into collaborative public discourse. Companies like H&M and Nike, previously criticized for labor practices, have since implemented public-facing reforms, including transparency reports and partnerships with civil society groups (BBC News 2021). These responses reflect constructive brand behavior—aimed at restoring trust and showing accountability—thus reframing consumer-brand relations through shared goals of justice, sustainability, and social healing (Scepkova and Zauskova 2024).

14.5 Limitations and Suggestions for Future Research

While this study offers meaningful insights, it is important to acknowledge several limitations that may guide future research. The study did not include actual purchasing data. Although intention-based constructs such as boycott intention and negative word-of-mouth are well-established proxies in consumer behavior literature, particularly when sales data are inaccessible (Akhtar et al. 2024; Klein et al. 1998), future research can benefit from integrating behavioral or transactional data to enhance external validity.

Another limitation is the homogeneity of the sample. The study used Filipino university students as respondents. While this demographic is appropriate given Generation Z's high digital media usage and social awareness (Ozkan and Solmaz 2017), this choice limits the generalizability of the findings. The overreliance on student samples, which is a common limitation in consumer research, may not adequately capture the perspectives of older, less digitally engaged consumers or those from different socioeconomic backgrounds. A more diverse sample would allow for comparative insights and a broader understanding of consumer animosity across demographic segments.

While the Philippines imports mostly industrial and raw materials from Russia—products that university students do not typically consume—consumer animosity research emphasizes symbolic rather than purely behavioral consumption. Boycotts are often expressive acts, motivated by moral stances, rather than direct economic impacts. This suggests that even without direct product interaction, consumer attitudes remain relevant, aligning with prior research (Klein et al. 1998).

The study also employed a cross-sectional design, which provides a snapshot of attitudes and behaviors but does not capture changes over time. A longitudinal approach could offer richer insights into how media exposure and animosity evolve. Nevertheless, cross-sectional designs remain valid for exploring theoretical relationships and constructing predictive models (Hair et al. 2019).

While the study exclusively used Structural Equation Modeling (SEM), future studies could incorporate qualitative or experimental methods to triangulate findings. SEM remains appropriate for investigating complex behavioral relationships, particularly when exploring indirect effects. However, integrating interviews or experimental validation could yield deeper contextual understanding (Harmeling et al. 2015; Leonidou et al. 2019).

Perceived intrusiveness was analyzed explicitly as an antecedent in our theoretical model, providing insights into how media exposure heightens consumer animosity, rather than as a moderator. The current analysis effectively captures the significant influence of perceived intrusiveness on consumer animosity as demonstrated in previous literature (Li et al. 2002; Smink et al. 2020). Although moderation analysis would enrich the understanding for future studies, the existing approach still robustly demonstrates critical theoretical pathways.

The current model did not include interaction terms, which could have added nuance to the relationships among variables. Although not essential to the study's core arguments, future work could examine moderation effects, particularly how contextual or personal factors influence the strength of the animosity-behavior pathway.

Further, although the study is grounded in media theory, it did not include a content analysis of digital media narratives. Existing literature provides strong evidence that media framing impacts consumer animosity (Fernandez-Ferrin et al. 2015), yet a content analysis could enhance the robustness of future models by directly measuring narrative tone and framing.

The research did not explicitly control for economic or cultural variables. While the sample's homogeneity provided a controlled cultural lens, future research could employ multi-country or multi-cultural designs to assess how these factors influence animosity expression and media interpretation.

14.6 Conclusion

Digital media plays a pivotal role in shaping consumer animosity during geopolitical conflicts. Findings from this study indicate that increased media intrusiveness correlates with heightened animosity toward products associated with hostile countries. This animosity, in turn, triggers consumer responses such as fear, negative word-of-mouth, and product boycotts. These results highlight the need for marketers, policymakers, and media platforms to account for emotional and ethical sensitivities when crafting communications in conflict-laden contexts (Ettenson and Klein 2005).

The study underscores the importance of monitoring the emotional tone and framing of digital content. Early cues in media coverage can significantly affect public perception, with implications for brand management and strategic planning. Companies such as McDonald's, Apple, and Ikea that strategically withdrew from the Russian market following the Ukraine invasion were perceived positively by global consumers (Akhtar et al. 2024). This reinforces the argument that aligning with public sentiment during crises can protect, and even enhance, brand reputation.

Moreover, this chapter revisits the Uses and Gratifications (U&G) theory, demonstrating that media engagement—while often satisfying psychological needs—can also lead to unintended negative outcomes such as fear and consumer animosity (Ducoffe 1995). Understanding this dual nature of media use is crucial for brands operating in politically sensitive markets.

From a conflict transformation perspective, the study explores how negative emotions such as fear and boycott can be reframed into constructive engagement.

Boycotts function as non-violent protests, enabling consumers to express dissent in ethical and peaceful ways. Anwar et al. (2018) support the view that boycott behaviors are aligned with strategic non-violence and peacebuilding. Recent global campaigns against unethical labor practices and authoritarian regimes exemplify the power of collective economic action to enforce accountability and uphold human rights (BBC News 2021; Business and Human Rights Resource Centre 2021; Reuters 2021).

In parallel, fear, often dismissed as a negative emotion, can serve as a catalyst for engagement. Brands that respond to consumer fear with transparent communication and ethical practice can build stronger, trust-based relationships. Companies like Patagonia, Everlane, and TotalEnergies have leveraged fear-based criticisms to deepen their sustainability commitments and foster consumer loyalty (Garcia 2019; Gross and Stelzenmuller 2024). These examples illustrate how fear, when constructively addressed, can lead to transformative business strategies that support peace and social good.

Ultimately, this study calls for greater sensitivity in marketing strategies during times of conflict and supports a vision of business as a proactive force for peace. Rather than merely mitigating harm, brands can play an active role in fostering reconciliation, ethical consumption, and collective well-being.

References

Akhtar, N.; Khan, H.; Siddiqi, U.; Islam, T. & Atanassova, I. (2024) Critical perspective on consumer animosity amidst Russia-Ukraine war. *Critical Perspectives on International Business,* 20 (1).49-70. https://doi.org/10.1108/cpoib-02-2023-0015

Anwar, F.; Fry, D.P. & Grigaityte, I. (2018). Aggression prevention and reduction in diverse cultures and contexts. *Current Opinion in Psychology,* 19(1). 49-54. https://doi.org/10.1016/j.copsyc.2017.03.029

Areiza-Padilla, J.A.; Moise, M.S. & Manzi-Puertas, M.A. (2021). Breaking the barriers of animosity: innovation in business models as a positioning strategy. *Heliyon,* https://doi.org/10.1016/j.heliyon.2021.e07545

Bahaee, M. & Pisani, M.J. (2009). Iranian consumer animosity and U.S. products: a witch's brew or elixir? *International Business Review,* 18(2).199-210. https://doi.org/10.1016/j.ibusrev.2009.02.002.

BBC News. (2021, March 25). *Nike, H&M face China fury over Xinjiang cotton 'concerns'.* https://www.bbc.com/news/world-asia-china-56519411

Bougie, R., Pieters, R., & Zeelenberg, M. 2003. Angry customers don't come back, they get back: The experience and behavioral implications of anger and dissatisfaction in services. *Journal of the Academy of Marketing Science*, 31(4): 377–393. https://doi.org/10.1177/0092070303254412

Business & Human Rights Resource Centre. (2021, March 24). *H&M faces boycott in China over previous decision not to use Xinjiang cotton due to concerns over alleged forced labor.* https://www.business-humanrights.org/en/latest-news/hm-faces-boycott-in-china-over-previous-decision-not-to-use-xinjiang-cotton-due-to-concerns-over-alleged-forced-labour/.

Charney, T., & Greenberg, B. (2002). *Uses and gratifications of the Internet.* In C. A. Lin & D. J. Atkin (Eds.), *Communication technology and society: Audience adoption and uses* (pp. 379–407). Hampton Press.

Diddi, A., & LaRose, R. (2006). Getting hooked on news: Uses and gratifications and the formation of news habits among college students in an internet environment. *Journal of Broadcasting & Electronic Media*, 50(2), 193–210. https://doi.org/10.1207/s15506878jobem5002_2R

Ducoffe, R. H. (1995). How consumers assess the value of advertising. *Journal of Current Issues & Research in Advertising*, *17*(1), 1-18. https://doi.org/10.1080/10641734.1995.10505022

Ettenson, R., & Klein, J. G. (2005). The fallout from French nuclear testing in the South Pacific: A longitudinal study of consumer boycotts. *International Marketing Review*, *22*(2), 199-224. https://doi.org/10.1108/02651330510593278

Ferguson, D. A., & Perse, E. M. (2000). The World Wide Web as a functional alternative to television. *Journal of Broadcasting & Electronic Media* 44(2), 155–174. https://doi.org/10.1207/s15506878jobem4402_1

Fernandez-Ferrin, P.; Bande-Vilela, B.; Klein, J.G. & Del Rio-Araujo,M.L, (2015). Consumer ethnocentrism and consumer animosity: antecedents & consequences. *International Journal of Emerging Markets,* 10(1). 73-88. https://doi.org/10.1108/IJOEM-11-2011-0102

Fornell, C. & Larcker, D. F. (1981). Evaluating Structural Equation Models with Unobservable Variables and Measurement Error. *Journal of Marketing Research*, 18(1), 39-50. https://doi.org/10.2307/3151312

Garcia, D. (2019, August 30). From Patagonia to Everlane: how to fit the fashion system into a sustainable DNA. *The Global Fashion Business Journal*. https://www.themds.com/back-stage/from-patagonia-to-everlane-how-to-fit-the-fashion-system-into-a-sustainable-dna.html

Grappi, S.; Romani, S. & Bagozzi, R. (2013). Consumer response to corporate irresponsible behavior: moral emotions and virtues. *Journal of Business Research,* 66 (10). 1814-1821. https://doi.org/10.1016/j.jbusres.2013.02.002.

Gross, S. & Stelzenmuller, C. (2024, June 18). *Europe's Messy Russian Gas Divorce*. Brookings. https://www.brookings.edu/articles/europes-messy-russian-gas-divorce/

Harmeling, C. M., Magnusson, P. & Singh, N. (2015). Beyond anger: A deeper look at consumer animosity. *Journal of International Business Studies*, 46(6), 676–693. http://www.jstor.org/stable/43653767

Hair, J. F., Black, W. C., Babin, B. J., & Anderson, R. E. (2019). *Multivariate Data Analysis* (8th ed.). Pearson

Haridakis, P., & Hanson, G. (2009). Social interaction and co-viewing with YouTube: blending mass communication reception and social connection. *Journal of Broadcasting & Electronic Media*, 53(2), 317–335. https://doi.org/10.1080/08838150902908270

Henseler, J., Ringle, C.M. & Sarstedt, M. (2015). A new criterion for assessing discriminant validity in variance-based structural equation modeling. *Journal of the Academy of Marketing Science*, 43(1).115-135. https://doi.org/10.1007/s11747-014-0403-8

Janiszewski, C., & Laran, J. (2024). Paradigmatic conformity blinds us from opportunity: a rejoinder. *Consumer Psychology Review*, 7(1), 127-130. https://psycnet.apa.org/doi/10.1002/arcp.1094

Katz, E., Blumler, J. G., & Gurevitch, M. (1973). Uses and gratifications research. *The Public Opinion Quarterly*, 37(4), 509-523. https://psycnet.apa.org/doi/10.1086/268109

Kaye, B. K. (1998). Uses and gratifications of the World Wide Web: From couch potato to Web potato. *The New Jersey Journal of Communication,* 6(1), 21–40. https://doi.org/10.1080/15456879809367333

Klein, J.G.; Ettenson, R. & Morris, M.D. (1998). The animosity model of foreign product purchase: an empirical test in The People's Republic of China. *Journal of Marketing,* 62 (1) 89-100. https://doi.org/10.1177/002224299806200108

Ko, H., Cho, C. H., & Roberts, M. S. (2005). Internet uses and gratifications: A structural equation model of interactive advertising. *Journal of Advertising*, 34(2), 57–70. https://www.jstor.org/stable/4189297

Lim, W.M. & Ting, D.H. (2012). E-shopping: an analysis of the uses and gratifications theory. *Modern Applied Science,* 6 (5). 48-63. https://doi.org/10.5539/mas.v6n5p48

Leonidou, L.C., Kvasova, O., Christodoulides, P. & Tokar, S. (2019). Personality traits, consumer animosity, and foreign product avoidance: the moderating role of individual cultural characteristics. *Journal of International Marketing*, (27) 2.76-96, https://doi.org/10.1177/1069031X19834936

Li, H., Edwards, S.M. & Lee, J.H. (2002). Measuring the intrusiveness of advertisements: scale development and validation. *Journal of Advertising*,31(2), 37-47, https://doi.org/10.1080/00913367.2002.10673665.

Nesse, R. M. (1990). Evolutionary explanations of emotions. *Human Nature*, 1(3): 261–289. https://doi.org/10.1007/BF02733986

Nunnally, J. C. (1978). *Psychometric Theory* (2nd ed.). McGraw-Hill.

Ozkan, M. & Solmaz, B. (2017). Generation Z - The global market's new consumers- and their consumption habits: generation Z consumption scale. *European Journal of Multidisciplinary Studies*, 2(5), 222-229.

Papacharissi, Z., & Rubin, A. M. (2000). Predictors of internet use. *Journal of Broadcasting & Electronic Media,* 44(2), 175–196. https://doi.org/10.1207/s15506878jobem4402_2

Reuters (2021, March 25). *Nike, Adidas join brands feeling Chinese social media heat over Xinjiang cotton.* https://www.reuters.com/world/china/after-hm-more-foreign-retail-brands-under-fire-china-xinjiang-fallout-2021-03-25/

Sebastian, C. (2024, September 22.) Ukraine wants to fire Western long-range weapons into Russia. But the battlefield gain is not clear-cut. *CNN.* https://edition.cnn.com/2024/09/22/europe/ukraine-western-long-range-weapons-russia-intl/index.html.

Scepkova, S. & Zauskova, A. (2024) Ecoproducts in the dynamic era of digital communications: a neuromarketing study of users' emotional reactions on social media. *Communication Today,* 15(2).84-110.

Siddiqi, U.I., Sun, J.& Akhtar, N. (2020). Ulterior motives in peer and expert supplementary online reviews and consumers' perceived deception. *Asia Pacific Journal of Marketing and Logistics*, 33(1)73-98. https://doi.org/10.1108/APJML-06-2019-0399.

Shimp, T.A. & Sharma, S. (1987). Consumer ethnocentrism: construction and validation of the CETSCALE. *Journal of Marketing Research,* 24 (1) 280-289. https://doi.org/10.2307/3151638

Smink, A.R., van Reijmersdal, E.A., van Noort, G. & Neijens, P.C. (2020) Shopping in augmented reality: the effects of spatial presence, personalization and intrusiveness on app and brand responses. *Journal of Business Research*, 118, 474-485, https://doi.org/10.1016/j.jbusres.2020.07.018

Stafford, T. F., & Stafford, M. R. (2001). Identifying motivations for the use of commercial web sites. *Information Resources Management Journal*, 14(1), 22–30.

Zolkepli, I. A., & Kamarulzaman, Y. (2011). Understanding social media adoption: The role of perceived media needs and technology characteristics. *World Journal of Social Sciences,* 1(1), 188–199.

Chapter 15
Foreign Aid and Failed Peace Dividends: A Critical Study of Development Aid in the Bangsamoro, Philippines

Raihan A. Yusoph

15.1 Introduction

For decades, the Bangsamoro in the southern Philippines has been the center of one of Southeast Asia's most protracted conflicts. Rooted in historical grievances, armed struggles for self-determination, and recurring cycles of violence, the region has faced economic and political marginalization. In response, a wide range of foreign aid, including bilateral partners such as Japan, Australia, and the United States, as well as multilateral institutions like the World Bank and the UNDP, has supported development interventions in coordination with the Government of the Philippines (GPH). These aid efforts have aimed to address underdevelopment and poverty in most conflict-affected regions worldwide (Riddell 2007; De Ree and Nillesen 2009; Asongu 2019; Ofoma 2023).

Two historic peace agreements, the 1996 Final Peace Agreement (FPA) with the Moro National Liberation Front (MNLF) and the 2014 Comprehensive Agreement on the Bangsamoro (CAB) with the Moro Islamic Liberation Front (MILF), marked significant political breakthroughs (Jeffery 2021; Kovács 2024). These political settlements were intended not only to end hostilities but also to usher in a post-conflict era supported by robust development aid. Programs such as the UN/Multi-Donor Program and the Mindanao Trust Fund–Reconstruction and Development Program (MTF-RDP) were launched to operationalize peace dividends by enhancing governance, improving livelihoods, and promoting social cohesion (Aguja 2000; Gabriel 2019).

R. A. Yusoph (✉)
Mindanao State University-Main Campus, Marawi City, Philippines

Hiroshima University, Higashihiroshima, Japan
e-mail: raihan.yusoph@msumain.edu.ph

A. Sharifi et al. (eds.), *Frameworks and Practices Towards a Sustainable and Peaceful Future*, World Sustainability Series,
https://doi.org/10.1007/978-981-95-2310-8_15

Yet, nearly three decades since the first major agreement, the Bangsamoro continues to face high levels of poverty, institutional fragility, and localized insecurity (World Bank 2021). This paradox, where development aid has been abundant but outcomes remain limited, raises critical questions about the effectiveness and sustainability of aid in fragile and post-conflict contexts. While much of the literature evaluates aid primarily through macroeconomic indicators and national-level data (Burnside and Dollar 2000; Easterly 2006; Moyo 2010; Deaton 2013; Miller 2014), fewer studies examine how aid is experienced and operationalized in subnational, historically marginalized regions, such as the Bangsamoro. Existing empirical work often overlooks the social, political, and institutional factors that shape aid effectiveness in such settings (Addison and McGillivray 2004; Dayanath and Ichihashi 2021).

This chapter seeks to fill that gap by critically examining how foreign-funded development aid introduced in the aftermath of peace agreements has influenced peacebuilding and development outcomes in the Bangsamoro. Focusing on two major programs, the UN/Multi-Donor Program and the MTF-RDP, it examines why, despite extensive aid interventions, the region remains underdeveloped. It further explores how donor-driven agendas, limited local ownership and participation, mismanagement, and resource misallocation, as well as the persistent disruption caused by *rido* (clan-based conflict), have contributed to a cycle of aid dependency, ultimately undermining the transformative potential of development aid.

15.2 Literature Review

The concept of "peace dividends" is central to post-conflict development theory, referring to the tangible socioeconomic benefits expected to follow the cessation of armed conflict (Campbell and Jankowitz 2024). These include improved livelihoods, education, governance, and infrastructure, which are intended not only as rewards for peace but also as mechanisms to prevent the resurgence of violence. In contexts like the Bangsamoro, the timely and equitable delivery of peace dividends is critical to legitimizing political settlements and sustaining public confidence in fragile peace processes. When these dividends fail to materialize—or are distributed unevenly—they can deepen mistrust and undermine both governance and long-term peacebuilding.

As Currie-Alder (2016) notes, development is both a field of scholarly inquiry and an arena of lived, political practice. Among its most contested instruments is foreign aid, which is widely debated in peace and development studies. While some scholars argue that aid can reduce conflict risks by addressing structural inequalities and enhancing state capacity (Azram and Mesnard 2003; Fearon 2004; Sachs 2006; Collier 2007; Rajan and Subramanian 2008; Savun and Tirone 2011), others caution that aid can have counterproductive effects, particularly when it is donor-driven, bypasses local institutions, or fosters elite capture and dependency (Chauvet and Guillaumont 2004; Easterly 2006; Moyo 2010; Deaton 2013; Miller 2014;

Steinwand 2015). However, much of this literature remains focused on macro-level, national outcomes and often neglects how aid is experienced and contested in sub-national, post-conflict regions.

In the case of the Bangsamoro, where development aid has been heavily deployed in the wake of formal peace agreements, there is a pressing need for research grounded in reality that moves beyond policy narratives and aggregate economic metrics. This chapter addresses that gap by critically examining how development aid is designed, implemented, and perceived at the community level, and why, despite decades of support, the promised peace dividends remain elusive.

15.3 Research Methodology

This chapter employed a qualitative research design to investigate the impact of development aid on peacebuilding outcomes in the Bangsamoro. Among the many aid initiatives implemented from 1997 to the present, this research focuses on two particularly significant programs: the SPCPD–NEDA–UN/Multi-Donor Program and the Mindanao Trust Fund–Reconstruction and Development Program (MTF-RDP). These cases were purposefully selected because they represent major donor-led interventions during two distinct phases of the peace process, the immediate aftermath of the MNLF agreement and the post-MILF normalization period. Their prominence in scale, duration, and visibility has made them influential in shaping both public perception and policy debates on aid effectiveness in the region, offering a meaningful lens through which broader structural challenges in aid delivery can be critically examined.

Data collection combined three methods: key informant interviews (KIIs), focus group discussions (FGDs), and document analysis. Reports and evaluation documents from agencies such as UNDP and the World Bank were analyzed to trace program objectives, implementation, and outcomes. To triangulate findings and enrich contextual depth, KIIs were conducted with MNLF and MILF commanders, local government officials, development agency personnel, and civil society actors. Two FGDs were also held in key beneficiary communities—Camp Busra Somiorang, Butig (Lanao del Sur), and Barangay Salbu, Datu Piang (Maguindanao del Sur).

Thematic analysis was conducted using Braun and Clarke's (2006) six-phase framework, supported by the 6Rs method by Naeem et al. (2023). Ethical approval was granted by the Research Ethics Committee of the Graduate School of Humanities and Social Sciences, Hiroshima University. All participants gave their informed consent, and strict measures were followed to ensure confidentiality throughout the study. This methodology, grounded in Creswell's (2015) standards for qualitative rigor, enables a nuanced and context-sensitive understanding of development aid dynamics in a historically marginalized post-conflict setting.

15.4 Results

This section presents the study's key findings based on an in-depth analysis of two significant aid initiatives: the SPCPD-NEDA-UN/Multi-Donor Program and the MTF-RDP, implemented between 1997 and 2017. Drawing on document reviews, key informant interviews, and focus group discussions, this chapter examines how these programs were designed, implemented, and experienced at the community level. The findings offer critical insights for enhancing the relevance, inclusivity, and sustainability of future post-conflict aid strategies in the Bangsamoro.

15.4.1 The SPCPD-NEDA-UN/Multi-Donor

On April 22, 1997, the Southern Philippines Council for Peace and Development (SPCPD), National Economic and Development Authority (NEDA), and the United Nations launched Project PHI/97/004—commonly known as the UN/Multi-Donor Program—to support the reintegration of MNLF combatants through livelihood creation, basic services, and job generation across Special Zone of Peace and Development (SZOPAD) areas (SPCPD/NEDA 1997). Significantly, this was the first time international development aid was coordinated through local bodies, SPCPD and NEDA Region XII, rather than centralized national agencies. The program initially received a grant of approximately USD 8.7 million, with contributions from Australia (USD 1.5 million), Belgium (USD 1.1 million), the Netherlands (USD 1 million), Spain, New Zealand, Sweden, Canada, Switzerland, and Norway. After a pilot phase, the program expanded in scope and funding, with an additional USD 7.3 million allocated to cover ten more MNLF-administered areas. By October 1998, 16.2% of funds had been spent during the pilot phase, and 62% was designated for the expanded phase, demonstrating substantial international support for post-conflict development in the Bangsamoro (SPCPD/NEDA 1998).

15.4.1.1 Program Construction

The structure and implementation of the UN/Multi-Donor Program can be best understood through the lens of converging interests among key stakeholders. For the MNLF, which had persuaded its constituents to accept autonomy rather than full independence, delivering the so-called "peace dividends" became crucial to maintaining its political legitimacy. Failure to do so risked not only diminishing its leadership among the Moro people but also jeopardizing the peace agreement itself. Thus, the MNLF's participation in the program, channeled through the SPCPD, was shaped by political pragmatism and the imperative of organizational survival (SPCPD/NEDA 1998; Aguja 2000).

For the GPH, particularly under the Ramos administration, the cessation of hostilities in Mindanao aligned with both political and economic priorities. Promoting peace enabled the government to attract foreign investment to the resource-rich but conflict-ridden region, which was central to the administration's Philippines 2000 economic strategy. However, given the state's fiscal constraints, mobilizing donor support was essential not only to fulfill its commitments under the peace agreement but also to demonstrate credibility to the international community, especially to members of the Organization of Islamic Cooperation (OIC).

15.4.1.2 Program Implementation

A highly centralized institutional arrangement characterized by the UN/Multi-Donor Program. The Southern Philippines Council for Peace and Development (SPCPD) served as the main executing agency, with UNDP providing overall coordination support (Aguja 2000). Sector-specific components were implemented by major UN bodies: the FAO led agricultural livelihood support, the ILO managed skills training and enterprise development, and the IOM handled information campaigns and confidence-building.

Strategic direction was provided by the Program Management Advisory Group (PMAG), a high-level oversight body comprising donor countries (e.g., Australia, Belgium, Canada, the Netherlands, New Zealand, Norway, Spain, Sweden, and Switzerland), UN agencies, SPCPD officials, and NEDA Region XII. Operational matters were overseen by the Executive Committee (EXECOM), which had authority over project approval, fund disbursement, and monitoring (SPCPD/NEDA 1998).

At the local level, five Area Management Teams (AMTs) were stationed in key urban centers—Palawan, Zamboanga, Iligan, Cotabato, and General Santos—to cover the SZOPAD. These AMTs, supported by UN Volunteers, acted as field coordinators and intermediaries with target communities (SPCPD/NEDA 1997).

15.4.2 Mindanao Trust Fund: Reconstruction and Development Project

The MTF-RDP commenced in December 2005 and officially concluded in 2020, having implemented seven interrelated initiatives. Beginning in 2016, the program intensified its activities within MILF territories, focusing particularly on the six officially recognized camps, as well as 25 additional base camps. The first phase had an eleven-year duration, which concluded in March 2017. The program benefited more than 638,000 individuals—52% of whom were women. It also completed 614 community-level projects across 332 barangays, spanning 114 municipalities in 19 provinces. Furthermore, the initiative facilitated the formation of over 300 community-based organizations (World Bank 2018; Gabriel 2019).

A follow-up phase to the program commenced in April 2018 under the title Mindanao Trust Fund—Reconstruction and Development Project Phase Two (MTF-RDP/2). This stage focused specifically on the six principal camps previously identified as being under MILF control. By the close of August 2019, the initial phase of RDP/2 had implemented 30 community-level projects, reaching a minimum of 42,321 beneficiaries—52% of whom were women. In response to both articulated and emerging needs within these MILF camps, as well as favorable developments in the program's rollout, supplementary funding was approved in late August 2019. This additional support enabled an expansion of the Project's scope and extended its operational timeline through December 31, 2020 (World Bank 2018).

15.4.2.1 Program Construction

Meanwhile, the architecture of MTF-RDP reflects a deliberate effort to blend peacebuilding with development practice. At its core was a community-driven development (CDD) strategy, which empowered local actors, barangay residents, indigenous leaders, and women's associations, to identify, prioritize, and implement development interventions. This design was informed by earlier lessons from MTF-RDP Phase I, particularly the need for stronger local participation, conflict sensitivity, and alignment with Bangsamoro self-governance aspirations. Unlike many top-down donor initiatives, MTF-RDP emphasized partnership and co-ownership through iterative consultations and participatory planning (World Bank 2018).

Donor agencies, including Australia, New Zealand, Canada, Sweden, the European Union, and the United States, channeled funds through the World Bank, which in turn provided technical guidance and fiduciary oversight. The Office of the Presidential Adviser on Peace, Reconciliation, and Unity (OPAPRU) functioned as the government counterpart, ensuring political alignment and coherence with the broader peace and normalization process. The Bangsamoro Development Authority (BDA), meanwhile, served as the principal implementing agency, tasked with managing project planning, disbursing funds, and coordinating with CSOs and LGUs. This tripartite structure ensured both vertical accountability (from donors and government) and horizontal responsiveness (to community needs) (World Bank 2018; Gabriel 2019).

15.4.2.2 Program Implementation

In contrast, the MTF-RDP Phase II, implemented after the CAB, emphasized community-led planning and participatory development. Project selection followed a structured process beginning with Participatory Community Assessments (PCAs) and the development of Community Development Plans (CDPs). Infrastructure outputs included solar streetlights, school buildings, access roads, and agricultural support facilities (Padua 2021).

To institutionalize transparency and governance, each barangay formed Community Volunteer Committees (CVC) responsible for procurement, monitoring, and reporting. These committees were supported by BDA field officers and trained CSO partners. Project evaluation reports showed that 86% of infrastructure outputs were technically rated as satisfactory, and 95% of surveyed beneficiaries expressed satisfaction with the services delivered (Padua 2021).

15.4.3 Key Challenges in the Implementation of the UN/Multi-donor Program and the MTF-RDP

This section identifies four interrelated challenges that have undermined the effectiveness of both the SPCPD-NEDA-UN/Multi-Donor Program and the MTF-RDP. These include donor-driven agendas, limited local ownership and participation, mismanagement and resource misallocation, and the persistent disruption caused by *rido* (clan-based conflict). Collectively, these challenges reveal deep-seated systemic issues that have constrained the sustainability and transformative impact of development aid interventions in the Bangsamoro.

15.4.3.1 Donor-Driven Priorities

A central limitation of the UN/Multi-donor program was its top-down orientation, which offered little space for genuine community engagement. Although framed as a post-conflict reconstruction initiative responsive to local needs, interviewees emphasized that most project components were pre-approved by higher authorities. As one MNLF commander in Madalum, Lanao del Sur, noted, "Most of the projects were pre-approved. We were just told to receive them. They didn't ask if those were really what the communities needed. At that time, we thought that no matter what projects we receive, it is our right to have it because it is [referring to the peace dividend] intended for us" (personal communication, March 25, 2025). This sentiment reflects a broader pattern in which aid is perceived as an entitlement rather than a negotiated outcome.

In contrast, the MTF-RDP was designed to follow a participatory model, incorporating tools such as PCAs and CDPs. However, FGD participants in Camp Busra expressed a disconnect between the planning process and the actual interventions delivered. As one MILF combatant explained, "We identified the irrigation system and water pump as our top priority, but the final Project that was delivered to us was farming inputs and a multipurpose building. Useful, yes, but not urgent. It felt like the decision was made elsewhere" (personal communication, April 2, 2025). This disjunction illustrates how even participatory frameworks can fall short when final decisions remain donor driven.

15.4.3.2 Limited Local Ownership and Participation

The externally driven nature of the UN/Multi-Donor Program significantly undermined local ownership across the design, planning, and implementation phases. While structures like the SPCPD and AMTs were intended to support decentralized coordination, respondents reported being excluded from meaningful decision-making. As the president of an MNLF cooperative in Cotabato City recalled, "We were told that this was our [MNLF] project, but the truth is, we were only invited once everything was finalized. There was no real consultation. As the president of the cooperative, I should have been consulted because my people keep on asking me what we will receive" (MNLF commander, personal communication, March 28, 2025). In the absence of genuine engagement, local actors were often reduced to passive implementers of predetermined agendas, weakening both their commitment to the Project and the sustainability of its outcomes.

Similarly, in the case of the MTF-RDP, formal mechanisms for community engagement, such as stakeholder workshops, did not necessarily translate into meaningful influence. Civil society partners and LGU representatives were included in planning sessions, but many described the process as performative. As one civil society representative based in Cotabato City remarked, "We sat in the meetings and workshops, yes. But the decisions had already been made. It felt more like a formality than real participation" (A. Mama, personal communication, March 28, 2025). This perception of tokenistic engagement eroded community ownership and limited the program's ability to strengthen local institutions and ensure a lasting development impact.

15.4.3.3 Mismanagement and Misallocation of Resources

Weak accountability mechanisms further undermined the effectiveness of the UN/Multi-Donor Program. Respondents cited unmonitored disbursements, missing equipment, and discrepancies between reported and actual deliveries. As one former MNLF combatant in Cotabato City explained, "Actually, we had funds for farming tools and equipment, but some of them never reached the community. They were recorded as delivered, but nothing arrived. There was no accountability for how those resources were spent. Some of us have accused our commander of selling it, but we do not have any evidence. Maybe there was no equipment at all because we didn't see it" (personal communication, March 28, 2025). Such mismatches between documentation and community experience reflect serious gaps in oversight and financial transparency, resulting in both resource loss and deepened mistrust among beneficiaries.

While the MTF-RDP introduced more stringent procurement and monitoring procedures, issues of poor implementation persisted. A resident from Camp Busra described a school infrastructure project: "They said we would get two classrooms for the *madrasah* (Islamic school), but only one was built. Even that was substandard—the walls cracked within months. It's hard to believe the full budget was allocated for that. What we heard is that it costs one million or more" (S. Abduljabbar, personal communication, April 2, 2025). These accounts point to ongoing weaknesses in contractor oversight and technical monitoring, particularly in remote, conflict-affected areas. Even in the absence of proven corruption, implementation failures compromised project outcomes and reduced community confidence in aid delivery.

15.4.3.4 Presence of Rido or Clan-based Conflict

Rido is often rooted in disputes over land, political rivalry, and inter-familial grievance and has long undermined local governance and development in the region (Torres III 2014; Durano Sr and Teodoro 2018; Yusoph and Ali 2022). A significant operational barrier for both programs was the persistent threat of *rido,* which created security risks in several project areas. In conflict-affected zones, active *rido* made it unsafe for teams and contractors to access sites, forcing delays, suspensions, or even cancellations of planned interventions. As one LGU officer in Madalum, Lanao del Sur, mentioned, "There were barangays we couldn't even reach because of active *rido*. The project timelines suffered, and some areas had to be dropped from the plan entirely" (M. Macarimbang, personal communication, March 25, 2025).

The MTF-RDP, though explicitly designed to promote peace through development, was not immune to the disruptive effects of unresolved *rido* in neighboring communities. While Camp Busra itself was relatively secure, escalating tensions nearby directly impacted implementation. A field staff member from the BDA recalled, "We had to halt construction of the multipurpose building inside the camp because of a brewing *rido* near our camp. Even the staff and laborers became targets, and some fled. There was nothing we could do until things settled down" (A. Johary, personal communication, April 2, 2025). These accounts emphasize how localized conflict, if unaddressed, can severely undermine even well-intentioned development programs.

15.5 Discussion

15.5.1 Aid Dependency and Internal Periphery

Although foreign-funded development aid in the Bangsamoro has aimed to improve socioeconomic conditions and support peacebuilding, it has largely failed to produce lasting outcomes. Rather than promoting resilience and local self-reliance, many of these interventions have fostered dependency, undermining the broader goals of sustainable peace and inclusive growth. Despite the volume of foreign aid, this dynamic remains underexplored in both academic literature and policy discussions.

Drawing on dependency theory, this study examines how external aid, while intended to spur development, can entrench structural inequalities and reinforce reliance on donors. Scholars such as Seers and Joy (1971), Lensink and Howard (1999), and Peet and Hartwick (2015) refer to this as "aid dependency," wherein development stagnates in the absence of continued external support. Lensink and Howard (1999), for instance, define aid dependence as a condition in which a recipient of aid is unlikely to achieve its development objectives without ongoing aid.

This perspective is particularly relevant to the Bangsamoro context, where aid, despite addressing urgent post-conflict needs, has often failed to overcome underlying structural challenges such as political instability, persistent insecurity, weak institutions, and poorly designed interventions. As a result, development aid has, in many cases, inadvertently perpetuated the very underdevelopment it seeks to resolve.

Dependency theory has long served as a critical framework for understanding the structural roots of underdevelopment in the Global South, with the Philippines often cited as a paradigmatic case. Scholars such as David (1980) and Nemenzo Jr. (2012) have examined the country's postcolonial condition, portraying it as a peripheral nation embedded in a global system dominated by core powers, particularly the United States. These early formulations focused on macro-level relations between states, where the Philippines was seen as a passive recipient of external economic and political influence.

While these insights remain foundational, they offer limited analytical traction in understanding how dependency manifests within the internal dynamics of peripheral states, especially in regions shaped by conflict, historical injustice, and domestic asymmetries. This study builds on and extends dependency theory by applying it to the subnational context of the Bangsamoro, where foreign aid interventions are influenced not only by foreign donor agendas but also by national governance shortcomings and elite politics. As Ellner (2024) notes, the contemporary value of dependency theory lies in its capacity to trace the persistence of asymmetrical relationships and constrained autonomy in peripheral societies. Likewise, Ahumada and Torres (2024) propose a localized evolution of the theory, suggesting that subnational areas often occupy a "double periphery", marginalized by both the global system and their central governments.

In this context, the Bangsamoro can be conceptualized as an "internal periphery within a peripheral state," where aid does not simply replicate center-periphery dynamics, but generates complex interactions among national agencies and donor institutions. These interactions unfold within fragile governance structures that have been shaped by decades of armed conflict and underdevelopment. Foreign aid, while often framed as a tool for peacebuilding and recovery, can in practice reinforce existing inequalities when implemented through top-down modalities.

This analysis aligns with scholars who emphasize that the impact of foreign aid is contingent upon the presence of enabling environments. Burnside and Dollar (2000), Chauvet and Guillaumont (2004), and Todaro and Smith (2015) argue that aid is more likely to generate developmental gains when aligned with strong institutions and inclusive policies. Deaton (2013) similarly cautions that without accountable governance structures, even well-intentioned interventions may yield minimal or adverse effects. As Muggah (2014) points out, development and security objectives are rarely achieved through external interventions alone; their success hinges on deep local engagement and the development of institutional capacity.

Despite nearly three decades of government pronouncements and international donor support, national statistics continue to depict a stark and persistent underdevelopment in the Bangsamoro. Provinces such as Sulu, Tawi-Tawi, Maguindanao, Lanao del Sur, and Basilan have consistently ranked among the poorest in the country, with poverty rates in some municipalities exceeding 70% as recently as 2015 (Philippine Statistics Authority 2015). Although the World Bank (2021) recorded a decline in poverty from 61.8% in 2018 to 37.2% in 2021, this reduction was largely attributed to temporary emergency aid and cash subsidies during the COVID-19 pandemic rather than the long-term effects of development programs. As of 2023, both the Philippine Statistics Authority and the Global Data Lab continue to identify the Bangsamoro region as one of the areas with the lowest poverty incidence and human development nationwide, with an HDI of 0.629 (Philippine Statistics Authority 2023; Global Data Lab 2022).

This persistent poverty stands in sharp contrast to the optimistic claims of aid effectiveness literature. Scholars such as Burnside and Dollar (2000), Sachs (2006), and Minoiu and Reddy (2009) argue that foreign aid can reduce poverty and help economic growth. Yet, the Bangsamoro experience defies this assumption: despite decades of peacebuilding and aid interventions from 1997 to the present, it remains one of the most impoverished regions in the Philippines.

While multiple structural and political factors have contributed to this stagnation, this study narrows its focus to a critical gap in the literature—namely, the design and implementation flaws of key development aid programs introduced in the wake of peace agreements. Rather than transforming long-standing inequalities, these interventions may have inadvertently reinforced them.

15.5.2 Limitations of Development Aid in the Bangsamoro

Development aid has been central to the Bangsamoro's post-conflict reconstruction efforts; however, its implementation reveals persistent structural limitations that compromise its long-term effectiveness. Through a triangulated analysis of key informant interviews, focus group discussions, and official program documents, this study surfaced four recurring challenges that critically shape how aid is designed, delivered, and perceived at the community level. By systematically comparing and cross-validating insights from these diverse sources, the study identified patterns that informed the core themes discussed. These findings highlight deeply embedded flaws in the development of aid architecture, flaws that ultimately hinder its ability to promote sustainable peace and inclusive development in the region.

Central among these challenges is the predominance of donor-driven priorities. Both the UN/Multi-Donor Program and the MTF-RDP exhibit a pattern in which aid interventions were largely predetermined by external actors, with minimal genuine consultation or decision-making power granted to local communities. Interviews revealed that local stakeholders were often relegated to passive roles, limiting the contextual relevance and responsiveness of development initiatives. Community inputs, even when solicited through participatory mechanisms like CDPs, frequently failed to influence actual project selection, which tended to reflect donor funding conditions over locally identified needs.

This dynamic reflects a broader critique of development aid, which serves the geopolitical, economic, and strategic interests of donor countries rather than the priorities of recipient populations. Scholars such as Fleck and Kilby (2010) and Bermeo (2017) emphasize how geopolitical considerations and alliance-building frequently influence the allocation of aid. In the Philippine context, Tadem (2008) and de Guzman and Montiel (2012) emphasize how tied aid, aid-conditional assistance that favors donors' firms or services, undermines national sovereignty and restricts the autonomy of local institutions in shaping their development trajectories.

A critical challenge in the Bangsamoro's development aid landscape is the misalignment between donor priorities and actual community needs, resulting in mismatched interventions. For example, under the UN/Multi-Donor Program, MNLF communities prioritized practical infrastructure, such as water pumps and irrigation systems, to address urgent livelihood concerns. Yet, these proposals were sidelined in favor of constructing multipurpose buildings or peace centers, which were chosen for their convenience within donor frameworks. While such facilities may offer symbolic value, they fall short of meeting pressing needs such as access to potable water.

This case exemplifies a broader pattern of externally driven, top-down aid initiatives that lack genuine local ownership and sustainability (Trajano 2020; Cardeño et al. 2023). Although participatory mechanisms may be employed, substantive decision-making often remains centralized with donors, limiting community empowerment. As Riddell (2007) and Aja-Eke (2023) caution, aid that is not carefully aligned with local contexts risks ineffectiveness or even fostering dependency.

While the literature affirms that development aid has the potential to foster socio-economic growth when embedded in authentic partnerships and responsive planning (Sachs 2006; Minoiu and Reddy 2009; Dayanath and Ichihashi 2021), such success hinges on donor commitment to the lived realities of recipient communities. In the Bangsamoro, recurring donor-centric patterns highlight the urgent need for development processes that genuinely empower local stakeholders, rather than positioning them as passive beneficiaries.

Second is the limited ownership and participation of target communities in the design, planning, implementation, and monitoring of development aid programs. When local actors—particularly LGUs and CSOs, crucial to peacebuilding—are excluded, interventions often fail to reflect socio-cultural and material realities, reducing their relevance and legitimacy. It was evident in the UN/Multi-Donor

Program, where decision-making was centralized among donors, the national government, and MNLF elites, while broader communities remained uninformed or marginalized. Evaluations noted that many perceived the program as an exclusive benefit for select MNLF members, with younger combatants receiving more advantages than long-time liberators still living in poverty (Adriano and Thomas 2013).

The widespread lack of sustainability planning and exit strategies in donor programs exacerbates this exclusion. Treating communities as passive recipients rather than empowered stakeholders fosters dependency, as most initiatives lack clear handover mechanisms, capacity-building, or benchmarks for long-term maintenance. Consequently, many projects deteriorate after donor withdrawal, reinforcing a cycle in which development gains are not sustained locally. This critical weakness in aid architecture underscores the urgent need for integrated sustainability frameworks and community-anchored exit plans to ensure lasting impact.

Third, both programs exhibited significant challenges related to mismanagement and weak resource accountability, with issues more pronounced in the UN/Multi-Donor Program. Participants reported experiencing ghost deliveries, incomplete infrastructure, and questionable procurement practices, reflecting gaps in auditing and monitoring that allowed funds to leak and undermined the program's credibility among beneficiaries. Although the MTF-RDP implemented stricter procurement protocols, technical shortcomings, such as substandard school buildings and incomplete classroom construction, persisted due to inadequate contractor supervision and limited local capacity for absorption.

These operational weaknesses highlight a broader issue: even with local safety, development aid remains vulnerable to inefficiency and corruption in contexts where governance structures remain fragile. Fernandez (2014) highlights that key regional institutions, including the SPCPD, SPDA, and ARMM Planning Office, have faced persistent criticism of incompetence and corruption. As a result, resources intended to assist conflict-affected communities frequently failed to reach their intended targets, exacerbating grievances and deepening underdevelopment, while missing critical opportunities to address the root causes of conflict.

These observations align with a broad scholarly consensus linking aid effectiveness to the quality of fund management, institutional transparency, and governance. Annen and Moers (2017) emphasize that donor success depends largely on embedded fiscal responsibility and transparency, while Findley et al. (2017) show how weak governance in aid-dependent contexts can exacerbate corruption and mismanagement. Asongu and Nwachukwu (2015) further argue that foreign aid may undermine institutional quality by weakening the rule of law and accountability mechanisms.

Nevertheless, aid retains significant potential. Scholars such as Muggah (2014), Deaton (2013), and Todaro and Smith (2015) contend that, when well-managed and aligned with inclusive policies, aid can promote growth and recovery in underdeveloped regions. In the Philippine context, Landingin (2008) highlights how large-scale foreign-funded projects have often suffered from overpricing and poor implementation, resulting in wasted resources and unmet development goals.

These findings underscore the urgent need for aid programs to adopt transparent management, robust monitoring, and inclusive governance. Without such frameworks and meaningful stakeholder engagement, development aid risks perpetuating the very inequalities and governance failures it aims to resolve. While aid plays a crucial role in post-conflict reconstruction, it will remain insufficient unless systemic mismanagement and corruption are effectively addressed.

Finally, one of the most underacknowledged yet critical barriers to effective development aid in the Bangsamoro is the persistent presence of *rido* fueled by disputes over land, political rivalries, and the widespread availability of firearms (Lara Jr. and de la Rosa 2023). In both the UN/Multi-Donor Program and the MTF-RDP, *rido* severely disrupted project implementation by delaying timelines, restricting access, and, at times, forcing suspension or abandonment of interventions. Beyond logistical challenges, these conflicts pose a threat to personnel and beneficiaries, ultimately undermining the delivery of meaningful peace dividends.

Despite its profound impact, neither program integrated localized conflict resolution or peacebuilding frameworks into its strategies, reflecting a critical oversight. By treating insecurity as a peripheral issue rather than a central, dynamic factor, aid efforts have failed to address the realities that shape access, participation, and sustainability on the ground.

This omission exemplifies a broader pattern in post-conflict reconstruction, which is often driven by short-term performance metrics and donor compliance rather than conflict-sensitive, locally grounded, and transformative peacebuilding. The findings of this chapter highlight systemic flaws in the design and implementation of aid in the Bangsamoro. Despite decades of foreign aid aimed at poverty reduction and recovery, these programs have largely fallen short of fostering sustainable development, instead contributing to a cycle of dependency where communities perceive aid as an entitlement rather than a pathway to self-reliance.

This dynamic encapsulates the core premise of dependency theory, which this chapter adapts from its traditional national-level critique of global economic hierarchies to a subnational analysis of the Bangsamoro as an "internal periphery" within the Philippine state. Marginalized by national structures and reliant on external aid, the region experiences entrenched unequal power relations that limit local institutions and community-driven development. Aid recipients often have little influence over program priorities or implementation, which remain largely dictated by donors. As a result, when aid concludes or shifts, communities lack the institutional and social infrastructure to sustain progress, perpetuating a cycle of dependence.

By applying dependency theory at the community level, this chapter critiques how foreign aid, rather than alleviating poverty, can reproduce dependency under the guise of development. In the Bangsamoro, development aid often serves as a tool of control, yielding short-term outcomes without promoting socioeconomic transformation. Without a fundamental shift toward development approaches that center on local conflict dynamics and community priorities, the promise of peace dividends will remain unmet, and the region's persistent developmental challenges will continue.

15.6 Conclusion

This chapter critically examined the role of foreign-funded development aid in shaping peacebuilding and development outcomes in the Bangsamoro following the formal signing of peace agreements. Through a comparative analysis of the UN/Multi-Donor Program and the Mindanao Trust Fund–Reconstruction and Development Project (MTF-RDP), the study identified four interrelated and persistent challenges: donor-driven agendas, limited local ownership and participation, mismanagement and resource misallocation, and the ongoing disruption caused by localized conflict, particularly *rido*. These factors have collectively undermined the effectiveness, legitimacy, and sustainability of aid interventions in the Bangsamoro.

Framed within dependency theory, the analysis positions the Bangsamoro as an internal periphery—marginalized not only by the central Philippine state but also by a global aid architecture that reproduces asymmetrical power relations. This framing reveals that the problem is not the volume of aid, but its top-down design, delivery mechanisms, and failure to cultivate genuine local participation. As a result, many interventions have produced limited outcomes in terms of institutional trust, livelihood improvement, and governance transformation. Community disillusionment has persisted, alongside a deepened dependency on external actors for both peacebuilding and development.

To address these structural limitations, future aid programs in the Bangsamoro must embed sustainability, conflict sensitivity, and local ownership as core principles. Interventions should be accompanied by clear exit strategies, robust capacity-building measures, and institutional handover mechanisms to ensure continuity beyond the project cycle. Centering community-driven planning processes and integrating localized conflict resolution—particularly in areas affected by *rido*—can increase both the legitimacy and security of development efforts. Donors and implementers must shift from technocratic, performance-based models toward transformative frameworks that empower local actors and institutions to take the lead in their own development projects.

While this chapter provides critical insights into the challenges of foreign aid in post-conflict Bangsamoro, it is not without limitations. It primarily focused on two major aid programs and did not conduct a comprehensive assessment of all foreign interventions across the region. Additionally, it did not extensively explore the role of Bangsamoro civil society or the emerging BARMM bureaucracy in mediating or resisting these aid dynamics. Future research could examine the long-term institutional impact of these interventions, the role of grassroots actors in shaping aid effectiveness, or conduct comparative studies with other post-conflict autonomous regions to further deepen the analysis of aid dependency and institutional reform.

Crucially, suppose development aid in the Bangsamoro is to yield sustainable peace. In that case, it must be reimagined not as a temporary, but as a transformative instrument grounded in justice, inclusion, and long-term resilience. Only through a fundamental shift in how aid is conceptualized and delivered can it support the broader vision of a self-determined, peaceful, sustained, and prosperous Bangsamoro.

References

Addison, T., & McGillivray, M. (2004). Aid to conflict-affected countries: Lessons for donors. *Conflict, Security & Development, 4*(3), 347–367. https://doi.org/10.1080/1467880042000319866

Adriano, F., & Thomas, P. (2013). *The contested corners of Asia: The case of Mindanao, Philippines*. The Asia Foundation.

Aguja, M. J. (2000). The aftermath of ethnic violence: Post-war reconstruction in the Southern Philippines: A preliminary assessment of the role of the international community. *Forum of International Development Studies,* Nagoya University.

Ahumada, J. M., & Torres, M. (2024). Dependency theory. In *the Global handbook of inequality* (227–247). Cham: Springer Nature Switzerland.

Aja-Eke, D. (2023). Foreign Aid and socioeconomic development: The perceptions of Aid in Nigeria. *Journal of Asian and African Studies*. https://doi.org/10.1177/00219096241270702

Annen, K., & Moers, L. (2017). Donor competition for aid impact and aid fragmentation. *World Bank Economic Review, 31*(3), 708–729. https://doi.org/10.1093/wber/lhw019

Asongu, Anna J. (2019). Foreign Aid, instability, and governance in Africa. *Politics & Policy* 47(4): 807–848. https://doi.org/10.1111/polp.12320

Asongu, S. A., & Nwachukwu, J. C. (2015). Foreign aid and governance in Africa. *International Review of Applied Economics,* 30(1), 69–88. https://doi.org/10.1080/02692171.2015.1074164

Azram, J., & Mesnard, A. (2003). Civil war and the social contract. *Public Choice, 115*(3), 455–475. https://doi.org/10.1023/A:1024265631528

Bermeo, S. B. (2017). Aid Allocation and Targeted Development in an Increasingly Connected World. *International Organization, 71*(4), 735–766. https://doi.org/10.1017/S0020818317000315

Braun, V., & Clarke, V. (2006). Using thematic analysis in psychology. *Qualitative Research in Psychology, 3*(2), 77–101. https://doi.org/10.1191/1478088706qp063oa

Burnside, C., & Dollar, D. (2000). Aid, policies, and growth. *American Economic Review, 90*(4), 847–868. https://doi.org/10.1257/aer.90.4.847

Campbell, P., & Jankowitz, S. (2024). Funding art in post-conflict society: a 'peace dividend'? *International Journal of Cultural Policy,* 30(7), 848–863. https://doi.org/10.1080/10286632.2024.2304297

Cardeño, C. E., Olivius, E., & Åkebo, M. (2023). 'Jihad is Planted in Our Hearts': International Aid, Rebel Institutions and Women's Participation in the Bangsamoro. *Civil Wars,* 27(1), 42–69. https://doi.org/10.1080/13698249.2023.2230824

Chauvet, L. and Guillaumont, P. (2004) *Aid and Growth Revisited: Policy, Economic Vulnerability and Political Instability.* Working Papers, CERDI.

Collier, P. (2007). *The Bottom Billion: Why the poorest countries are failing and what can be done about it.* Oxford University Press.

Creswell, J. (2015). *30 Essential Skills for the Qualitative Researcher.* SAGE Publications.

Currie-Alder, B. (2016). The state of development studies: Origins, evolution and prospects. *Canadian Journal of Development Studies / Revue canadienne d'études du développement, 37*(1), 5–26. https://doi.org/10.1080/02255189.2016.1135788

David, R. S. (1980). Philippine underdevelopment and dependency theory. *Philippine Sociological Review*, 28(1–4), 81–87. http://www.jstor.org/stable/23892107

Dayanath, G. and Ichihashi, M. (2021). Does Foreign Aid Work Well in Developing Countries? A Mediating Effect Approach, *Journal of International Development and Cooperation,* Vol. 27, No. 1 & 2.

De Guzman, J. M., & Montiel, C. J. (2012). Social Representations of Foreign Aid: Exploring Meaning-Making in Aid Practice in Sulu, Southern Philippines. *Journal of Pacific Rim Psychology*, 6(1), 1-17. https://doi.org/10.1017/prp.2012.2

De Ree, J., & Nillesen, E. (2009). Aiding violence or peace? The impact of foreign aid on the risk of civil conflict in Sub-Saharan Africa. *Journal of Development Economics,* 88(2), 301-313. https://doi.org/10.1016/j.jdeveco.2008.03.005

Deaton, A. (2013). *The Great Escape: Health, wealth, and the origins of inequality*. Princeton University Press.

Durano Sr, C. R., & Teodoro, C. G. (2018). Dynasticism as an Indirect Driver of Civil Wars: Clan Feuding in the Philippines and Its Impact on Intrastate Political Violence. *An Unforgetting Wrath: Dynastic Conflicts, Vendetta Politics, and Civil War Violence*, *1001*, 126.

Easterly, W. (2006). *The white man's burden: Why the West's efforts to aid the rest have done so much ill and so little good*. Penguin Press.

Ellner, S. (2024). Dependency theory and its revival in the twenty-first century. *Latin American Research Review*, 59(3), 744 - 757. https://doi.org/10.1017/lar.2024.12

Fearon, J. D. (2004). Why do some civil wars last so much longer than others? *Journal of Peace Research, 41*(3), 275–303. https://doi.org/10.1177/0022343304043770

Fernandez, M. C. (2014, September). *Implementing peace and development in the Bangsamoro: Potentials and constraints of socioeconomic programs for conflict-affected areas in Southern Philippines (1913–2015)* [Conference session]. APISA Conference, Chiang Mai, Thailand. https://icafernandez.com/wp-content/uploads/2017/06/fernandez-pd-in-the-bangsamoro-2017-06-20.pdf

Findley, M. G., Harris, A. S., Milner, H. V. & Nielson, D. (2017). Who controls foreign aid? Elite versus public perceptions of donor influence in aid-dependent Uganda. *International Organization, 71*(4), 633–663. https://doi.org/10.1017/S0020818317000273

Fleck, R. K., & Kilby, C. (2010). Changing aid regimes? U.S. foreign aid from the Cold War to the War on Terror. *Journal of Development Economics, 91*(2), 185–197. https://doi.org/10.1016/j.jdeveco.2009.09.011

Gabriel, C. S. (2019). *Does aid really help? The nexus between development aid and state-society resilience in fragile situations* [Doctoral dissertation, Kennesaw State University]. Kennesaw State University Digital Commons. https://digitalcommons.kennesaw.edu/incmdoc_etd/28

Global Data Lab. (2022). *Subnational Human Development Index (SHDI) database*. https://globaldatalab.org/shdi/

Jeffery, R. (2021). *Negotiating Peace: Amnesties, Justice, and Human Rights*. Cambridge University Press. https://doi.org/10.1017/9781108937184

Kovács, B. Á. (2024). *From entry points to sustainable action: Equipping peace processes for accountability and integrity. The case of the Bangsamoro*. Berghof Foundation.

Landingin, R. (2008, February 12). Perils, pitfalls of aid. *Philstar.com*. https://www.philstar.com/headlines/2008/02/12/44129/perils-pitfalls-aid

Lara, F. J., Jr., & de la Rosa, N. P. C. (Eds.). (2023). Conflict's *long game: A decade of violence in the Bangsamoro*. Bughaw.

Lensink, R., & Howard, W. (1999). *Aid dependence: Issues and indicators* (EGDI Studies). Almqvist & Wiksell International.

Miller, R. (2014). *Aid as peacemaker: Canadian development assistance and Third World conflict*. McGill–Queen's University Press.

Minoiu, C., & Reddy, S. (2009). *Development aid and economic growth: A positive long-run relation*. International Monetary Fund. https://doi.org/10.1016/j.qref.2009.10.004

Moyo, D. (2010). Dead Aid: Why aid is not working and how there is a better way for Africa. *Canadian Journal of African Studies / Revue Canadienne des Études Africaines, 44*(2), 422–424. http://www.jstor.org/stable/23236833

Muggah, R. (2014). Deconstructing the Fragile City: Exploring Insecurity, Violence, and Resilience. *Environment & Urbanization,* 26(2), 345-358. https://doi.org/10.1177/0956247814533627

Naeem, M., Ozuem, W., Howell, K., & Ranfagni, S. (2023). A step-by-step process of thematic analysis to develop a conceptual model in qualitative research. *International Journal of Qualitative Methods, 22*, 1–18. https://doi.org/10.1177/16094069231205789

Nemenzo Jr., F. (2012). Dependency and Liberation: A Focus on the Third World. *Philippine Political Science Journal, 5*(7), 120–128. https://doi.org/10.1080/01154451.1978.9753999

Ofoma, C. (2023). Foreign Aid in Nigeria: A departure from economic development to social development impact. *Journal of Management Sciences and Applications,* 2(2): 219–232. https://www.ceeol.com/search/article-detail?id=1203924

Padua, M. L. (2021). *Implementation Completion and Results Report (ICR) Document - Mindanao Trust Fund-Reconstruction and Development Project Phase II - P164019* Washington, D.C.: World Bank Group. Retrieved from. http://documents.worldbank.org/curated/en/680101634507508998 Accessed on January 7, 2025.

Peet, R., & Hartwick, E. R. (2015). *Theories of development: Contentions, arguments, alternatives.* Guilford Press.

Philippine Statistics Authority. (2015). *Population of the Autonomous Region in Muslim Mindanao: Based on the 2015 Census of Population.* PSA. Retrieved from https://psa.gov.ph/statistics/population-and-housing/node/57706 Accessed on January 2, 2025.

Philippine Statistics Authority. (2023). *Official poverty statistics of the Philippines.* PSA. Retrieved from https://psa.gov.ph/statistics/poverty Accessed on January 2, 2025.

Rajan, R., & Subramanian, A. (2008). Aid and growth: What does the cross-country evidence show? *Review of Economics and Statistics, 90*(4), 643–665. https://doi.org/10.1162/rest.90.4.643

Riddell, R. C. (2007). *Does foreign aid really work?* Oxford University Press.

Sachs, J. D. (2006). *The end of poverty: Economic possibilities for our time.* Penguin Press.

Savun, B., & Tirone, J. (2011). Foreign Aid, Democratization, and Civil Conflict: How does democracy aid affect civil conflict? *American Journal of Political Science, 55*(2), 233–246. https://doi.org/10.1111/j.1540-5907.2010.00501.x

Seers, D., & Joy, L. (Eds.). (1971). *Development in a divided world.* Harmondsworth, UK: Penguin Books.

SPCPD/NEDA. (1997). *Special Zone of Peace and Development, Southern Philippines: An Integrated Development Framework.* UNDP.

SPCPD/NEDA. (1998). *Taking the initiative: Terminal report on the SPCPD-UN-NEDA Program.* UNDP.

Steinwand, M. C. (2015). Compete or Coordinate? Aid Fragmentation and Lead Donorship. International Organization, 69(2), 443–472. https://doi.org/10.1017/S0020818314000381

Tadem, E. C. (2008). Japan, the U.S., and official development assistance to the Philippines. *Kasarinlan: Philippine Journal of Third World Studies, 5*(4). https://journals.upd.edu.ph/index.php/kasarinlan/article/view/986

Todaro, M. P., & Smith, S. C. (2015). *Economic development* (12th ed.). Pearson Education.

Torres, W. M. III (Ed.). (2014). *Rido: Clan feuding and conflict management in Mindanao.* Ateneo de Manila University Press.

Trajano, J. C. I. (2020). Bottom-up peacebuilding: Role of grassroots and local actors in the Mindanao peace process. *Asian Journal of Peacebuilding, 8*(2), 357–372. https://doi.org/10.18588/202011.00a097

World Bank. (2018). *Multi-donor facility: Reconstruction and development: Implementation completion and results report.* Retrieved from https://projects-beta.worldbank.org/en/projects-operations/document-detail/P095173 Accessed on January 7, 2025.

World Bank. (2021) *Philippines Economic Update - Beyond the Numbers: Sustaining Poverty Reduction in BARMM (English).* Washington, D.C.: World Bank Group. Retrieved from http://documents.worldbank.org/curated/en/099060324081514520 Accessed on January 7, 2025.

Yusoph, R. A., & Ali, A. A. (2022). Peace Reconciliation: In the Context and Realities of the Bangsamoro Autonomous Region in Muslim Mindanao. *International Journal of Social Sciences: Current and Future Research Trends, 15*(1), 188-206. https://ijsscfrtjournal.isrra.org/index.php/Social_Science_Journal/article/view/1190

MIX
Papier aus verantwortungsvollen Quellen
Paper from responsible sources
FSC® C105338

FSC
www.fsc.org

If you have any concerns about our products,
you can contact us on
ProductSafety@springernature.com

In case Publisher is established outside the EU,
the EU authorized representative is:
Springer Nature Customer Service Center GmbH
Europaplatz 3, 69115 Heidelberg, Germany

Printed by Libri Plureos GmbH
in Hamburg, Germany